Radio Science and Engineering

Radio Science and Engineering

Edited by Claude McMillan

CLANRYE
INTERNATIONAL
www.clanryeinternational.com

Clanrye International,
750 Third Avenue, 9th Floor,
New York, NY 10017, USA

ISBN: 978-1-63240-586-9

Cataloging-in-publication Data

Radio science and engineering / edited by Claude McMillan.
p. cm.
Includes bibliographical references and index.
ISBN 978-1-63240-586-9
1. Radio. 2. Telecommunication. 3. Communication in engineering. I. McMillan, Claude.
TK6550 .R33 2017
621.384--dc23

For information on all Clanrye International publications
visit our website at www.clanryeinternational.com

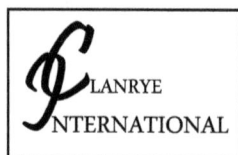

℄LANRYE
INTERNATIONAL

Printed in the United States of America.

Contents

Preface

Radio science is defined as the study of range and travel of radio waves in various environments. This book on radio science discusses the new scientific procedures that employ radio sciences and the different technologies that are developed for its induction, transmission and reception. Statistics and data that could predict atmospheric changes as well as long-term changes in the earth's climate and biosphere can be calculated using analyses and methodology used in radio science. This book is a valuable compilation of topics ranging from the basic to the most complex advancements in the field of radio science. The aim of this text is to present researches that have transformed this discipline and aided its advancement. This book will prove helpful for students and researchers in the fields of radio science and related fields like astronomy, radiation and nuclear physics.

The researches compiled throughout the book are authentic and of high quality, combining several disciplines and from very diverse regions from around the world. Drawing on the contributions of many researchers from diverse countries, the book's objective is to provide the readers with the latest achievements in the area of research. This book will surely be a source of knowledge to all interested and researching the field.

In the end, I would like to express my deep sense of gratitude to all the authors for meeting the set deadlines in completing and submitting their research chapters. I would also like to thank the publisher for the support offered to us throughout the course of the book. Finally, I extend my sincere thanks to my family for being a constant source of inspiration and encouragement.

Editor

A novel ZePoC encoder for sinusoidal signals with a predictable accuracy for an AC power standard

T. Vennemann[1], **T. Frye**[1], **Z. Liu**[1], **M. Kahmann**[2], and **W. Mathis**[1]

[1]Institut für Theoretische Elektrotechnik (TET), Gottfried Wilhelm Leibniz Universität , Hannover, Germany
[2]Physikalisch-Technische Bundesanstalt (PTB), Braunschweig, Germany

Correspondence to: T. Vennemann (vennemann@tet.uni-hannover.de)

Abstract. In this paper we present an analytical formulation of a Zero Position Coding (ZePoC) encoder for an AC power standard based on class-D topologies. For controlling a class-D power stage a binary signal with special spectral characteristics will be generated by this ZePoC encoder for sinusoidal signals. These spectral characteristics have a predictable accuracy within a separated baseband to keep the noise floor below a specified level. Simulation results will validate the accuracy of this novel ZePoC encoder. For a real-time implementation of the encoder on a DSP/FPGA hardware architecture a trade-off between accuracy and speed of the ZePoC algorithm has to be made. Therefore the numerical effects of different floating point formats will be analyzed.

1 Introduction

ZePoC was invented and initially implemented for audio coding by the Institut für Theoretische Elektrotechnik (TET). The main advantages of a complete digital class-D power amplifier using ZePoC are the low switching rate and the separated baseband (Streitenberger, 2005). A lot of effort was made to develop a prototype of the ZePoC audio power amplifier resulting in a number of publications. A good overview can be found in the white paper from Texas Instruments (Texas Instruments, 2005).

ZePoC is also ideal for an AC power standard based on class-D topologies (Wellmann, 2010). An AC power standard consists of two channels. One channel provides a highly accurate sinusoidal voltage and the other one a highly accurate sinusoidal current. The mathematical methods presented in this contribution are suitable for both channels. Figure 1 represents the block diagram of the voltage channel.

The ZePoC encoder is implemented on a digital signal processor (DSP). Depending on some encoding parameters the duty cycle of a pulse width modulated (PWM) signal is computed and transferred to a field programmable gate array (FPGA) via an I^{2}S interface. A pulse shaper implemented on the FPGA converts the received duty cycles into a bit stream. This bit stream is a time discrete form of the binary PWM signal. A gigabit serializer inside the FPGA allows a transmission of the binary signal with a very high time resolution.

To amplify the binary PWM signal a class-D power stage is used. An inverting driver controls a P-channel and a N-channel MOSFET. At the output of the power stage the high voltage of the PWM signal equals V$_{DD}$ and the low voltage equals V$_{SS}$. If V$_{DD}$ and V$_{SS}$ are derived from natural constants using a physical effect (e.g. Josephson effect), the amplitude of the amplified PWM signal is highly accurate. The problem of harmonic distortions caused by non-ideal switching transients of the power stage and the static drain-to-source on-resistance of the MOSFETs is not content of this paper.

At the end of the signal chain an analogue low pass filter (LPF) is used to suppress the disturbances caused by the switching. This is only possible because of the separated baseband which ensures a spectral gap between the AC signal and the switching disturbances at higher frequencies. The absolute value of the LPF frequency response has to be equal to one at the frequency of the AC signal.

2 Pulse shaper

An example for the time function of the PWM signal used for the AC power standard is displayed in Fig. 2. The periodic time is T_{sw} and the switching frequency results in $f_{sw} = \frac{1}{T_{sw}}$.

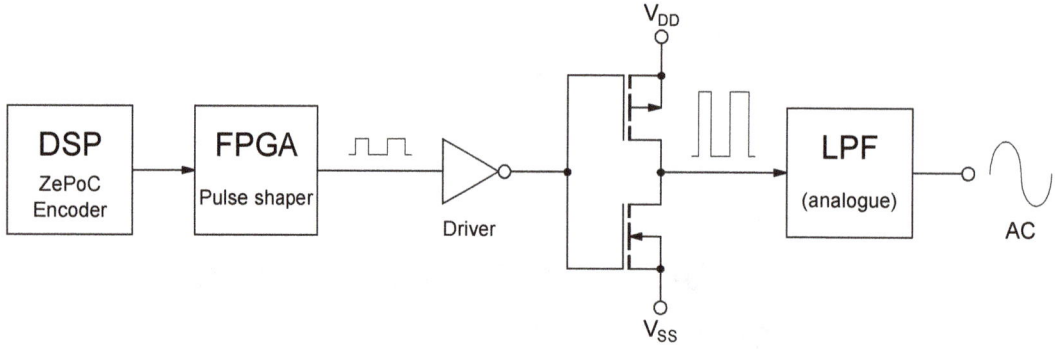

Figure 1. AC power standard: block diagram of the voltage channel.

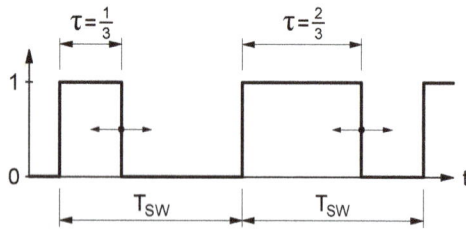

Figure 2. Time function of the PWM signal.

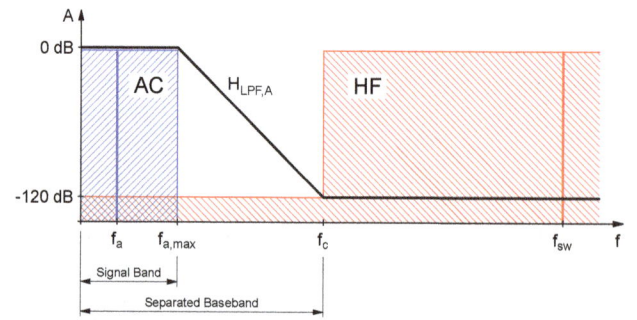

Figure 3. Specified spectral characteristics of the PWM signal.

All rising edges are equidistant in time, while the falling edges are modulated. The modulation is done by controlling the switch-on time of each period. For this purpose the duty cycle $\tau \in (0, 1)$ is used. $\tau \leq 0$ or $\tau \geq 1$ violate the switching condition (overmodulation) and have to be avoided by the ZePoC encoder.

For generating the binary PWM signal a digital pulse shaper is used. The main characteristic of a pulse shaper is its time resolution. In a synchronous digital circuit the change of output signals is only possible in a discrete time grid determined by the clock frequency. Therefore a very high time resolution is necessary to guarentee high accuracy.

The DSP used for this application is an ADSP-21369KBPZ-3A manufactured by Analog Devices. This DSP contains a digital PWM generator supporting a maximum clock frequency of 200 MHz (Analog Devices, 2013). If the binary signal is directly generated by the DSP, the time resolution is only 5 ns.

To increase the performance of the PWM generator inside the DSP a special module containing delay lines and analogue multiplexers was developed (Weber, 2014). This module is able to enhance the time resolution up to approximately 313 ps without increasing any of the clock frequencies.

A better time resolution and jitter performance could be achieved with a FPGA. The Altera Arria V GX 5AGXFB3H4F35C4N contains 24 transceivers supporting clock frequencies up to 6.5536 GHz (Altera, 2015). A serializer is part of each transceiver and is able to generate a binary signal with a time resolution up to approximately 153 ps.

2.1 Spectral characteristics of the PWM signal

The noise floor inside the separated baseband has to be lower than -120 dB. For this a minimum amount of $10^6 = 1\,000\,000$ possible positions for the falling edge between two enframing rising edges are necessary. In digital systems it is often useful to deal with powers of two. $2^{20} = 1\,048\,576$ is chosen which leaves a margin of 4.8 % to define an adequate minimum pulse width. With a serializer running at 6.5536 GHz the switching frequency of the PWM signal is

$$f_{sw} = \frac{6.5536\,\text{GHz}}{2^{20}} = 6.25\,\text{kHz}. \tag{1}$$

Figure 3 shows the spectral characteristics of the binary PWM signal. The frequency of the carrier signal $f_c = \frac{1}{2} f_{sw}$ defines the separated baseband $0 \leq f < f_c$. One sinusoidal AC signal with frequency $f_a = \frac{1}{c} f_c$ and amplitude a is inside the signal band $0 < f < f_{a,max}$. The upper limit $f_{a,max}$ depends on the frequency response $H_{LPF,A}$ of the analogue LPF and c is the frequency factor between f_a and f_c.

3 ZePoC encoder

The block diagram of the ZePoC encoder is displayed in Fig. 4. Real and imaginary part of complex signals are processed separately and indexed with "R" and "I".

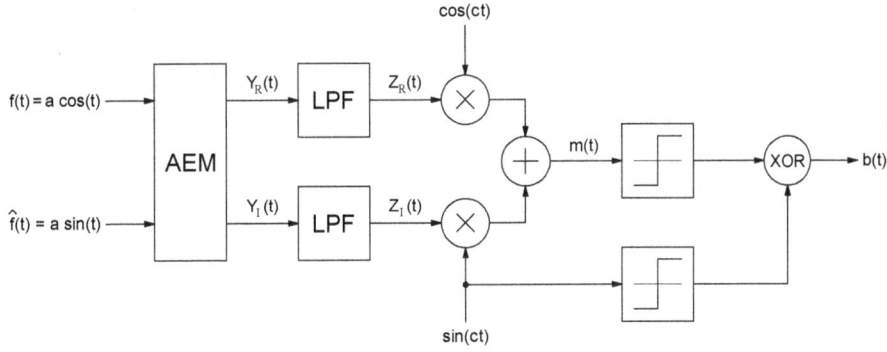

Figure 4. Block diagram of the ZePoC encoder.

Only the open-loop structure allows to find an analytical formulation for the modulated signal $m(t)$. In the following subsections the ZePoC encoder and its analytical formulation are described in detail.

3.1 Input signal

The input signal must be an analytical sinusoidal signal with an adjustable amplitude a. To avoid overmodulation the factor a is limited to γ which defines also the minimum pulse width of the binary output signal $b(t)$. To simplify all equations the time variable t will be defined as $t := \omega T$. The time function

$$f(a,t) = a\cos(t), \quad 0 \leq a \leq \gamma < 1, \quad a, \gamma \in \mathbb{R}, \tag{2}$$

and its Hilbert transform

$$\hat{f}(a,t) = \mathcal{H}\{f(a,t)\} = a\sin(t) \tag{3}$$

results in the analytical input signal

$$F(a,t) = f(a,t) + j\,\hat{f}(a,t). \tag{4}$$

3.2 Analytical exponential modulation (AEM)

The AEM is defined as

$$X(a,t) = \underbrace{\text{Re}\{X(a,t)\}}_{=:X_R(a,t)} + j\underbrace{\text{Im}\{X(a,t)\}}_{=:X_I(a,t)} = e^{-jF(a,t)} \tag{5}$$

which could be separated into its real and imaginary part

$$\Leftrightarrow \begin{cases} X_R(a,t) = e^{\hat{f}(a,t)}\cos\left(f(a,t)\right) \\ X_I(a,t) = -e^{\hat{f}(a,t)}\sin\left(f(a,t)\right). \end{cases} \tag{6}$$

The cosine and the sine term describe a vector in the complex plane. The complex signal $X(a,t)$ is limited to 1 in polar notation. $X(a,t)$ is not depicted in Fig. 4 because it is only the definition of the AEM.

To expand all possible positions of the vector described in Eq. (6) to a half circle in the complex plane the arguments have to be multiplied with $\frac{\pi}{2}$. Because of the complex input signal $F(a,t)$ the argument of the exponential term has to be multiplied with the same factor. Finally the constant factor $e^{-\frac{\pi}{2}}$ must be appended to limit the output signals of the AEM $Y_R(a,t)$ and $Y_I(a,t)$ to the closed interval $[-1,1]$ with

$$\Rightarrow \begin{cases} Y_R(a,t) = e^{-\frac{\pi}{2}} \cdot e^{\frac{\pi}{2}\hat{f}(a,t)} \cdot \cos\left(\frac{\pi}{2}f(a,t)\right) \\ Y_I(a,t) = -e^{-\frac{\pi}{2}} \cdot e^{\frac{\pi}{2}\hat{f}(a,t)} \cdot \sin\left(\frac{\pi}{2}f(a,t)\right). \end{cases} \tag{7}$$

In the next part of the ZePoC encoder the signals $Y_R(a,t)$ and $Y_I(a,t)$ have to pass a low pass filter (LPF). For an analytical formulation of the LPF it is essential to know the frequency component of these signals. Therefore the sine, cosine and the exponential function will be substituted by Taylor polynomials

$$\Rightarrow \begin{cases} Y_R(a,t) = e^{-\frac{\pi}{2}} \cdot \text{EXP}_N(y) \cdot \text{COS}_N(x) \\ Y_I(a,t) = -e^{-\frac{\pi}{2}} \cdot \text{EXP}_N(y) \cdot \text{SIN}_N(x). \end{cases} \tag{8}$$

3.3 Approximation of the AEM by Taylor polynomials

The exponential function e^y and the trigonometric functions sine and cosine are approximated by Taylor polynomials where N determines the degree. The Taylor expansions are performed at the point 0 because of the symmetrical range of the arguments around this point.

Beginning with the definition of the Taylor polynomial $\text{EXP}_N(y)$ for e^y the variable y is set to the argument of the exponential function which has to be approximated. This gives a sum of $\sin^n(t)$ terms with constant coefficients. Expanding these terms results in a sum of cosine terms with frequency factors $n = 0, 1, \ldots, N$ and constant coefficients $A_{N,n,s}$. The index N is the degree of the Taylor polynomial, n is the frequency factor and $s = 0$ means there is no alternating sign in the sum. The Taylor polynomial for $y = \frac{\pi}{2}\hat{f}(a,t)$ is defined as

$$\text{EXP}_N\left(y = \frac{\pi}{2}\hat{f}(a,t)\right) = \sum_{n=0}^{N} \frac{y^n}{n!}$$

$$= \sum_{n=0}^{N} \frac{(\pi a)^n \sin^n(t)}{2^n\,n!} \tag{9}$$

$$= \sum_{n=0}^{N} A_{N,n,0}(a) \cos\left(n(t - \frac{\pi}{2})\right), \ N \in \mathbb{N}_0.$$

Similar to Eq. (9) the sine and cosine function are approximated by Taylor polynomials. The index $s = 1$ of the amplitude coefficients $A_{N,k,s}(a)$ indicates that there are alternating signs $(-1)^n$ in the sums COS_N and SIN_N with

$$\mathrm{COS}_N\left(x = \frac{\pi}{2}f(a,t)\right) = \sum_{n=0}^{\lfloor \frac{N}{2} \rfloor} (-1)^n \frac{x^{2n}}{(2n)!}$$

$$= \sum_{n=0}^{\lfloor \frac{N}{2} \rfloor} (-1)^n \frac{(\pi a)^{2n} \cos^{2n}(t)}{2^{2n}(2n)!} \quad (10)$$

$$= \sum_{n=0}^{\lfloor \frac{N}{2} \rfloor} (-1)^n A_{N,2n,1}(a) \cos(2nt), \ N \in \mathbb{N}_0$$

and

$$\mathrm{SIN}_N\left(x = \frac{\pi}{2}f(a,t)\right) = \sum_{n=0}^{\lfloor \frac{N-1}{2} \rfloor} (-1)^n \frac{x^{2n+1}}{(2n+1)!}$$

$$= \sum_{n=0}^{\lfloor \frac{N-1}{2} \rfloor} (-1)^n \frac{(\pi a)^{2n+1} \cos^{2n+1}(t)}{2^{2n+1}(2n+1)!} \quad (11)$$

$$= \sum_{n=0}^{\lfloor \frac{N-1}{2} \rfloor} (-1)^n A_{N,2n+1,1}(a) \cos((2n+1)t), \ N \in \mathbb{N}.$$

COS_N contains only even powers of the argument x $\left(0, 2, \ldots, 2 \cdot \lfloor \frac{N}{2} \rfloor\right)$ while SIN_N contains only odd powers $\left(1, 3, \ldots, 2 \cdot \lfloor \frac{N-1}{2} \rfloor\right)$. The floor functions ensure that every degree $N \in \mathbb{N}$ can be used in Eqs. (10) and (11).

The constant amplitude coefficients $A_{N,k,s}(a)$ can be written in a compact sigma notation. Three different cases have to be considered: $k = 0$, k is even and k is odd.

1st case: $k = 0$, $N \in \mathbb{N}$, $s \in \{0, 1\}$: $\quad$ (12)

$$A_{N,k,s}(a) = \sum_{n=0}^{\lfloor \frac{N}{2} \rfloor} (-1)^{s \cdot n} \binom{2n}{n} \frac{(\pi \cdot a)^{2n}}{2^{4n}(2n)!}$$

2nd case: $k \in \mathbb{N}$ and even, $k \leq N \in \mathbb{N}$, $s \in \{0, 1\}$: $\quad$ (13)

$$A_{N,k,s}(a) = \sum_{n=\frac{k}{2}}^{\lfloor \frac{N}{2} \rfloor} (-1)^{s(n-\frac{k}{2})} \binom{2n}{n - \frac{k}{2}} \frac{(\pi \cdot a)^{2n}}{2^{4n-1}(2n)!}$$

3rd case: $k \in \mathbb{N}$ and odd, $k \leq N \in \mathbb{N}$, $s \in \{0, 1\}$: $\quad$ (14)

$$A_{N,k,s}(a) = \sum_{n=\frac{k-1}{2}}^{\lceil \frac{N}{2} \rceil - 1} (-1)^{s(n-\frac{k-1}{2})} \binom{2n+1}{n - \frac{k-1}{2}} \frac{(\pi \cdot a)^{2n+1}}{2^{4n+1}(2n+1)!}$$

Table 1. Accuracy of the approximated AEM

Degree N	$|\mathrm{Error}|_{\max}$	Accuracy
10	1.23×10^{-5}	98 dB
11	1.60×10^{-6}	115 dB
12	1.91×10^{-7}	134 dB
13	2.13×10^{-8}	153 dB
14	2.22×10^{-9}	173 dB

Next the products consisting of the sums $\mathrm{EXP}_N \cdot \mathrm{COS}_N$ and $\mathrm{EXP}_N \cdot \mathrm{SIN}_N$ have to be expanded. After sorting all terms by frequency and summarizing all constants, $Y_\mathrm{R}(a,t)$ and $Y_\mathrm{I}(a,t)$ can be written in a compact sigma notation as

$$\Rightarrow \begin{cases} Y_\mathrm{R}(a,t) = e^{-\frac{\pi}{2}} \sum_{n=0}^{2N-1} M_{N,n}(a) \cos\left(n(t - \frac{\pi}{2})\right) \\ Y_\mathrm{I}(a,t) = e^{-\frac{\pi}{2}} \sum_{n=1}^{2N-1} M_{N,n}(a) \sin\left(n(t - \frac{\pi}{2})\right). \end{cases} \quad (15)$$

The constant coefficients $M_{N,k}(a)$ are given by a double sum over the products of two amplitude coefficients $A_{N,k,s}(a)$ and defined as

$$M_{N,k}(a) = \frac{1}{2} \sum_{\substack{m=0 \\ m+2n=k}}^{N} \sum_{n=0}^{\lfloor \frac{N}{2} \rfloor} A_{N,m,0}(a) \cdot A_{N,2n,1}(a) \quad (16)$$

$$+ \frac{1}{2} \sum_{\substack{m=0 \\ |m-2n|=k}}^{N} \sum_{n=0}^{\lfloor \frac{N}{2} \rfloor} A_{N,m,0}(a) \cdot A_{N,2n,1}(a).$$

All $M_{N,k}(a)$ have to be recalculated if the amplitude a of the input signal is modified.

Table 1 shows the absolute error and the accuracy of the AEM approximated by Taylor polynomials of degree N. It can be seen that for the specified accuracy of 120 dB a degree of $N = 12$ is sufficient.

3.4 Low pass filter (LPF)

In consequence of nonlinear mathematical operations inside the AEM the signals $Y_\mathrm{R}(a,t)$ and $Y_\mathrm{I}(a,t)$ are no longer band-limited. All signal components with frequencies above $f_c = c \cdot f_a$ have to be suppressed by the low pass filters (LPFs). To apply the LPFs only the upper bound of summation $2N - 1$ has to be replaced by $\lfloor c \rfloor$. No frequency components with a frequency factor greater than c can pass the LPFs. The output signals of the LPFs are

$$\Rightarrow \begin{cases} Z_\mathrm{R}(c,a,t) = e^{-\frac{\pi}{2}} \sum_{n=0}^{\lfloor c \rfloor} M_{N,n}(a) \cos\left(n(t - \frac{\pi}{2})\right) \\ Z_\mathrm{I}(c,a,t) = e^{-\frac{\pi}{2}} \sum_{n=1}^{\lfloor c \rfloor} M_{N,n}(a) \sin\left(n(t - \frac{\pi}{2})\right). \end{cases} \quad (17)$$

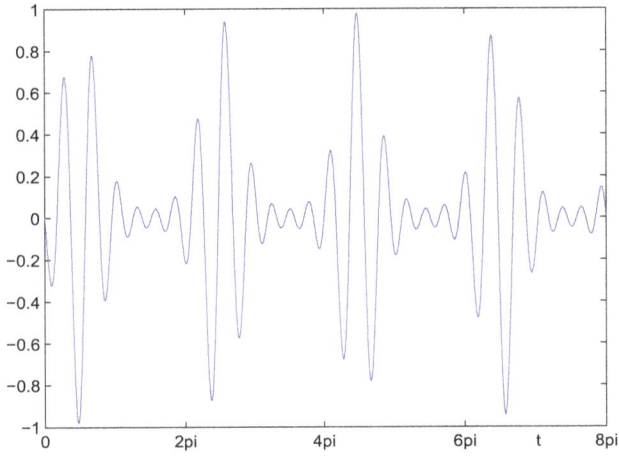

Figure 5. One periode of $m(c, a, t)$ for $a = 0.99$ and $c = 6.25$.

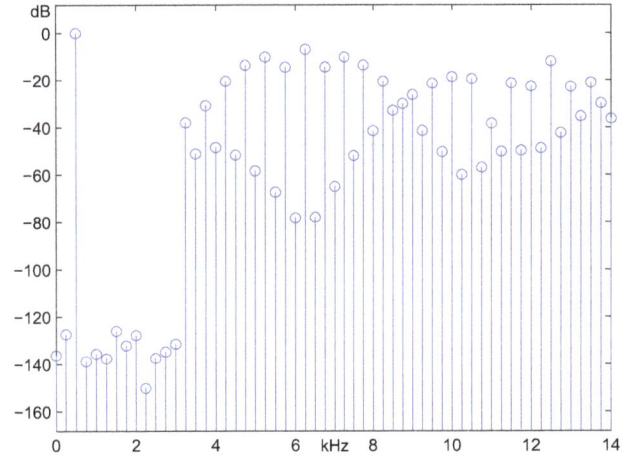

Figure 6. ZePoC spectrum for $a = 0.99$, $f_a = 500\,\text{Hz}$, $c = 6.25$, $f_c = 3.125\,\text{kHz}$, $f_{\text{sw}} = 6.25\,\text{kHz}$, serializer@6.5536 GHz.

If the LPFs are disabled, ZePoC encoding becomes equal to a natural pulse width modulation (NPWM).

3.5 Single-sideband phase modulation

The time function $m(c, a, t)$ is built by mixing the complex signal $Z(c, a, t) = Z_R(c, a, t) + j Z_I(c, a, t)$ with the complex carrier signal $C = \cos(ct) + j \sin(ct)$ as follows:

$$m(c, a, t) = Z_R(c, a, t) \cdot \cos(ct) + Z_I(c, a, t) \cdot \sin(ct). \quad (18)$$

For $a = 0.99$ and $c = 6.25$ this function is periodic with period $t = 8\pi$. Figure 5 shows a plot of $m(c, a, t)$ over one full period.

3.6 Generating the bit stream

To generate the bit stream for the serializer each bit of the stream must be set to a defined value. Therefore the function $m(c, a, t)$ must be evaluated at the center of each bit. Depending on this result (greater or less than zero) these bits are set. In Fig. 4 this operation is depicted as sign function.

For generating the bit stream in an efficient way, the zeros of $m(c, a, t)$ must be found. It is not necessary to find the exact positions of the zeros because of the bit stream's discrete time domain.

To find the zeros with little effort, a binary search algorithm is used. Figure 8 illustrates the concept of this algorithm for $n = 3$. The search area is exactly between two zeros of the carrier signal $\sin(ct)$, enframing one zero of $m(c, a, t)$.

The search starts in the middle of the search area. At the current position, the sign of $m(c, a, t)$ will be evaluated. Depending on the sign the direction of the jump is chosen to be left or right. After each jump the jump distance will be halved. For a search area of 2^n bits n jumps have to be executed to find the bit Z where the zero crossing of $m(c, a, t)$ is.

All bits in the search area left of Z are set to logic one and all bits right of Z are set to logic zero. To find out if the bit Z

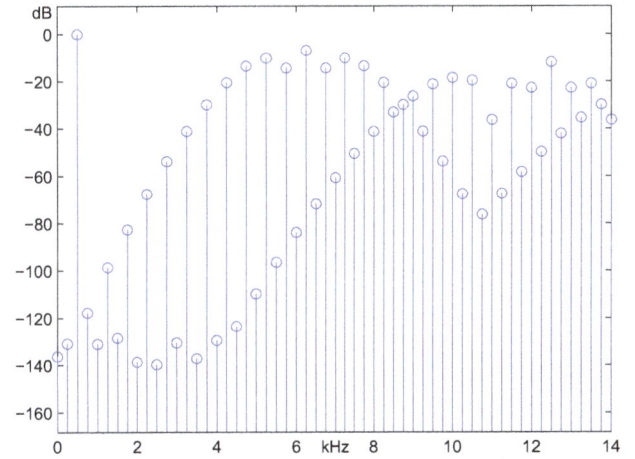

Figure 7. NPWM spectrum for $a = 0.99$, $f_a = 500\,\text{Hz}$, $c = 6.25$, $f_c = 3.125\,\text{kHz}$, $f_{\text{sw}} = 6.25\,\text{kHz}$, serializer@6.5536 GHz.

is a logic one or zero, an additional jump must be executed. If this last jump was to the right, bit Z is set to logic one, else Z is set to zero. The additional jump is similar to a rounding function and helps to reduce the DC component of the binary signal $b(t)$.

3.7 Simulation results

Figure 6 displays the spectrum for $a = 0.99$ and $c = 6.25$ of the binary signal $b(t)$ generated by the ZePoC encoder. The disturbances within the separated baseband are below $-120\,\text{dB}$. The spectral gap between the AC signal and the disturbances allows the use of an analogue LPF to suppress the switching noise at the output of the voltage channel.

If the LPFs inside the ZePoC encoder are disabled, ZePoC encoding becomes equal to NPWM. Figure 7 shows the NPWM spectrum with the same parameters a and c. Here it is not possible to separate the AC signal and the disturbances

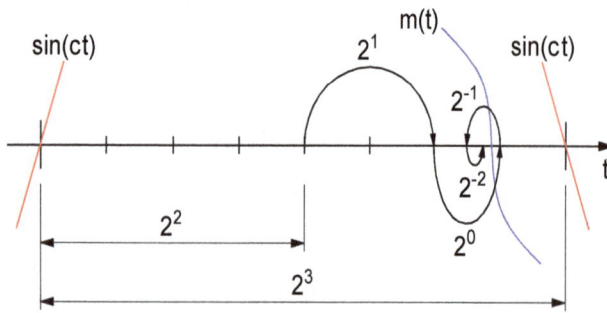

Figure 8. Concept: binary search for zeros of $m(c, a, t)$ for $n = 3$.

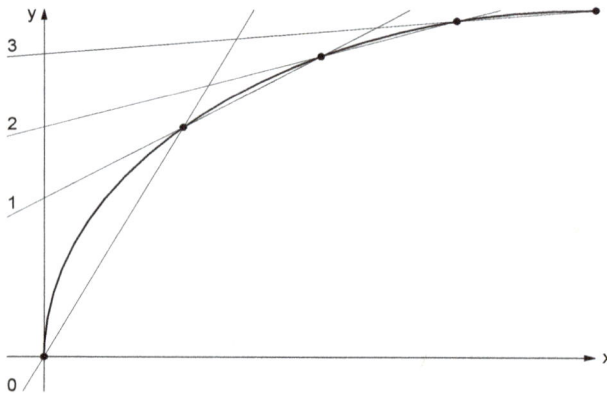

Figure 9. Linear interpolation of a discretized function.

with an analogue LPF. For all simulations the time base is assumed to be ideal.

4 Real-time implementation

During simulation a frequency factor $c \in \mathbb{Q}$ is used which results in a periodic binary signal $b(t)$. If $b(t)$ is periodical then one period of $b(t)$ can be computed and stored in memory. No real-time processing is necessary in this case.

For an arbitrary frequency of the AC signal $c \in \mathbb{R}$ is required. Now $b(t)$ is no longer periodical and a real-time implementation of the ZePoC encoder is required.

4.1 Interpolation of sine and cosine functions

For the numerical evaluation of sine and cosine functions often the CORDIC algorithm is used. CORDIC stands for COordinate Rotation DIgital Computer and is an iterative algorithm (Gupta, 2010). Iterative algorithms often need too much computing time and are not suitable for a real-time implementation of this ZePoC encoder on a DSP.

A very fast way to evaluate functions is to use a look-up table (LUT) combined with linear interpolation. The argument of the function will be equidistant discretized. For each discrete argument the exact function value will be evaluated. Every two function values next to each other define a straight line as shown in Fig. 9. The slope and offset of each line are stored in a table.

To compute a function value the correct line must be located depending on the argument. Then the slope and offset of the selected straight line are read from the table. The resulting linear equation returns the interpolated function value.

For an accuracy of 120 dB the absolute error must be less than 10^{-6}. Linear interpolation of $\sin(x)$ and $\cos(x)$ with $x \in [-\pi, \pi]$ using tables with $2^{13} = 8192$ entries has a maximum error of 0.3×10^{-6}. Each table contains 4096 slope- and 4096 offset-coefficients. Both tables fit in one internal 0.75 Mbit SRAM block of the DSP.

The size of the tables can be reduced using the symmetries $\sin(-x) = -\sin(x)$ and $\sin\left(\frac{\pi}{2} + x\right) = \sin\left(\frac{\pi}{2} - x\right)$. Also the cosine function can be expressed as a $\frac{\pi}{2}$ phase shifted sine function. If a reduced table is used a case analysis is necessary which needs valuable computing time in a real-time system.

4.2 Effects of different floating point formats

Floating point (FP) numbers are stored in a special binary format. Any FP number consists of a sign S, an exponent E and a mantissa M. For normalized signals in the closed interval $[-1, 1]$ the length of the mantissa M is the most important factor for accuracy (Muller, 2009).

Table 2 lists the properties of the used FP formats single, extended single and double. Always one bit is used for the sign whereas the number of bits representing exponent and mantissa varies. Only single and double precision numbers are standardized by IEEE 754 and natively supported by MATLAB®. All simulations in the last section were done using 64 bit double precision.

Double precision numbers are not natively supported by the DSP. To use these 64 bit numbers on the DSP software emulation is required. This software emulation needs a lot of computing power and slows down the ZePoC algorithm dramatically. Therefore it is very important to use only FP formats natively supported by the DSP. The computation time for algorithms using 40 bit instead of 32 bit FP numbers is the same.

For generating an AC signal with a frequency $f_a = 50$ Hz at a switching frequency of $f_{sw} = 2f_c = 2cf_a = 6.25$ kHz a factor $c = 62.5$ is required. An amplitude of $a = 0.99$ avoids overmodulation. The binary PWM signal $b(t)$ is computed by the real-time ZePoC encoder implemented on the DSP. All three FP formats are used to generate $b(t)$ with different accuracies. Figures 10, 11 and 12 show the resulting spectra of the PWM signal for the different FP number formats.

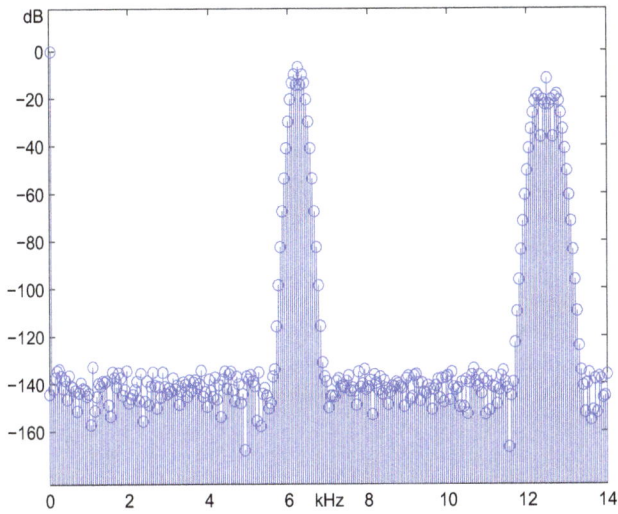

Figure 10. Spectrum of PWM signal for double precision $a = 0.99$, $f_a = 50\,\text{Hz}$, $c = 6.25$, $f_{\text{sw}} = 6.25\,\text{kHz}$, serializer@6.5536 GHz

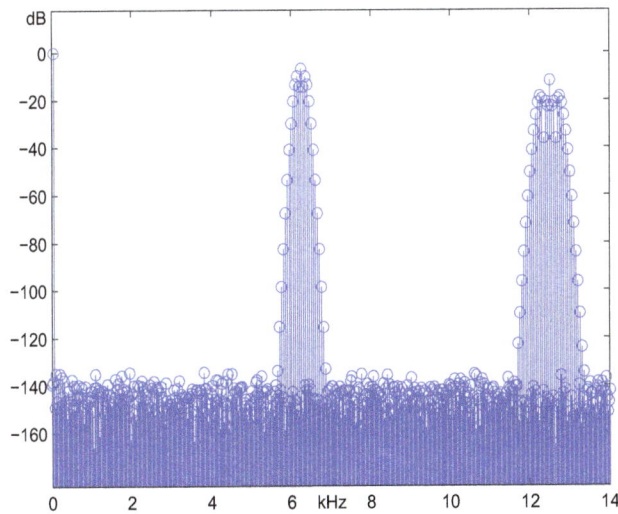

Figure 12. Spectrum of PWM signal for single precision $a = 0.99$, $f_a = 50\,\text{Hz}$, $c = 6.25$, $f_{\text{sw}} = 6.25\,\text{kHz}$, serializer@6.5536 GHz

Table 2. Floating point formats.

	Single 32 bit	Ext. Single 40 bit	Double 64 bit
Sign S	1	1	1
Exponent E	8	8	11
Mantissa M	23	31	52
Standardized (IEEE 754)	X		X
Supported by MATLAB®	X		X
Supported by DSP	X	X	

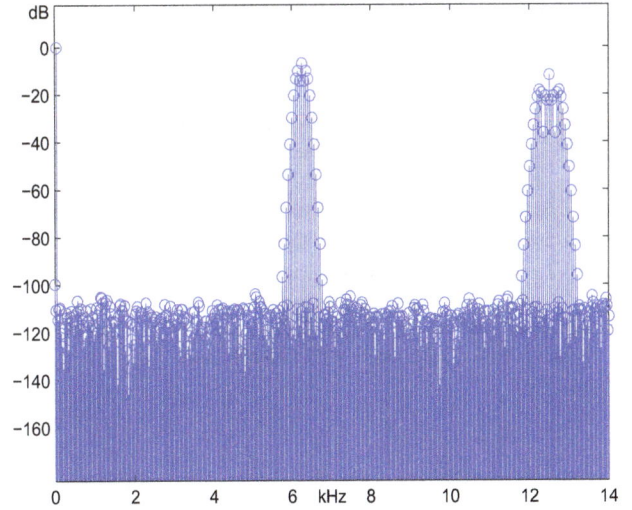

Figure 11. Spectrum of PWM signal for extended single precision $a = 0.99$, $f_a = 50\,\text{Hz}$, $c = 6.25$, $f_{\text{sw}} = 6.25\,\text{kHz}$, @6.5536 GHz

5 Conclusions

This contribution shows that an analytical formulation of a ZePoC encoder is possible for sinusoidal input signals. The accuracy can be determined by the degree of the Taylor polynomials used to approximate the AEM.

A new pulse shaper with a very high time resolution based on a FPGA will be used for the voltage channel of the AC power standard. Therefore an easy and fast algorithm to generate a bit stream for the GHz serializer inside the FPGA is presented.

For the real-time implementation on a DSP the effects of different floating point number formats are analyzed. The accuracy of extended single precision floating point numbers is sufficient for the specified AC power standard.

The main difference between the spectra is the noise floor. For double precision every bin is below $-130\,\text{dB}$ and many bins are not visible because they are below the plot range of $-180\,\text{dB}$. When using extended single precision the noise floor contains many more visible bins but also here no bin exceeds $-130\,\text{dB}$. Only for single precision numbers the noise floor reaches $-100\,\text{dB}$ which is outside the specification of $-120\,\text{dB}$.

References

Altera: Arria V Device Overview, 23.01.2015, available at: www.altera.com (last access: 20 February 2015), 2015.

Analog Devices: Data Sheet SHARC Processor ADSP-21367/ADSP-21368/ADSP-21369, Rev. F, available at: www.analog.com (last access: 20 February 2015), 2013.

Gupta, A.: CORDIC Implementation of Sine-Cosine Functions. Algorithm and Implemantation Floating-Point Format, LAP LAMBERT Academic Publishing, Saarbrücken, Germany, 72 pp., 2010.

Muller, J.-M., Brisebarre, N., and de Dinechin, F.: Handbook of Floating-Point Arithmetic, Birkhäuser, Basel, Switzerland, 572 pp., 2009.

Streitenberger, M.: Zur Theorie digitaler Klasse-D Audioleistungsverstärker und deren Implementierung, VDE-Verlag, Berlin, Germany, 228 pp., 2005.

Texas Instruments: A New Audio File Format for Low-Cost, High Fidelity, Portable Digital Audio Amplifiers, White Paper, Texas Instruments, Dallas, TX, USA, 2005.

Weber, M., Vennemann, T., and Mathis, W.: Increasing the time resolution of a pulse width modulator in a class D power amplifier by using delay lines, Adv. Radio Sci., 12, 91–94, doi:10.5194/ars-12-91-2014, 2014.

Wellmann, J., Kahmann, M., and Mathis, W.: TET-Watt – An AC Power Standard based on Class-D Topologies using ZePoC-Coding, Conference on Precision Electromagnetic Measurements, Daejeon, Korea, 13–18 June, 2010.

Application of transmission-line super theory to classical transmission lines with risers

R. Rambousky[1]**, J. Nitsch**[2]**, and S. Tkachenko**[2]

[1]Bundeswehr Research Institute for Protective Technologies and NBC Protection (WIS), Munster, Germany
[2]Otto-von-Guericke University Magdeburg, Magdeburg, Germany

Correspondence to: R. Rambousky (ronald.rambousky@ieee.org)

Abstract. By applying the Transmission-Line Super Theory (TLST) to a practical transmission-line configuration (two risers and a horizontal part of the line parallel to the ground plane) it is elaborated under which physical and geometrical conditions the horizontal part of the transmission-line can be represented by a classical telegrapher equation with a sufficiently accurate description of the physical properties of the line. The risers together with the part of the horizontal line close to them are treated as separate lines using the TLST. Novel frequency and local dependent reflection coefficients are introduced to take into account the action of the bends and their radiation. They can be derived from the matrizant elements of the TLST solution. It is shown that the solution of the resulting network and the TLST solution of the entire line agree for certain line configurations. The physical and geometrical parameters for these corresponding configurations are determined in this paper.

1 Introduction

Transmission-Line Super Theory (TLST) was introduced by Haase and Nitsch (2001, 2003) more than one decade ago. In this theory Maxwell's equations are represented for a system of lossless nonuniform thin transmission lines in a system of equations which have the same structure as the telegrapher equations. In particular, the TLST equations take into account all field modes and physical effects that might occur, including radiation losses. It surpasses the classical transmission-line theory, which is a special case. Their complex parameters are local and frequency dependent and are obtained by the solution of integral equations.

In the paper a practical classical transmission-line (cTL) is regarded. The considered TL consists of a finite part parallel to the ground plane and two vertical risers connecting the horizontal part to the conducting ground plane at the ends.

In Sect. 2 the classical analysis of the TL is briefly described and the classical reflection coefficients are introduced. In Sect. 3 the fundamentals of TLST are presented and the finite transmission line with risers is analyzed using the numerical TLST procedure. The local and frequency dependent parameter matrix elements representing the per unit length inductance and capacitance values are discussed. A procedure is shown where the whole TL can be separated in uniform and nonuniform parts. Only for the nonuniform parts TLST has to be used and the asymptotic part can be handled classically. In Sect. 4 novel local and frequency dependent reflection coefficients are introduced and it is shown how they are related to the matrizant elements of the TLST. Finally it is shown how the current on the TL can be calculated using the reflection coefficients. Numerical results for the finite uniform TL with risers are shown and discussed in Sect. 5. Calculated results for the novel reflection coefficients and current values on the TL are shown for several geometrical constitutions of the TL. Finally the results are discussed and summarized in Sect. 6.

2 A finite TL in classical transmission-line theory

Classical transmission-line theory (cTLT) does not take care of effects of finite open ends or of risers to the ground plane. The TL is regarded physically infinite and mathematically a total length is designated to meet the right resonance frequencies. The dominating TEM mode of such a TL is commonly

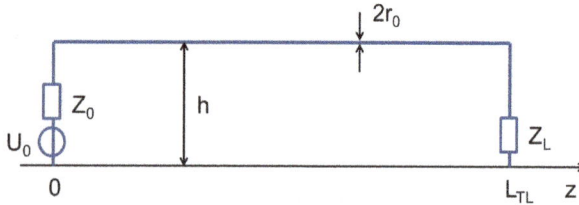

Figure 1. Geometry of a horizontal finite TL (length L_{TL}) with height h over PEC ground including two risers.

described by the classical transmission-line equations

$$\frac{dU(z)}{dz} + j\omega L'_{cTL} I(z) = 0$$

$$\frac{dI(z)}{dz} + j\omega C'_{cTL} U(z) = 0, \tag{1}$$

where z is the axial orientation of the TL (see Fig. 1) and $U(z)$ and $I(z)$ are the complex voltage and current distributions on the line. The classical (and constant) per unit length (p.u.l) inductance and capacitance are named as L'_{cTL} and C'_{cTL}, respectively.

The finite classical TL with risers over a conducting ground plane (PEC) regarded in this work is shown in Fig. 1. To simplify the presentation the following parameters for the TL are chosen, although the outlined method works in general: wire radius $r_0 = 0.5$ mm; height over ground $h = 5$ cm; length of the horizontal part $L_{TL} = 2$ m; total arc length of the TL $L = L_{TL} + 2h = 2.1$ m. At the beginning the line is fed by a lumped source with voltage $U_0 = 1$ V and source impedance $Z_0 = 50\,\Omega$. The line is terminated with a load impedance $Z_L = 50\,\Omega$. The formulas for the classical p.u.l. inductance L'_{cTL} and capacitance C'_{cTL} are

$$L'_{cTL} = \frac{\mu_0}{2\pi} \ln\left(\frac{2h}{r_0}\right) = 1.06 \times 10^{-6} \frac{\text{V s}}{\text{A m}} \tag{2}$$

$$C'_{cTL} = \frac{2\pi\epsilon_0}{\ln\left(\frac{2h}{r_0}\right)} = 1.05 \times 10^{-11} \frac{\text{A s}}{\text{V m}}, \tag{3}$$

resulting in a characteristic line impedance of $Z_C = \sqrt{L'_{cTL} C'^{-1}_{cTL}} = 318\,\Omega$.

A current wave originating at $+\infty$, traveling on the horizontal part of the TL in $-z$ direction and being reflected at the beginning of the TL at $z = 0$ can be expressed using the classical left-hand current reflection coefficient R^{class}_+ as

$$I(z) = I_1\left(e^{jkz} + R^{class}_+ e^{-jkz}\right). \tag{4}$$

with I_1 being an appropriate constant. From Eq. (1) the expression for the voltage $U(z) = -\frac{1}{j\omega C'}\frac{dI(z)}{dz}$ can be deduced and together with Eq. (4) the result for the left-hand current reflection coefficient is calculated as

$$R^{class}_+ = e^{2jkz}\frac{Z_C I(z) + U(z)}{Z_C I(z) - U(z)} = \frac{Z_C - Z_0}{Z_C + Z_0}. \tag{5}$$

For the final step in Eq. (5) the general classical solutions for $U(z)$ and $I(z)$ for a uniform TL

$$\begin{pmatrix} U(z) \\ I(z) \end{pmatrix} = \begin{pmatrix} \cos(kz) & -jZ_C\sin(kz) \\ -\frac{j}{Z_C}\sin(kz) & \cos(kz) \end{pmatrix} \cdot \begin{pmatrix} U(0) \\ I(0) \end{pmatrix}, \tag{6}$$

was used. It is obvious that in cTLT the reflection coefficients are constant.

Corresponding considerations using a current wave traveling on the horizontal part of the line originating at $-\infty$ lead to the (constant) classical right-hand current reflection coefficient R^{class}_- as

$$R^{class}_- = e^{-2jk(z-L_{TL})}\frac{Z_C I(z) - U(z)}{Z_C I(z) + U(z)} = \frac{Z_C - Z_L}{Z_C + Z_L}. \tag{7}$$

3 TLST analysis of a finite TL with risers

3.1 Fundamentals of TLST

Transmission-line super theory (Haase and Nitsch, 2001; Haase et al., 2003; Haase, 2005; Nitsch et al., 2009; Nitsch and Tkachenko, 2010) is a full wave description of Maxwell's equations cast into the form of telegrapher's equations. For a single wire system (with return conductor or ground plane) the super theory transmission-line equation for lumped sources or loads at the line ends in the potential-current representation states (Rambousky et al., 2012)

$$\frac{\partial}{\partial l}\begin{bmatrix} \varphi(l,f) \\ i(l,f) \end{bmatrix} + j\omega\overline{\mathbf{P}}^{*(1)}(l,f)\begin{bmatrix} \varphi(l,f) \\ i(l,f) \end{bmatrix} = \begin{bmatrix} 0 \\ 0 \end{bmatrix}. \tag{8}$$

The potential on the transmission-line is denoted by $\varphi(l,f)$ and the current by $i(l,f)$. The best choice for the line parameter is the (natural) arc length l of the line and f is the frequency. The super matrix $\overline{\mathbf{P}}^{*(1)}$ is the transmission-line parameter matrix. In the case of a one wire system $\overline{\mathbf{P}}^{*(1)}$ is a 2 by 2 matrix. In contrast to cTLT the transmission-line parameter matrix $\overline{\mathbf{P}}^{*(1)}(l,f)$ now is complex valued and both local (l) and frequency (f) dependent. This parameter matrix is calculated by an iteration process starting with a low frequency approximation in the zeroth iteration step resulting in a frequency independent but already local parameter matrix $\overline{\mathbf{P}}^{*(0)}(l)$ (Nitsch et al., 2009; Rambousky et al., 2012). In previous work we could show that already the first iteration step results in an acceptable accuracy (Rambousky et al., 2013a). The general solution of the super theory transmission-line equation (8) for the one wire case can be written as

$$\begin{bmatrix} \varphi(l,f) \\ i(l,f) \end{bmatrix} = \mathcal{M}^l_{l_0}\left\{-j\omega\overline{\mathbf{P}}^{*(1)}\right\}\begin{bmatrix} \varphi(l_0,f) \\ i(l_0,f) \end{bmatrix}, \tag{9}$$

where the expression $\mathcal{M}^l_{l_0}$ is the so called matrizant or product integral (Gantmacher, 1984), and l_0 and $l > l_0$ represent two spatial positions on the TL. Regarding only lumped

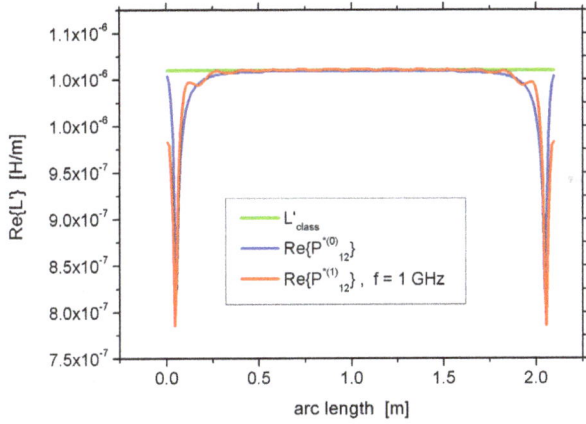

Figure 2. Parameter matrix elements representing the real part of the p.u.l. inductance for TLST analysis of the TL configuration.

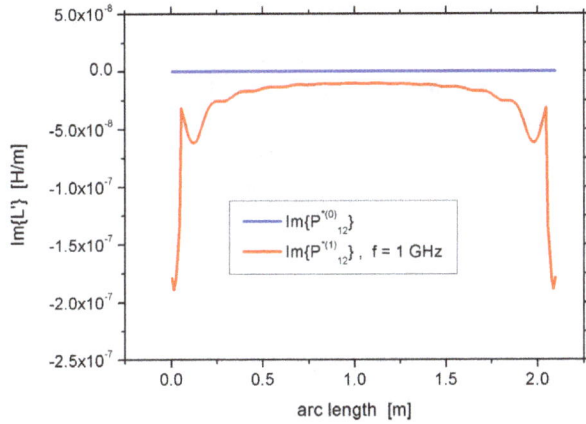

Figure 3. Parameter matrix elements representing the imaginary part of the p.u.l inductance for TLST analysis of the TL configuration.

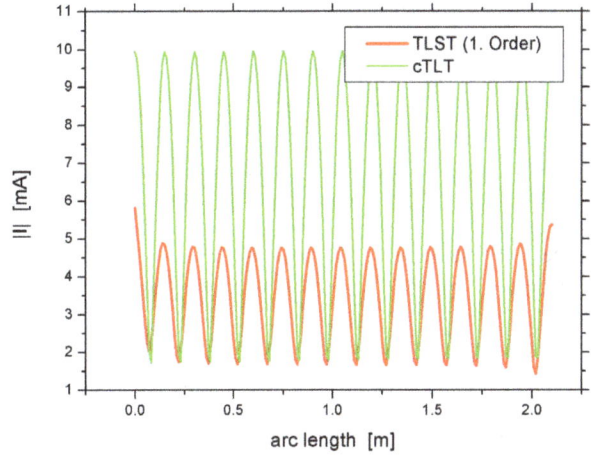

Figure 4. Current on the TL calculated using TLST approach for $f = 1\,\text{GHz}$ compared to classical TL theory.

creases again. Also the imaginary part of L' deviates significantly from the classical value (zero) at the risers, indicating the most radiative parts of the TL.

The current on the TL calculated using TLST with first order iteration parameter matrix is shown in Fig. 4 for a frequency of 1 GHz. To have a comparable arc length, the total length of the classical TL was also set to 2.1 m. It is clearly seen that the real current distribution deviates significantly from the classical theory, mainly because of the radiating losses at the used frequency.

3.3 Decomposition of the TL based on the group property of the matrizant

In the example of Fig. 1 the TL can be decomposed in the left-hand riser part, the uniform middle part (asymptotic region) and the right-hand riser part. Attention has to be paid that the junctions are located where the composed TL shows almost classical behavior (see Fig. 2). Therefore the junctions have to be sufficiently far away from the riser, like at z_1 and z_2 as shown in Fig. 5. Now, the TLST parameter matrices for the single parts of the decomposed TL can be calculated. Because even the tail end of an otherwise classical TL shows significant deviation of the line parameters in TLST, the elements of the parameter matrix have to be adjusted due to the junction. For the used TL with risers the asymptotic region (part II) was defined as a classical TL with constant line parameters using Eqs. (2) and (3). The riser parts I and III were adjusted for the junctions by hand to ignore the tail ends and to meet the classical values at the junctions. This is shown in Fig. 6 with the dashed curves.

For the whole TL current and potential in the load Z_L at the end of the line ($l = L$) can be calculated using the first

sources or lumped loads at the ends of the wires, Eq. (9) can be calculated using the appropriate boundary conditions of the TL model.

3.2 TLST parameter elements for a finite TL with risers

The parameter matrices $\overline{\mathbf{P}}^{*(0)}(l)$ and $\overline{\mathbf{P}}^{*(1)}(l, f)$ resulting from the TLST iteration process are independent of the lumped sources and loads. The $\mathbf{P}^*_{12}$ elements representing the p.u.l. inductance are shown in Fig. 2 (real part) and in Fig. 3 (imaginary part).

Figure 2 indicates that in the TLST the classical p.u.l. inductance L'_{cTL} is reached at a certain distance away from the risers for the configuration presented in Fig. 1. The graph of the inductance bends when approaching the ends of the horizontal part of the line and reaches its minima at the ends of the horizontal parts. A reason for this behavior can be found in Nitsch et al. (2009). When passing through the risers it in-

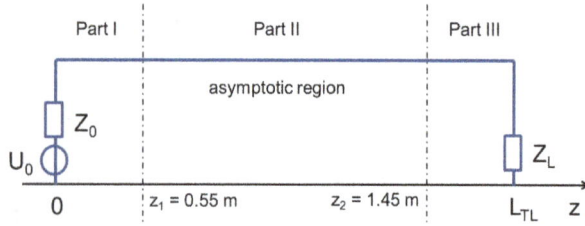

Figure 5. Partitioning of the nonuniform TL of Fig. 1 into an asymptotic region (part II) and the two riser regions (parts I and III).

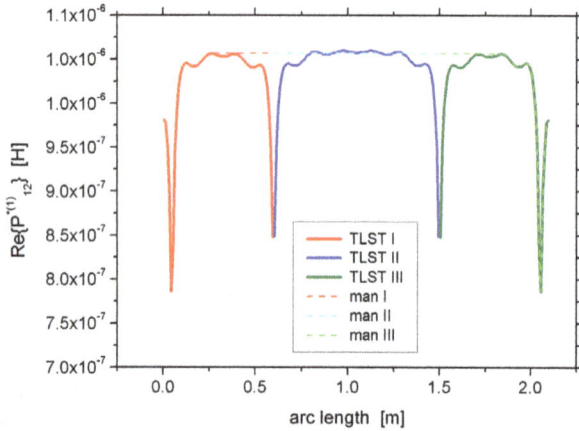

Figure 6. Real part of the parameter matrix element $\overline{P}_{12}^{*(1)}$ for the defined three parts of the nonuniform TL and their manual adjustment at the junctions.

order parameter matrix $\overline{\mathbf{P}}^{*(1)}$

$$\begin{bmatrix} \varphi(L) \\ i(L) \end{bmatrix} = \mathcal{M}_0^L \left\{ -j\omega \overline{\mathbf{P}}^{*(1)} \right\} \begin{bmatrix} \varphi(0) \\ i(0) \end{bmatrix} = \mathbf{M}_L \begin{bmatrix} \varphi(0) \\ i(0) \end{bmatrix}. \quad (10)$$

The matrix $\mathbf{M}_L$ is the matrizant over the whole arc length of the TL using the parameter matrix $\overline{\mathbf{P}}^{*(1)}$ of the whole TL.

On the other hand, current and potential in the load Z_L can be calculated using the matrizants $\mathbf{M}_I^{man}$, $\mathbf{M}_{II}^{man}$ and $\mathbf{M}_{III}^{man}$ of the single parts I, II and III of the TL with the manually (at the junctions) adapted parameter matrices as

$$\begin{bmatrix} \varphi(L) \\ i(L) \end{bmatrix} = \mathbf{M}_{III}^{man} \cdot \mathbf{M}_{II}^{man} \cdot \mathbf{M}_{I}^{man} \begin{bmatrix} \varphi(0) \\ i(0) \end{bmatrix}. \quad (11)$$

For example $\mathbf{M}_I^{man}$ is the matrizant covering the arc length from $l = 0$ to $l = z_1 + h = 0.6\,\mathrm{m}$ using the manually (at the junction) adapted first order parameter matrix $\overline{\mathbf{P}}_{I,man}^{*(1)}$ resulting in

$$\mathbf{M}_I^{man} = \mathcal{M}_0^{z_1+h} \left\{ -j\omega \overline{\mathbf{P}}_{I,man}^{*(1)} \right\}. \quad (12)$$

To validate the equivalence of Eqs. (10) and (11) the current in Z_L at the end of the TL was calculated in both ways. The result is shown in Fig. 7 and gives very good agreement. It

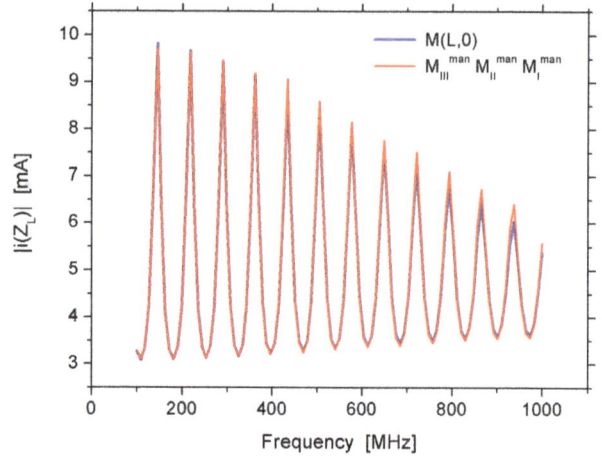

Figure 7. Current in the load Z_L at the end of the TL of Fig. 1.

has to be mentioned again that in the assembled solution the middle part (part II) of the TL was regarded as a pure classical TL. The results so far show that the current distribution on a real TL with risers can be calculated by dividing the line in uniform and nonuniform parts. The nonuniform parts have to be calculated using an advanced TLT, like TLST. The uniform parts can be handled as classical TL. The overall matrizant of the TL can be assembled by multiplying the single matrizants of the TL parts in correct order.

4 Novel local and frequency dependent current reflection coefficients and amplitude functions

The idea now is to transfer the concept of current reflection coefficients from cTLT to a realistic finite transmission-line with risers at both ends.

4.1 Derivation of the novel current reflection coefficients using TLST

In TLST voltage $U(z)$ is replaced by the potential $\varphi(l)$ and current $I(z)$ by $i(l)$. Again l is the natural parameter of the TL (arc length) including the risers. As an extension of Eqs. (5) and (7) the now l and frequency dependent current reflection coefficients can also be defined as the quotient of an incoming and outgoing current wave as

$$\widetilde{R}_+(l) := e^{2jkl} \frac{Z_C i(l) + \varphi(l)}{Z_C i(l) - \varphi(l)} \quad (13)$$

and

$$\widetilde{R}_-(l) := e^{-2jk(l-L)} \frac{Z_C i(l) - \varphi(l)}{Z_C i(l) + \varphi(l)}. \quad (14)$$

The advanced current reflection coefficients $\widetilde{R}_+(l)$ and $\widetilde{R}_-(l)$ now are expressed using the results of TLST calculations. Due to the group feature of the resulting matrizants (see

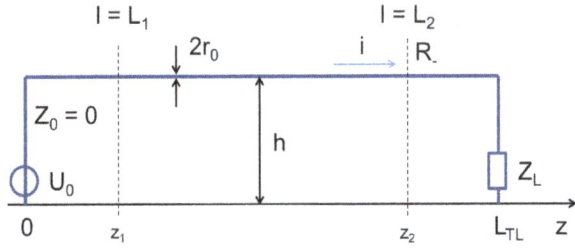

Figure 8. Transmission-line configuration for derivation of the right-hand current reflection coefficient $\widetilde{R}_-(l)$ of the TL with risers.

Sect. 3.3), the matrizant of the whole TL can be composed as a matrix product of matrizants representing parts of the TL. Therefore, actually only the riser parts I and III have to be calculated using TLST and the classical matrizant can be used for the asymptotic region (part II). A theoretical restriction for our approach is that no radiation coupling between the two riser parts of the TL is allowed. This is assured if the horizontal length L_{TL} of the TL is large compared to the height h over ground.

4.1.1 The right-hand reflection coefficient $\widetilde{R}_-(l)$

In Fig. 8 the TL configuration for derivation of the right-hand current reflection coefficient $R_-(l)$ is depicted. Imagine a current wave coming from $-\infty$ travels in positive z direction, gets reflected at the right-hand riser and travels back to $-\infty$. The classical telegrapher's equations are valid in the asymptotic region, defined by $L_1 \leq l \leq L_2$. The matrizant $\mathcal{M}_L^l\left\{-j\omega\overline{\mathbf{P}}^{*(1)}\right\} \equiv \mathcal{M}(l,L)$ can be decomposed in

$$\mathcal{M}(l,L) = \mathcal{M}(l,L_2) \cdot \mathcal{M}(L_2,L), \tag{15}$$

where the second factor on the right side of Eq. (15) is independent of l. The l dependence is restricted to the asymptotic region. For $L_{TL} \gg h$ the radiation coupling is negligible and $\widetilde{R}_-(l)$ is independent of $\widetilde{R}_+(l)$.

In a next step the quotient in Eq. (14) has to be expressed by matrizants of the TLST calculation. Generally the relation

$$\begin{bmatrix} \varphi(l_2) \\ i(l_2) \end{bmatrix} = \mathcal{M}(l_2,l_1)\begin{bmatrix} \varphi(l_1) \\ i(l_1) \end{bmatrix} \qquad \forall l_1, l_2 \in [0,L] \tag{16}$$

holds. Setting $l_2 = l$ and $l_1 = L$ and using the boundary condition $\varphi(L) = U_L = Z_L i(L)$ the potential-current vector at arc length l can be expressed as

$$\begin{bmatrix} \varphi(l) \\ i(l) \end{bmatrix} = \mathcal{M}(l,L)\begin{bmatrix} \varphi(L) \\ i(L) \end{bmatrix}$$

$$= i(L)\begin{bmatrix} \mathcal{M}_{11}(l,L) & \mathcal{M}_{12}(l,L) \\ \mathcal{M}_{21}(l,L) & \mathcal{M}_{22}(l,L) \end{bmatrix} \cdot \begin{bmatrix} Z_L \\ 1 \end{bmatrix}. \tag{17}$$

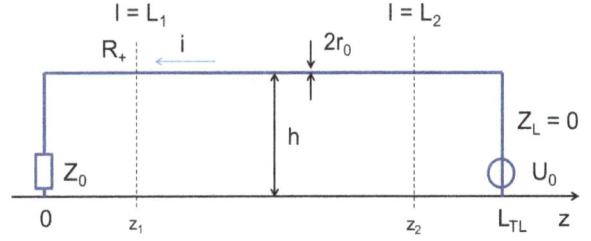

Figure 9. Transmission-line configuration for derivation of the left-hand current reflection coefficient $\widetilde{R}_+(l)$ of the TL with risers.

Inserting the results for $\varphi(l)$ and $i(l)$ in Eq. (14) leads to the advanced expression for the right-hand current reflection coefficient $\widetilde{R}_-(l)$ which is now local and frequency dependent.

$$\widetilde{R}_-(l) = e^{-2jk(l-L)}$$
$$\cdot (Z_L[-\mathcal{M}_{11}(l,L) + Z_C\mathcal{M}_{21}(l,L)]$$
$$+ (-\mathcal{M}_{12}(l,L) + Z_C\mathcal{M}_{22}(l,L)))$$
$$\cdot (Z_L[\mathcal{M}_{11}(l,L) + Z_C\mathcal{M}_{21}(l,L)]$$
$$+ \mathcal{M}_{12}(l,L) + Z_C\mathcal{M}_{22}(l,L))^{-1}. \tag{18}$$

4.1.2 The left-hand reflection coefficient $\widetilde{R}_+(l)$

The TL configuration for the derivation of $\widetilde{R}_+(l)$ is depicted in Fig. 9. It is assumed that a current wave traveling in $-z$ direction gets reflected at the left-hand side of the TL. Using the same concept as before the matrizant can be decomposed in a non l dependent part and an l dependent part (asymptotic region), that is $\mathcal{M}(l,0) = \mathcal{M}(l,L_1) \cdot \mathcal{M}(L_1,0)$. The second one includes the essential physical property of the reflection process.

Using the same derivation method as before (now setting $l_1 = 0$ and $l_2 = l$) one gets the following advanced expression for the left-hand current reflection coefficient $\widetilde{R}_+(l)$ which is again local and frequency dependent.

$$\widetilde{R}_+(l) = e^{2jkl}$$
$$\cdot (-Z_0[\mathcal{M}_{11}(l,0) + Z_C\mathcal{M}_{21}(l,0)]$$
$$+ \mathcal{M}_{12}(l,0) + Z_C\mathcal{M}_{22}(l,0))$$
$$\cdot (-Z_0[-\mathcal{M}_{11}(l,0) + Z_C\mathcal{M}_{21}(l,0)]$$
$$- \mathcal{M}_{12}(l,0) + Z_C\mathcal{M}_{22}(l,0))^{-1}. \tag{19}$$

4.2 Derivation of the amplitude function $\widetilde{C}_+(l)$

The configuration for calculating the amplitude function of a forward ($+z$-direction) traveling current wave is depicted in Fig. 5. It is assumed that the load impedance is ideal ($Z_L = Z_C$) for all used frequencies and the traveling current wave is not reflected at the end of the TL. With $i(l) = \widetilde{C}_+(l)e^{-jkl}$ and $\varphi(l) = Z_C\widetilde{C}_+(l)e^{-jkl}$ the following expression for $\widetilde{C}_+(l)$ can be derived by summation:

$$\widetilde{C}_+(l) = e^{jkl}\frac{i(l)Z_C + \varphi(l)}{2Z_C}. \tag{20}$$

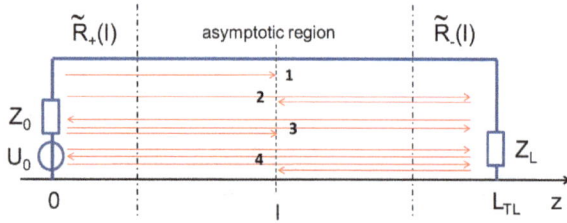

Figure 10. Transmission-line configuration for derivation of the left-hand amplitude function $\widetilde{C}_+(l)$ and the description of multiple reflections.

The quotient in Eq. (20) can be expressed again using the matrizants of the TLST calculation. With the relation

$$\begin{bmatrix} \varphi(l) \\ i(l) \end{bmatrix} = \mathcal{M}(l,0) \begin{bmatrix} U_0 - Z_0 i(0) \\ i(0) \end{bmatrix}, \tag{21}$$

leading to the expressions

$$\varphi(l) = i(l) Z_C \tag{22}$$
$$= \mathcal{M}_{11}(l,0)\,(U_0 - Z_0 i(0)) + \mathcal{M}_{12}(l,0) i(0)$$
$$i(l) = \mathcal{M}_{21}(l,0)\,(U_0 - Z_0 i(0)) + \mathcal{M}_{22}(l,0) i(0), \tag{23}$$

a result for the characteristic impedance Z_C can be received by division:

$$Z_C = \frac{\mathcal{M}_{11}(l,0)\,[U_0 - Z_0 i(0)] + \mathcal{M}_{12}(l,0) i(0)}{\mathcal{M}_{21}(l,0)\,[U_0 - Z_0 i(0)] + \mathcal{M}_{22}(l,0) i(0)}. \tag{24}$$

Solving Eq. (24) for $i(0)$ yields

$$i(0) = U_0 [\mathcal{M}_{21}(l,0) Z_C - \mathcal{M}_{11}(l,0)] [Z_0 Z_C \mathcal{M}_{21}(l,0)$$
$$- \mathcal{M}_{22}(l,0) Z_C - \mathcal{M}_{11}(l,0) Z_0 + \mathcal{M}_{12}(l,0)]^{-1}. \tag{25}$$

The intermediate result Eq. (25) has to be insertet into Eqs. (22) and (23). Then using Eq. (20) and considering that the determinant of $\mathcal{M}(l,0)$ is always 1, results in the final expression for $\widetilde{C}_+(l)$:

$$\widetilde{C}_+(l) = U_0 e^{jkl} [-Z_0 Z_C \mathcal{M}_{21}(l,0)$$
$$+ \mathcal{M}_{22}(l,0) Z_C + \mathcal{M}_{11}(l,0) Z_0 - \mathcal{M}_{12}(l,0)]^{-1} \tag{26}$$

Inserting the classical matrix elements from Eqs. (6) into (26) the cTLT expression for the amplitude function, $\widetilde{C}_+^{class} = U_0/(Z_0 + Z_C)$, is received.

4.3 Calculation of the TL current using novel reflection coefficients and amplitude function

In the last step the current on the TL has to be determined using the previously derived reflection coefficients $\widetilde{R}_-(l)$, $\widetilde{R}_+(l)$ and the amplitude function $\widetilde{C}_+(l)$. Therefore the configuration of Fig. 10 is used with arbitrary loads Z_0 and Z_L. A forward traveling outgoing current wave $i_1(l) = \widetilde{C}_+(l) e^{-jkl}$ would be reflected at the end and the current wave $i_2(l) =$

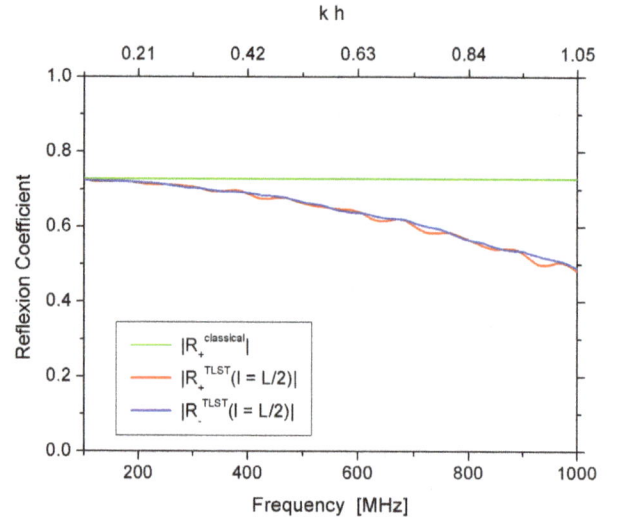

Figure 11. Reflection coefficients $|\widetilde{R}_+|$, $|\widetilde{R}_-|$ at the center point for the classical TL with risers of Fig. 1

$\widetilde{C}_+(l) e^{-jkl} \widetilde{R}_- e^{-jk(L-l)}$ would travel back. This reflected wave would be reflected again at the beginning of the TL and the current wave $i_3(l) = \widetilde{C}_+(l) e^{-jkl} \widetilde{R}_- e^{-jk(L-l)} \widetilde{R}_+ e^{-jkl}$ would travel also again to the end of the TL. Theoretically this procedure would be repeated endlessly leading to an expression for the current wave with two infinite sums

$$i(l) = \widetilde{C}_+(l) \sum_{n=0}^{\infty} \left(e^{-jkL} \widetilde{R}_- e^{-jkL} \widetilde{R}_+ \right)^n e^{-jkl}$$
$$+ \widetilde{C}_+(l) e^{-jkL} \widetilde{R}_- \sum_{n=0}^{\infty} \left(e^{-jkL} \widetilde{R}_+ e^{-jkL} \widetilde{R}_- \right)^n e^{-jk(L-l)} .. \tag{27}$$

The two sums in Eq. (27) represent geometrical series and can be simplified leading to the final result for the current on the TL

$$i(l) = \frac{\widetilde{C}_+(l) \left(e^{-jkl} + \widetilde{R}_- e^{-2jkL} e^{jkl} \right)}{1 - \widetilde{R}_- \widetilde{R}_+ e^{-2jkL}}. \tag{28}$$

It has to be mentioned that for the asymptotic region, that is $l \in [L_1, L_2]$, the reflection coefficients are constant.

When the current on the TL is known for example from a full wave simulation the current reflection coefficients can be calculated. This is shown for $\widetilde{R}_+(l)$. For the asymptotic region a backward traveling wave can be expressed as $i(l) = \widetilde{I}_1 \left(e^{jkl} + \widetilde{R}_+ e^{-jkl} \right)$. A straight forward calculation results in

$$\widetilde{R}_+(l) = \left(\frac{jki(l) - \frac{di(l)}{dl}}{jki(l) + \frac{di(l)}{dl}} \right) e^{2jkl}. \tag{29}$$

5 Numerical calculations for classical TL with risers

First, the TL from Fig. 1 is regarded with the before mentioned line parameters $L_{TL} = 2\,\mathrm{m}$, $L = 2.1\,\mathrm{m}$, $h = 5\,\mathrm{cm}$ and

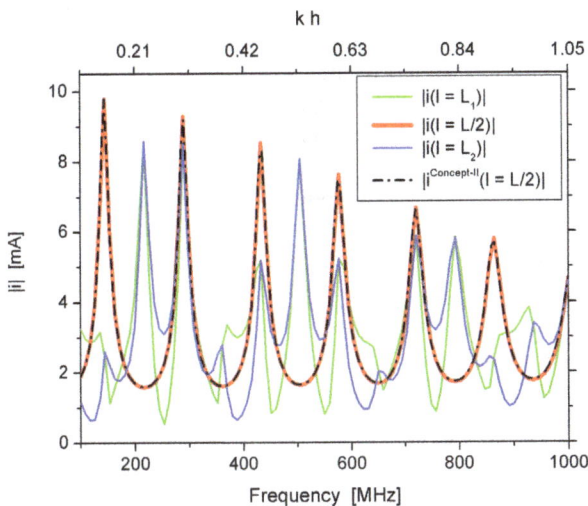

Figure 12. Current for different positions on the TL calculated using the reflection coefficients.

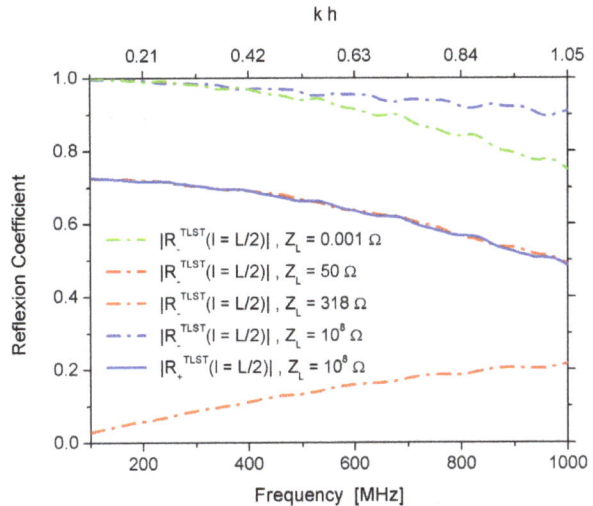

Figure 13. Reflection coefficients at $l = L/2$ for different load impedances Z_L.

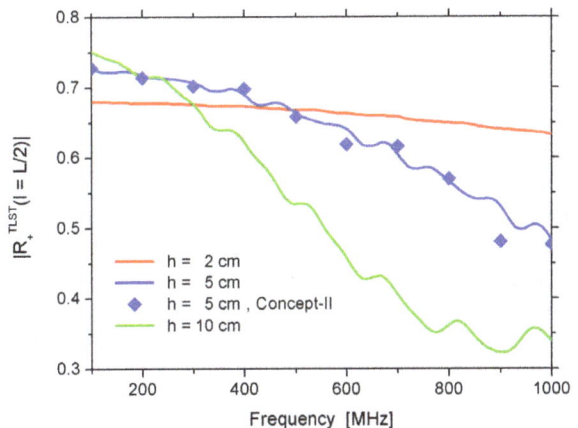

Figure 14. Reflection coefficients at $l = L/2$ for different heights h of the TL over PEC ground.

$r_0 = 0.5\,\text{mm}$. The TL is driven by a voltage source $U_0 = 1\,\text{V}$ with source impedance $Z_0 = 50\,\Omega$ and terminated by a load $Z_L = 50\,\Omega$. Elements of the parameter matrix $\overline{\mathbf{P}}^{*(1)}$ (TLST) are shown in Figs. 2 and 3. Using formulas Eqs. (19) and (18) the current reflection coefficients $|\widetilde{R}_+|$ and $|\widetilde{R}_-|$ were calculated for the center position on the TL at $l = L/2$ in the frequency range from 100 MHz to 1 GHz as shown in Fig. 11. It is clearly seen that the current reflection coefficients deviate from their classical value significantly with rising frequency because of the radiated energy losses. Because source and load impedance are the same in this configuration, $|\widetilde{R}_+|$ and $|\widetilde{R}_-|$ have the same value.

Using Eq. (28) the current can be calculated for different positions and frequencies. Fig. 12 shows the results for the positions L_1, $L/2$ and L_2. Also shown (black dash-dotted line) is the current $|I(L/2)|$ resulting from a MoM calculation using the Concept-II code (Brüns et al., 2011). The correspondence between full wave analysis and calculation using novel local and frequency dependent reflection coefficients is excellent.

In Fig. 13 the reflection coefficients $|\widetilde{R}_-|$ are shown for the TL from Fig. 1 with $h = 5\,\text{cm}$ for different loads Z_L (short circuited, $50\,\Omega$, matched and open). The source impedance remains at $Z_0 = 50\,\Omega$, so $|\widetilde{R}_+|$ is the same for all Z_L values and is explicitly shown for the open case. The classical current reflection coefficient for an open TL is negative because of the necessary phase shift of the current wave. In Fig. 13 the absolute value of $\widetilde{R}_-$ is presented, but of course for $\omega \to 0$ the real part of $\widetilde{R}_-$ would tend to the value -1. For a matched load the classical reflection coefficient is zero. That means the current wave would completely be absorbed in the load and no reflected wave would be produced. In a real TL with risers there is no fixed matched load for all frequencies any longer (Rambousky et al., 2013b). The nonuniformity of a

real TL is responsible for the scattering of the current wave at the local (e.g. bends) or distributed (e.g. varying height over ground) scattering centers. With a classical matched load impedance the current reflection coefficient $|\widetilde{R}_-|$ rises with frequency as can clearly be seen in Fig. 13.

Another interesting fact is the influence of the height h of the TL over ground on the reflection coefficient. With decreasing height h a TL should show increasingly classical behavior. This can be seen also in the gradient of the reflection coefficients. In Fig. 14 it is shown that for decreasing height h the frequency dependence of $|\widetilde{R}_+|$ decreases also and would approach the constant classical value for $h \to 0$. The curve with the blue diamonds in Fig. 14 was produced by calculating the current on the TL with the MoM code Concept-II and using Eq. (29) to calculate $\widetilde{R}_+(l)$. There is again a good correspondence between MoM and TLST results.

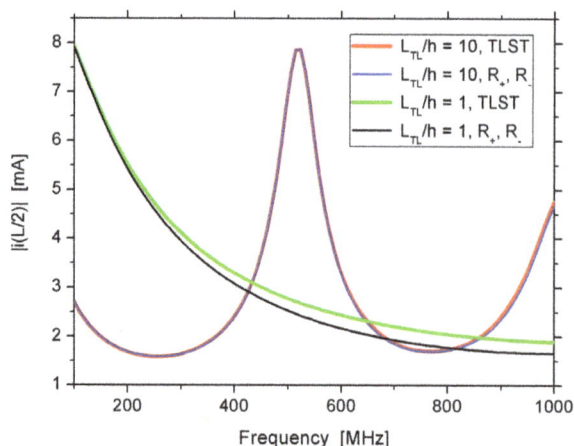

Figure 15. Current at the middle of the TL ($l = L/2$) for a constant height $h = 5$ cm and different horizontal length L_{TL}.

As mentioned before the original TLST calculation of the current on the TL is an exact solution while the calculation using the novel reflection coefficients is still an approximation because the mutual influence of the risers is neglected. When the ratio of the horizontal part of the TL and the height above the conducting ground plane, L_{TL}/h, is large enough there will be no significant influence due to the risers. The current on the TL then should be the same as for a pure TLST calculation and a current calculation using the above reflection coefficient method at least for the asymptotic region. This can be seen in Fig. 15 for the ratio $L_{TL}/h = 50$ cm$/5$ cm $= 10$. The correspondence is nearly perfect. Reducing the ratio dramatically to $L_{TL}/h = 5$ cm$/5$ cm $= 1$ where the length of the horizontal part is equal to the height of the risers, there is a distinct mismatch between the two current calculation procedures (see also Fig. 15). But from a practical point of view the differences are not crucial so that for practical applications the reflection coefficient method can be used even with smaller L_{TL}/h ratios.

6   Conclusion

In this paper it was shown that cTLT is not sufficient for a finite classical TL with risers at high frequencies. For efficient analysis the TL can be separated into the two riser parts and the asymptotic region. The latter can be handled with cTLT while the riser parts have to be calculated using an advanced TLT, like TLST. The product of matrizants for the three parts finally gives the matrizant for the original whole TL.

Novel reflection coefficients were defined according to the concept of the constant classical ones which are now local and frequency dependent. The current on the TL was calculated using these novel reflection coefficients. For TLs where the horizontal part is significantly larger than the height of the risers the so calculated current fits very well to the ex-

act solution using only TLST or a full wave method. Numerical results were shown for several configurations of load impedance or heights of the risers. The result for the current determined via the novel reflection coefficients leads even for small L_{TL}/h ratios to practically usable values.

The presented method for the analysis of nonuniform TLs is essential for a network theory where such TLs are to be handled. It could be shown that junctions are allowed in regions which show nearly classical behavior for an otherwise nonuniform TL. The formulas for the novel reflection coefficients possess exactly those type of poles, which are necessary for the singularity expansion method (SEM) analysis of the basic frequencies of a TL system. The extension of SEM for nonuniform TL will be of interest for future work.

References

Brüns, H., Freiberg, A., and Singer, H.: CONCEPT-II Manual of the Program System, user manual, Technische Universität Hamburg-Harburg, 2011.

Gantmacher, F.: The theory of matrices, Chelsea Publishing Company, New York, 1984.

Haase, H.: Full-Wave Interactions of Nonuniform Transmission Lines, in: Res Electricae Magdeburgenses (MAFO Vol.9), edited by: Nitsch, J. and Styczynski, Z., Magdeburg, 2005.

Haase, H. and Nitsch, J.: Full-wave transmission-line theory (FWTLT) for the analysis of three dimensional wire-like structures, in: Proc. 14th International Zurich Symposium and Technical Exhibition on Electromagnetic Compatibility, 235–240, Zurich, Switzerland, 2001.

Haase, H., Nitsch, J., and Steinmetz, T.: Transmission-Line Super Theory: A New Approach to an Effective Calculation of Electromagnetic Interactions, The Radio Science Bulletin, 307, 33–60, 2003.

Nitsch, J. and Tkachenko, S.: High-Frequency Multiconductor Transmission-Line Theory, Found. Phys., 40, 1231–1252, 2010.

Nitsch, J., Gronwald, F., and Wollenberg, G.: Radiating Nonuniform Transmission-Line Systems and the Partial Element Equivalent Circuit Method, Wiley, Chichester, West Sussex, UK, 2009.

Rambousky, R., Nitsch, J., and Garbe, H.: Analyzing Simplified Open TEM-Waveguides using Transmission-Line Super Theory, in: International Symposium on Electromagnetic Compatibility, EMC EUROPE 2012, 1–6, Rome, Italy, 2012.

Rambousky, R., Nitsch, J., and Garbe, H.: Application of the Transmission-Line Super Theory to Multiwire TEM-Waveguide Structures, IEEE Trans. EMC, 55, 1311–1319, 2013a.

Rambousky, R., Nitsch, J., and Garbe, H.: Matching the termination of radiating non-uniform transmission-lines, Adv. Radio Sci., 11, 259–264, doi:10.5194/ars-11-259-2013, 2013b.

3

Design investigation to improve voltage swing and bandwidth of the SiGe driver circuit for a silicon electro-optic ring modulator

A. Fatemi[1], H. Gaul[2], U. Keil[2], and H. Klar[1]

[1] Institute of Microelectronics, Technical University of Berlin, Berlin, Germany
[2] FCI Deutschland GmbH, Berlin, Germany

Correspondence to: A. Fatemi (adel.fatemi@tu-berlin.de)

Abstract. This paper reports on a new SiGe driver IC to address the low breakdown voltage level of modern BiCMOS transistors. An optical modulator driver IC in SiGe 250 nm technology with a supply voltage of 4.5 V is presented. This driver IC consists of pre- and main driver stages where a newly modified cascode topology and capacitance degeneration technique is employed to meet current application requirements; high voltage swing at high datarate. The simulation results show a differential output voltage swing of 3.9 Vp-p at 14 Gbps data rate, according to the FDR Infiniband standard.

1 Introduction

In recent years, many studies have worked to solve the bandwidth limitations of wire interconnection and the resulting cross talk in data communication. Photonics could be one way to solve this issue, even at higher transmission distances. By integrating optics into electronics, silicon phonic technology promises to offer a higher level of integrity, greater bandwidth and lower energy consumption in data transmission (Zuffada, 2012).

The optical modulator is a key element to implement the electrical-to-optical data conversion chain on the transmitter side. A micro-ring resonator, fabricated on a silicon substrate, is a high speed optical modulator (Zuffada, 2012). It maps voltage swing across its terminal to a specific level of optical output power. The proper Extinction Ratio (ER) is one of the main quality factors for optical modulation and is therefore dependent on the amplitude of the swing across the ring. Measurement results indicate (Giesecke et al., 2014) that voltage modulation must be in the range of 3–4 V to provide an acceptable ER. Since a back plane electrical signal could not directly fulfill this requirement (InfiniBand Architecture, 2012), an inter-stage broadband amplifier (here known as a driver IC) is needed to complete the electrical-to-optical conversion chain.

In broadband amplifier, Current Mode Logic (CML) configuration is typically used as the last stage of amplifier. Nowadays, with increasing transient frequency of modern BiCMOS transistors, their breakdown voltage capability is going to lower values (Mandegaran and Hajamiri, 2004). Therefore, obtaining high voltage swing in high transmission data rate is not straightforward.

In Sect. 2 of the present paper, a new modified cascode topology that divides all of the voltage stress among three transistors is explored, and a capacitance degenerative technique that extends bandwidth is also reviewed. Driver IC designed in SiGe 250 nm technology will be explored in Sect. 3.

2 Design concept

2.1 Modified breakdown voltage doubler of bipolar transistors

In a CML configuration, the output transistor has to sustain the entire output voltage swing. Cascode topology is a conventional solution to divide voltage stress between two transistors. A number of techniques have been reported to tune the amount of voltage drop on each transistor (Mandegaran and Hajamiri, 2004; Li et al., 2005a, b; Li and Tsai, 2006; Rakowski et al., 2012). All of these techniques are based on pushing the base of the upper transistor, Q2, to follow the output voltage as shown in Fig. 1.

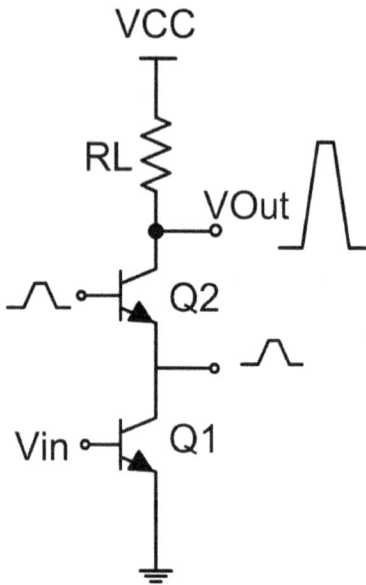

Figure 1. The idea behind modified cascode topology.

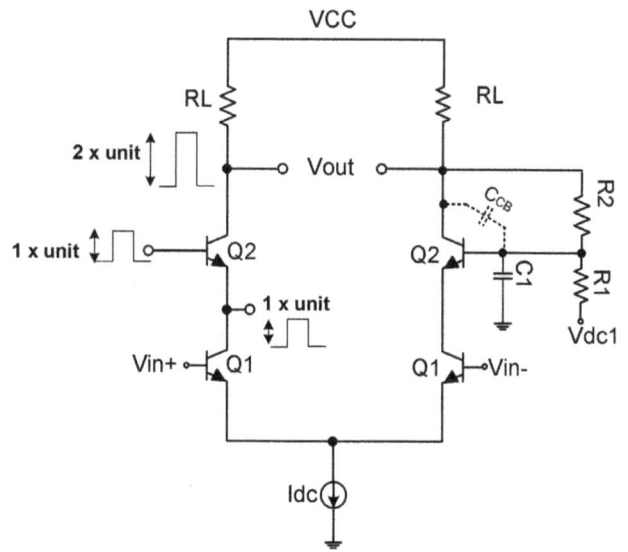

Figure 2. Modified cascode topology with feedback network and proper amplitude voltage ratio.

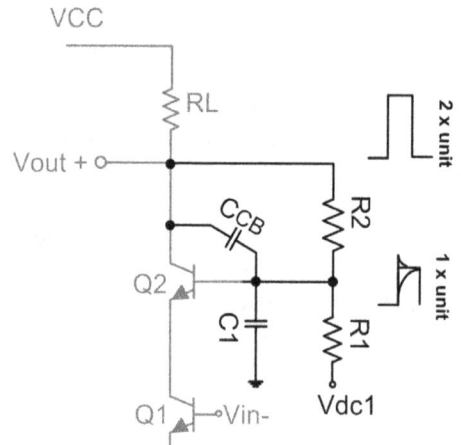

Figure 3. Simplified feedback network in half circuit differential mode.

If the output voltage is at its maximum, it is desired to evenly divide the stress across two transistors in the case of similar devices in the stacks. A possible solution reported (Li et al., 2005a) is to use the intrinsic collector-base capacitance (CCB) of the upper transistor, Q2, to feed part of the output signal back to its base. In this way, both the base of the upper transistor and its emitter follow the collector as shown in Fig. 2.

The precise analysis of the feedback network is not straightforward, as the influence of the feedback network and transistor, Q2, must be taken into account as is explained here (Li and Tsai, 2006). The simplified feedback network, with neglecting the effects of large base-collector ohmic resistance and other parasitic capacitances, can be modeled by C1, CCB, R1, and R2 as is highlighted in Fig. 3.

A resistive feedback network (R2 and R1) defines low frequencies of amplitude voltage of base of transistor, Q2, and C1 with collector-base capacitance provides voltage drop for higher frequencies. Choosing right value for C1 is quite important as smaller capacitance brings some spikes at the base of the transistor and higher values of C1 limits the bandwidth as is shown in Fig. 3. It should be considered that R1 and R2 are chosen comparably larger than RL to reduce power consumption and have less influence on 50 ohm load impedance (Li and Tsai, 2006). Vdc controls the DC voltage of the base and emitter of Q2. Maximum single-ended output swing is calculated using the following equation:

$$\Delta V_{\text{out}} = V_{\text{out,OFF}} - V_{\text{out,ON}} = \left(2 \times V_{\text{CE,max}} + V_{\text{CE,sat}}\right)$$
$$- 3 \times V_{\text{CE,sat}}. \tag{1}$$

In the present paper, the idea of using the intrinsic collector-base capacitance for the feedback network is extended into a three-stage cascode topology. This method can increase the maximum allowable output voltage swing while breakdown voltage is becoming less for upcoming downsized technologies. Now, maximum single-ended output swing is calculated using the following equation:

$$\Delta V_{\text{out}} = V_{\text{out,OFF}} - V_{\text{out,ON}} = \left(3 \times V_{\text{CE,max}} + V_{\text{CE,sat}}\right)$$
$$- 4 \times V_{\text{CE,sat}}. \tag{2}$$

For an identical output swing, it can be proofed that the transistors with lower breakdown voltage, as almost two-thirds of the breakdown voltage in previous method, can be employed.

In the new implementation, the entire output swing must be divided equally among three transistors, so that the base of the two upper transistors must be pushed with the same phase as the output node (collector of Q3) with proper am-

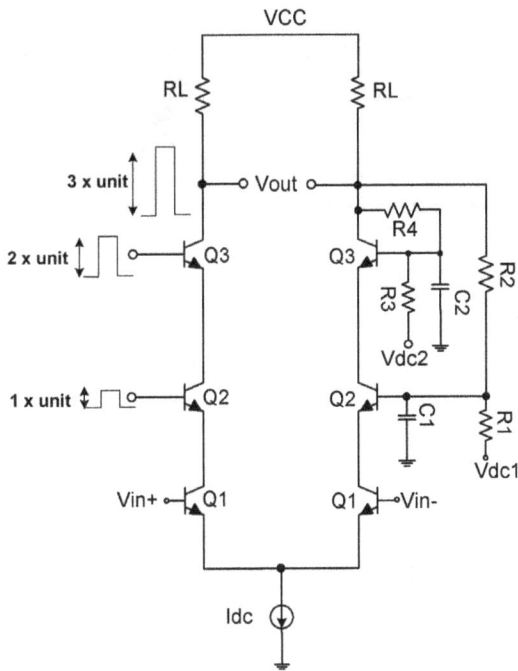

Figure 4. New modified multi-cascode topology with feedback network and corresponding voltage swing.

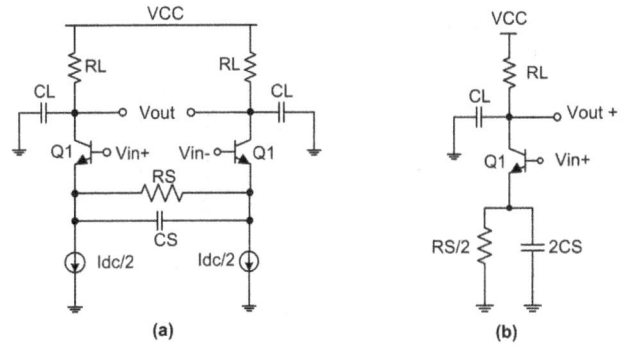

Figure 5. (a) Capacitance degeneration technique, (b) half circuit in differential mode.

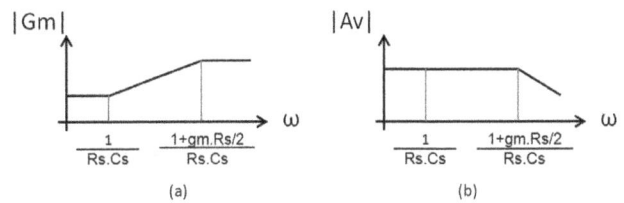

Figure 6. (a) Bandwidth of Gm, (b) overall system frequency response.

plitude. As is shown in Fig. 4, the feedback network must convey two-thirds, and one-third of the output signal for the base of Q2 and Q3 respectively, so as to drop one third of the whole output swing over each transistor. Here, the intrinsic collector-base capacitances, together with C1 and C2, provide a high frequency feedback network for Q1 and Q2.

2.2 Capacitance degenerative technique

In a differential configuration pair, a capacitance degeneration technique could improve bandwidth by adding a pole and a zero to the system (Razavi, 2012). The zero is designed to coincide with and cancel the pole of the system, and the 3dB bandwidth is determined by the second pole, which is away from the first one.

Figure 5 shows how the pole and the zero are added to the CML in a differential configuration. For half of the circuit in Fig. 5b, conductance of the common-emitter circuit is (Razavi, 2012):

$$G_m = \frac{g_m}{1 + g_m \times \left(\frac{R_s}{2} \| \frac{1 + g_m \times \frac{R_s}{2}}{R_S \times C_S} \right)}. \tag{3}$$

If one assumes that the output pole of the differential pair is dominated by $R_L \times C_L$, it would coincide with the added zero to cancel each other out, only if:

$$R_S \times C_S = R_L \times C_L. \tag{4}$$

Therefore, overall bandwidth of the system is extended to the second pole of the system, $(1 + g_m \times R_S/2)/(R_S \times C_S)$, as shown in Fig. 6.

3 Design and implementation

Using the concepts discussed in Sect. 2, a driver IC was implemented in SiGe 250 nm technology. This driver IC consists of two parts: a pre-driver stage and a main driver stage.

3.1 Pre-driver stage

Back-plane electrical signals fall within a certain standard range. This range is normally not strong enough to completely switch the current from one differential side to another one, a requirement for maximizing output swing and minimizing power consumption. In addition, a high amplitude input signal could bring the transistors into a deep saturation region where switching speed is reduced. Therefore, the pre-driver stage is placed before the main driver stage in order to prepare both the amplitude and the DC level of the input signal for the main stage driver IC.

As is shown in Fig. 7, the pre-driver stage is in a differential configuration, which consists of one common-emitter amplifier, followed by a common-collector stage. The common-emitter stage is designed to transfer low and high amplitude input signal to the specific level appropriate for driving the main stage driver IC. The main role of the common-collector stage is to provide proper impedance

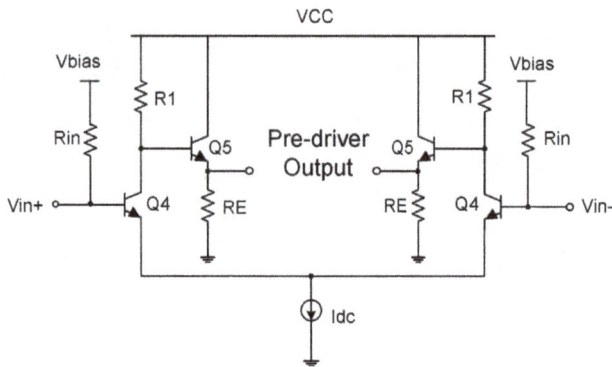

Figure 7. Proposed pre-driver stage.

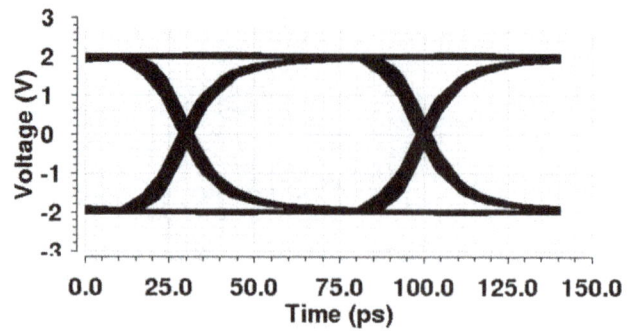

Figure 8. Proposed three stage cascode configuration.

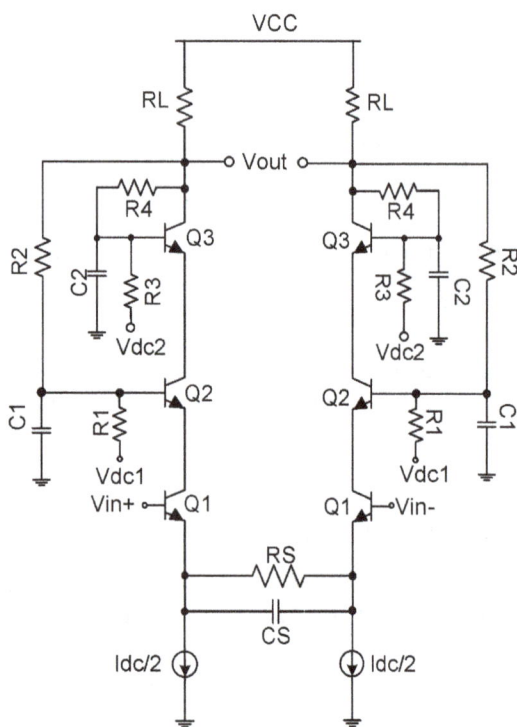

Figure 9. Simulated eye diagram of output signal at 14 Gpbs.

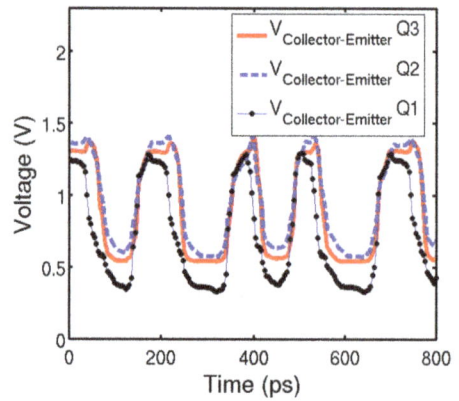

Figure 10. Simulated collector-emitter voltage stress in multi-cascode topology.

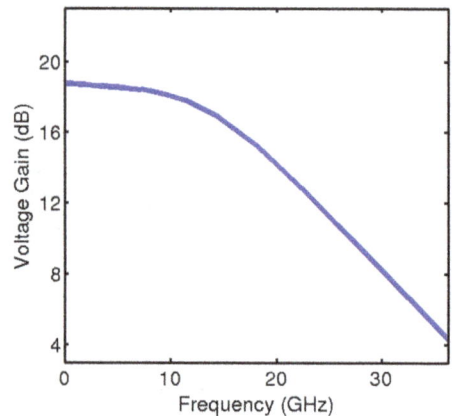

Figure 11. Simulated small signal voltage gain of driver IC.

for the common-emitter amplifier and cascade stage. In the present paper, the pre-driver output is designed to generate 500 mVp-p for the main driver stage.

3.2 Main driver stage

As target differential output swing is 4 Vp-p, supply voltage is increased to 4.5 V. Figure 8 shows how the main stage driver IC is constructed, it consists of a three-stage cascode and capacitance degeneration technique. A simulated eye diagram is depicted in Fig. 9, where its eye opening is 3.9 V at 14 Gbps.

Figure 10 shows how much voltage stress is on each device. This figure indicates that stress is less than 1.5 V, which is less than the breakdown voltage of 1.9 V in this technology, and it is shared quite evenly among devices as well. Figure 11 shows the small signal simulation of driver IC that indicates low frequency gain of 17 dB and corresponding 3 dB bandwidth of 14 GHz.

4 Conclusions

Silicon photonic technology offers a way to fully integrate optics into electronics in a single chip. A micro-ring resonator requires a comparatively high input voltage swing for proper performance (minimum ER). In the present paper, some techniques have been presented to obtain a high voltage swing in multi-Gbps data rate transmission. The technique that is introduced in this paper makes it possible to employ the transistors with lower breakdown voltage levels as two-third for high voltage swing application. Finally, the SiGe driver IC in 250 nm technology was investigated. Simulation results claim an output voltage swing of 3.9 Vp-p at 14 Gbps.

Acknowledgements. This work is funded by the Federal ministry of education (BMBF) in the SHyWA project under Grant No. 16BP1103. The Author would like to thank Friedel Gerfers for his valuable discussions and guidance.

References

Giesecke, A. L., Prinzen, A., Bolten, J., Porschatis, C., Chmielak, B., Matheisen, C., Wahlbrink, T., Lerch, H., Waldow, M., and Kurz, H.: Add-drop microring resonator for electro-optical switching and optical power monitoring, in: Proceedings of the 34th Conference on Lasers and Electro-Optics, 08–13 June 2014, San Jose, USA, 1–2, 2014.

InfiniBand Architecture, Volume 2: https://cw.infinibandta.org/document/dl/7141, last access: November 2012.

Li, D. U. and Tsai, C. M.: 10-Gbps modulator drivers with local feedback networks, IEEE J. Solid-St. Circ., 1, 1025–1030, 2006.

Li, D. U., Haung, L. R., and Tsai, C. M.: Low power consumption 10-Gbps SiGe modulator drivers with 9V$_{PP}$ differential output swing using intrinsic collector base capacitance feedback network, IEEE Rad. Freq. Integr., 1, 317–320, 2005a.

Li, D. U., Haung, L. R., and Tsai, C. M.: 10 Gbps CMOS laser driver with 3.3 V output swing, IEEE Cust. Integr. Cir., 1, 333–336, 2005b.

Mandegaran, S. and Hajimiri, A.: A breakdown voltage doubler for high voltage swing drivers, IEEE Cust. Integr. Cir., 1, 103–106, 2004.

Rakowski, M., Ryckaert, J., Pantouvaki, M., Yu, H., Bogaerts, W., De Meyer, K., Steyaert, M., Absil, P. P., and Van Campenhout, J.: Low power, 10-Gbps 1.5-Vpp differential CMOS driver for a silicon electro-optic ring modulator, IEEE Cust. Integr. Cir., 1, 9–12, 2012.

Razavi, B.: Capacitance Degeneration, in: Design of Integrated Circuit for Optical Communications, Second Edition, John Wiley and Sons, Hoboken, New Jersey, 140–143, 2012.

Zuffada, M.: The industrialization of the Silicon Photonics: Technology road map and applications, in: Proceedings of the 42th European Solid-State Device Research Conference, 17–21 September 2012, Bordeaux, France, 7–13, 2012.

4

Investigations on the magnetic field coupling of automotive high voltage systems to determine relevant parameters for an EMR-optimized designing

David Krause[1], Werner John[2], and Robert Weigel[3]

[1]AUDI AG, 85045 Ingolstadt, Germany
[2]SiL System Integration Laboratory GmbH, Technologiepark 32, 33100 Paderborn, Germany
[3]Lehrstuhl für Technische Elektronik, Friedrich-Alexander-Universität Erlangen-Nürnberg, Cauerstraße 9, 91058 Erlangen, Germany

Correspondence to: David Krause (david.krause@audi.de)

Abstract. The implementation of electrical drive trains in modern vehicles is a new challenge for EMC development. This contribution depicts a variety of investigations on magnetic field coupling of automotive high-voltage (HV) systems in order to fulfil the requirements of an EMR-optimized designing. The theoretical background is discussed within the scope of current analysis, including the determination of current paths and spectral behaviour. It furthermore presents models of shielded HV cables with particular focus on the magnetic shielding efficiency. Derived findings are validated by experimental measurements of a state-of-the-art demonstrator on system level. Finally EMC design rules are discussed in the context of minimized magnetic fields.

1 Introduction

The development of EMC concepts for electric vehicles requires a thorough understanding of electromagnetic coupling mechanisms of the high-voltage (HV) system. Due to the limited freedom of modification of vehicle drive-systems, available resources needs to be used optimally, which further demands identification and evaluation of EMC-related design parameters. Comparing the wavelengths of dominant electromagnetic interferences from voltage source inverters (VSI) with the geometrical dimension of a regularly sized vehicle enables a quasi-steady approach, where electric and magnetic fields can be investigated separately. In this con-

tribution the focus is laid on the magnetic field coupling of automotive HV systems.

The pulse-wide modulated voltage switching of power electronics generates parasitic currents apart from operational ones. They are primary caused by distributed capacitances present in HV components like the electric motor. In general current levels are scaling with the size of the components, because larger conductive structures mean an increased capacitive coupling. In other words, engines with more electric power generate more parasitic current that consequently causes higher magnetic fields. Apart from the current strength, magnetic field coupling on antennas, wires or other electric devices is significantly dependent on the geometry of cable routes and the characteristics in frequency domain. Related investigations therefore require identifying major current paths as well as the respective spectral behaviour, which had been the focus of former contributions e.g. Kempski et al. (2006), Grandi et al. (2004) and Jettanasen (2010), but were not considered in the context of magnetic fields.

Research work presents selected experimental EMI measurements of a state-of-the-art HV system demonstrator, focusing on quasi-steady magnetic fields with frequencies from 9 kHz until the lower MHz range. The theoretical background includes a current analysis of the AC-network, where common-mode (CM) and differential-mode (CM) circuit models are used to determine the underlying spectral functions. As fully shielded systems are commonly used in automotive HV systems (Hohloch et al., 2012), shielded HV cable models are introduced and used to analytically describe the magnetic shielding efficiency. In a further step they are verified by numerical field simulations and measurements.

With the help of a state-of-the-art system demonstrator derived findings are experimentally proved. Loop-antenna measurements are analysed in the scope of developed circuit models and spectral functions. It is demonstrated, that different operating conditions of the VSI can show either CM- or DM dominant magnetic coupling. Slight adjustments at the test setup made it possible to investigate unshielded and shielded conditions with reference to EMC-requirements (GB/T 18387, 2008), which are relevant for the type-approval of electric vehicles. Finally this contribution concludes with a discussion about EMC design rules and intents to support electric power train layout designers in dimensioning EMC-optimized HV topologies.

2   Current analysis

The application of VSIs into electric vehicles leads to a broadband spectrum of challenging EMC issues. One important aspect, discussed in this chapter, is the investigation of currents with their respective propagation paths and spectral functions. Based on the requirements of automotive EMC norms, which are defined particularly for electric vehicles, the relevant frequencies are ranging from 9 kHz until 30 MHz (GB/T 18387, 2008).

Due to reasons of personal safety automotive HV systems are always built up as IT(isole terra)-networks and are therefore isolated from the vehicle chassis, representing the common ground system. This consideration is only valid from the DC perspective and needs to be reversed, when investigating higher frequencies. From the EMC point of view the system then becomes more complex and parasitic properties (e.g. stray capacitances) need to be minded (see Fig. 1).

The basic task of the VSI is to generate changeable rotational speed for the electric engine by offering sinusoidal pulse-modulated voltages. These periodic voltage pulses are causing currents, which can be represented by a series of harmonics in frequency domain. For EMC analyses in general it is useful to distinguish between common-mode (CM) and differential-mode (DM). First one is describing the electrical behaviour to the ground system that in terms of automotive HV systems primary has a capacitive nature. Hence, time varying voltages like pulses lead to a flow of displacement current proportional to the value of systemic parasitical capacitances:

$$i_C(t) = C \frac{\mathrm{d}u_C(t)}{\mathrm{d}t}. \tag{1}$$

The second one, representing the symmetrical line-to-line behaviour, additionally shows major inductive influence especially for the very low frequency range. Current then is caused by the integration of the time-varying voltage pulses

and is indirectly proportional to the inductance value:

$$i_L(t) = \frac{1}{L} \int u_L(t)\mathrm{d}t. \tag{2}$$

The asymmetrical control of inverters leads to both CM and DM voltages and thus forms respective current paths with their own spectral characteristics that will be in detail discussed with focus on the AC-network below. Derivated linear equations are using variable spectral functions and are therefore independent from the underlying modulation technique. However, in the literature there are analytical closed-form expressions for all modulation methods, which are currently applied in automotive power electronics (SVPWM and Flat Top). The most native method, the sine-triangular PWM or SPWM, is used for the following SPICE-based calculations and for example can be analytically represented by Double Fourier analysis as a sum of carrier and side-band harmonics (Kempski et al., 2006).

2.1   Common-Mode current

CM currents, commonly also known as leakage currents, are always parasitical and therefore not part of the actual function of electric power trains. They are generated by alternating voltage drops at capacitively coupled reference systems that in automotive terms are not restricted to vehicle chassis, but also can be low-voltage (LV) networks or large conductive structures like motor shafts. In this contribution the basic CM behaviour of the AC-network is described, where some simplifications are performed.

For IT-based HV systems the CM path is significantly determined by its capacitive properties and is therefore modelled by discrete Y-capacitances (see Fig. 2). In real application they are distributed and shaped primary between the stator and the housing on the motor side and between the IGBT modules and the metallic cooling unit on the inverter side (Kempski et al., 2002). Inductances are mainly determined by HV cables, whereas for CM considerations the inductive influence of the motor coils can be neglected, because of bypassing winding capacitances. Since oscillating circuits are formed, the model additionally needs to include DC-resistances to avoid unrealistically sharp resonances.

The EMC-relevant CM source is the switched voltage drop between the phase- and the reference system. For a three-phase inverter the respective voltage is given by the vector sum of the instantaneous pulsed phase voltages (e.g., Cacciato et al., 1999),

$$\underline{U}_{CM} = \frac{\underline{U}_{u,\mathrm{gr}} + \underline{U}_{v,\mathrm{gr}} + \underline{U}_{w,\mathrm{gr}}}{3} \tag{3}$$

which is never zero, because of temporary electrical asymmetry (Kempski et al., 2002). According to the circuit, there is a characteristic resonance frequency at:

$$\omega_{CM,r} = \frac{1}{\sqrt{\frac{C_{Y,\mathrm{vsi}} \cdot C_{Y,\mathrm{mot}}}{C_{Y,\mathrm{vsi}} + C_{Y,\mathrm{mot}}} \cdot \frac{L_{wr}}{3}}}. \tag{4}$$

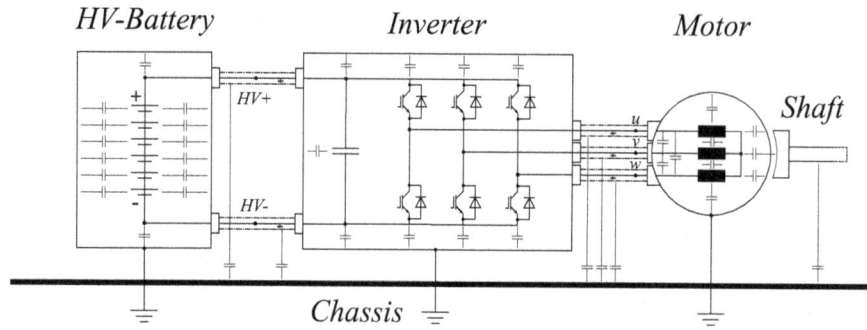

Figure 1. Distributed parasitic capacitances in automotive HV systems.

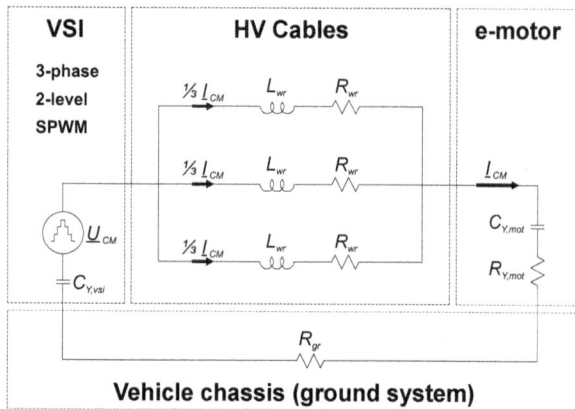

Figure 2. Simplified CM circuit model of the AC-/phase-network.

Table 1. Relevant parameters for calculating CM voltages and currents.

Parameter	Description	Value
U_{DC}	DC-voltage VSI	300 V
f_c	Clock frequency VSI	10 kHz
M	Modulation index SPWM	50 %
R_{wr}, R_{gr}	Resistance HV cable, ground	0.5 mΩ
L_{wr}	Self-inductance 2 m HV cable	2 μH
$R_{Y,mot}$	Y-path resistance e-Motor	3 Ω
$C_{Y,mot}$	Y-path capacitance e-Motor	6 nF
$C_{Y,vsi}$	Y-path capacitance VSI	100 nF

Hence, the spectral function of the CM current can be piecewise defined with:

$$\underline{I}_{CM} = \begin{cases} \underline{U}_{CM} \cdot \dfrac{j\omega C_{Y,mot} \cdot C_{Y,vsi}}{C_{Y,vsi} + C_{Y,mot}} & \text{if } \omega \ll \omega_{CM,r} \\ \underline{U}_{CM} \cdot \dfrac{1}{R_{wr}/3 + R_{Y,mot}} & \text{if } \omega = \omega_{CM,r} \\ \underline{U}_{CM} \cdot \dfrac{3}{j\omega L_{wr}} & \text{if } \omega \gg \omega_{CM,r} \end{cases} \quad (5)$$

Typical CM parameters are summarized in Table 1 and used to calculate the circuit model with SPICE (see Fig. 3). In time domain the voltage pulses are formed as step functions with three positive and negative steps. During the slopes, capacitances are excited and leakage current peaks are generated. According to the inverter's clock frequency pulses are periodically repeated, that consequently leads to an integer multiple of clock harmonics in frequency domain. Besides, there are additional side band harmonics, following the fundamental frequency of the modulated sine waves. For the observed frequency range until 30 MHz, the voltage source offers rectangular signals, hence harmonic levels are constantly decreasing with 20 dB per decade. Since Y-capacitances of VSIs are typically much higher than those of electric en-

gines, the current level is mainly restricted by the capacitive coupling at the motor side (see Eq. 5). According to Eq. (4) there is a resonance point at approximately 2.5 MHz, where the reactance is zero and the current level is determined by the overall DC resistance. The current spectrum behaves flat for frequencies far below, since simultaneously harmonics of the pulsed voltage are decreasing and the admittance of the CM path is rising with 20 dB per decade. Following again Eq. (5) the CM impedance for the clock frequency at 10 kHz equals 69 dBΩ, which expectably is the same result as the difference of voltage and current levels in the logarithmic frequency domain plot. The self-inductances of the HV cables become decisive, when frequencies far above the resonance point are considered. Here, currents are well attenuated and therefore less relevant from EMC perspective.

2.2 Differential-Mode current

When referring to DM currents of automotive HV systems, the following separation is useful:

- operational load currents
- current ripples
- capacitive displacement currents

Operational load currents represent the 120° phase shifted rotary current at the motor coils. Frequencies are equal to the

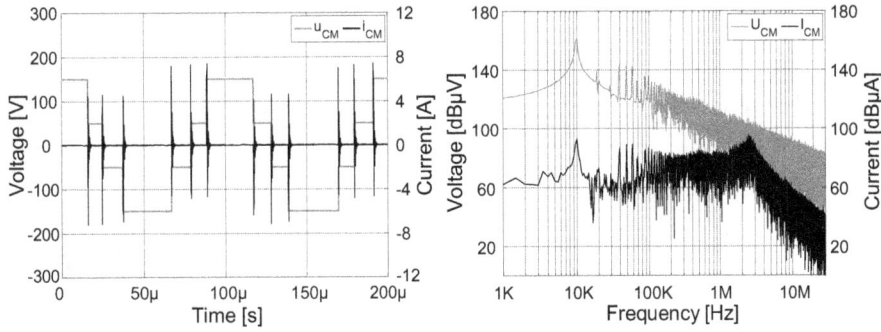

Figure 3. CM voltage and current in time and frequency domain.

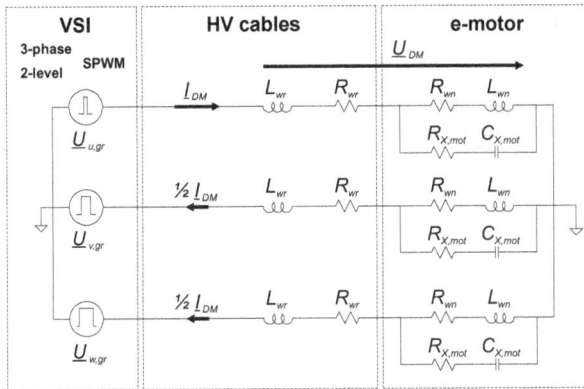

Figure 4. DM circuit model of the AC-network.

fundamental component of the respective modulation technique and in automotive applications are typically not higher than 1 kHz. As a functional part of the electric power train these currents will not be investigated, since their properties are outside the boundary conditions of this contribution. Current ripples and capacitive displacement currents contrarily lie in a relevant frequency area, have parasitic origins and are going to be discussed in this section.

To analyse them for a three-phase VSI, a symmetrical circuit model is used, where the electric machine is represented by ohmic-inductive loads in star connection (see. Fig. 4).

Parasitic properties are modelled by parallel X-capacitances, which are primary formed at the coil windings and the connector panel. The effective DM source is an arbitrary phase leg voltage given by:

$$\underline{U}_{DM} = \underline{U}_{u,gr} - \underline{U}_{CM} = \frac{2 \cdot \underline{U}_{u,gr} - \underline{U}_{v,gr} + \underline{U}_{w,gr}}{3} \quad (6)$$

According to the circuit model there are two characteristic frequencies. The first one is defined by the parallel circuit resonance of self-inductances and parallel X-capacitances, where the motor impedance changes from inductive to capacitive behaviour.

$$\omega_{DM,r1} = \frac{1}{\sqrt{C_{X,mot} \cdot L_{wn}}} \quad (7)$$

The second one is formed by the series circuit of wire inductances and X-capacitances and is simultaneously defining a resonance point of the overall DM path:

$$\omega_{DM,r2} = \frac{1}{\sqrt{C_{X,mot} \cdot L_{wr}}} \quad (8)$$

Assuming reasonable circumstances, where $L_{wn} > L_{wr}$ and consequently $\omega_{DM,r1} < \omega_{DM,r2}$, DM currents can be piecewise defined with:

$$\underline{I}_{DM} = \begin{cases} \underline{U}_{DM} \cdot \dfrac{1}{R_{wn} + j\omega L_{wn}} & \text{if } \omega \ll \omega_{DM,r1} \\[2mm] \underline{U}_{DM} \cdot j\omega C_{X,mot} & \text{if } \omega_{DM,r1} \ll \omega \ll \omega_{DM,r2} \\[2mm] \underline{U}_{DM} \cdot \dfrac{1}{R_{wr} + R_{x,mot}} & \text{if } \omega = \omega_{DM,r2} \\[2mm] \underline{U}_{DM} \cdot \dfrac{1}{j\omega L_{wr}} & \text{if } \omega \gg \omega_{DM,r2} \end{cases}$$

$$(9)$$

For typical DM parameters summarized in Table 2 respective voltages and currents are calculated with SPICE (see Fig. 5). Again periodically pulsed voltages are present that can be represented by Fourier series representations as a set of harmonics starting from the clock frequency. For frequencies below the first edge frequency $\omega_{DM,r1}$ the impedance is still ohmic-inductive (see Eq. 9) and currents are attenuated between 20 and 40 dB decade^{-1}. In time domain they can be identified as current ripples, superpositioning the operational sine currents. Considering even higher frequencies capacitive displacement currents outbalance current ripples, because of parasitic X-capacitances parallel to the motor coils. From now on the current spectrum behaves equally to the CM case with a resonance point at $\omega_{DM,r2}$.

3 Magnetic field coupling of shielded HV cables

Nowadays automotive HV systems used in series electric vehicles are fully shielded (Hohloch et al., 2012). In addition

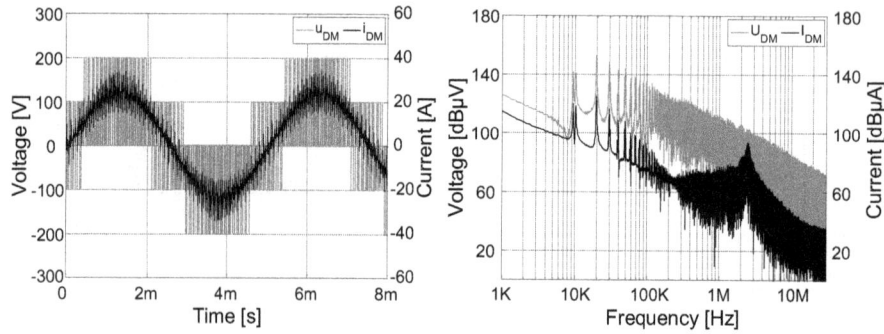

Figure 5. DM voltage and current in time and frequency domain.

Table 2. Relevant parameters for calculating DM voltages and currents.

Parameter	Description	Value
U_{DC}	DC-voltage VSI	300 V
f_c	Clock frequency VSI	10 kHz
f_i	Fundamental frequency VSI	200 Hz
M	Modulation Index SPWM	50 %
R_{wr}	Resistance HV cable	1 mΩ
L_{wr}	Self-inductance HV cable	0.5 μH
$R_{X,mot}$	X-path resistance e-Motor	3 Ω
$C_{X,mot}$	X-path capacitance e-Motor	2 nF
R_{wn}	Resistance coil windings	3 Ω
L_{wn}	Self-inductance coil windings	200 μH

to the metallic enclosures of HV components it implies, that also cable screens ensure a complete conductive surrounding of parts under (high) voltage. According to Ampère's law, an arbitrary closed-loop line integral of the magnetic field surrounding a shielded cable equals the vectorial sum of its inner wire current $\underline{I}_{in}$ and shield current $\underline{I}_{sh}$:

$$\oint_C \underline{H} ds = \underline{I}_{in} + \underline{I}_{sh} = \underline{I}_{sum} \tag{10}$$

For coaxial cables used in telecommunication systems, it is assumed that $\underline{I}_{in} = -\underline{I}_{sh}$ and the outer magnetic field is completely compensated. However, in the scope of power electronics applications this condition is not fulfilled, since operation modes of the cables as well as the considered frequency range are significantly different. Here, the residual sum current $\underline{I}_{sum}$ is causing an EMC-relevant magnetic field. In case of single shielded wire, which is routed straight and far away from its ground plane, the magnitude of the magnetic field vector can be calculated with the common formula:

$$\underline{H}(r) = \frac{\underline{I}_{sum}}{2\pi \cdot r} \tag{11}$$

In the following subsections this basic approach is used to find analytical expressions for calculating the magnetic

shielding efficiency (SE) of HV cables based on equivalent circuit models. In a further step they are validated by numerical field simulation and laboratory measurements.

3.1 Shielded cable models for quasi-steady currents

As shown in Sect. 2 the significant spectral content of currents is ranging until several Megahertz. In this context shielded HV cables can be modeled by their ohmic and inductive behaviour and capacitive displacement currents are neglectable (see Fig. 6).

Shielding against quasi-steady magnetic fields is dominantly driven by the mutual inductance between inner wire and the shield. Induced voltages lead to shield currents, that are dependent on the screen impedance and cause compensating fields (see Eq. 10). For the CM case the ground system also needs to be included. In contrast to the cables it shows minor inductive influence and is therefore modelled as a resistive path. Since pure symmetrical conditions are defined for the DM case, a potential ground path would carry no current and is thus not considered.

3.2 Calculating the magnetic shielding efficiency

Remembering the above mentioned definition, where the sum current $\underline{I}_{sum}$ is directly proportional to the outer magnetic field (see Eq. 11), the magnetic SE can be calculated with the following expressions:

$$SE_H = 20\log_{10}\left|\frac{\underline{I}_{in}}{\underline{I}_{sum}}\right| = 20\log_{10}\left|\frac{\underline{I}_{in}}{\underline{I}_{in} + \underline{I}_{sh}}\right| \tag{12}$$

$$= -20\log_{10}\left|1 + \frac{\underline{I}_{sh}}{\underline{I}_{in}}\right|$$

According to the CM circuit model (Fig. 6, left) the current ratio between cable shield and inner wire is given by:

$$\frac{\underline{I}_{CM,sh}}{\underline{I}_{CM,in}} = -\frac{R_{CM,gr}}{R_{CM,gr} + R_{CM,sh}} \cdot \frac{1 + j\omega\frac{M_{CM}}{R_{CM,gr}}}{1 + j\omega\frac{L_{CM,sh}}{R_{CM,gr} + R_{CM,sh}}} \tag{13}$$

This formula is likewise valid for a multi-conductor system, when each of the wire has the same properties and its

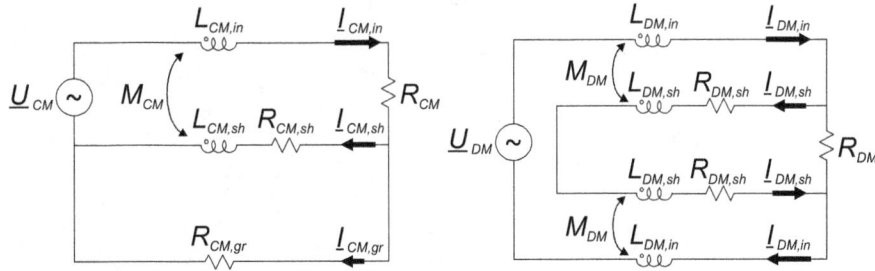

Figure 6. Simplified circuit models of shielded HV cables (left: CM; right: DM).

own ground path. In case n conductors are sharing a single ground path, the DC-resistance needs to be reduced by R_{gr}/n.

Now considering a DM voltage source feeding two or more conductors (see Fig. 6, right). Since pure symmetrical conditions are assumed, ground currents compensate and the ground path can be discarded. Setting $R_{gr} \to 0$ into Eq. (13) leads to the DM current ratio

$$\frac{\underline{I}_{DM,sh}}{\underline{I}_{DM,in}} = -\frac{j\omega M_{DM}}{R_{DM,sh} + j\omega L_{DM,sh}}. \tag{14}$$

In the literature there are closed analytical expressions to calculate the necessary inductances (Paul, 2006). For wires, parallel to a perfectly conducting wall (CM), the l.p.u. inductances is given by

$$L_{CM,in} = \frac{\mu_0}{2\pi} \cdot \ln\left(\frac{2d}{a}\right), \tag{15}$$

where a is the inner wire radius and d is the distance of current carrying conductors. Analogously the l.p.u. inductance of a single wire, as part of two parallel symmetrical wires (DM), can be calculated with:

$$L_{DM,in} = \frac{\mu_0}{2\pi} \cdot \ln\left(\frac{d}{a}\right) \tag{16}$$

Because current paths of inner wires and shields are approximately the same, $L_{in} \approx L_{sh} \approx M$ can be assumed.

3.3 Validation with simulations and measurements

In the following numerical simulations and laboratory measurements (see Fig. 7) are performed to verify the calculation method presented in the previous section.

The CM setup is a one-cable-configuration, which is located 50 mm above a ground plane. Both, the source and the load are positioned between the inner wire and the ground potential. DM simulation and measurement setups are built up as symmetrical two-cable-configurations with an inter-cable distance of 50 mm. The current is set to flow in-between the wires, where one cable carries the load and the other one the return current. Since pure symmetrical conditions are ensured, the ground plane carries no current and can

either be connected or isolated. To allow a flow of shield current, cable screens are connected at both ends to conductive structures and thus sharing their potentials with the ground plane. Since the magnetic SE is independent of the current strength and the load impedance, both are arbitrarily chosen. The resistance of the overall shield path is 2 mΩ, including the intrinsic screen resistance and transfer resistances of the shield connections. The ground path has only a minor resistive influence of 0.1 mΩ, which is the general case for automotive applications.

Numerical field simulations are performed by the LF Frequency Domain Solver of Computer Simulation Technology (CST) EM Studio (CST EM Studio Manual, 2014), based on the finite integration technique (FIT). Using the magnetoquasistatic (MQS) condition, where displacement currents are discarded, helps to significantly accelerate simulation speed and accuracy. At an arbitrary point $P(x_0, y_0, z_0)$, outside the shielded cable, the absolute value of the magnetic field vector is calculated for shielded and unshielded conditions. Hence, the magnetic SE is determined with:

$$SE_{H,sim} = 20\log_{10}\left|\frac{|\boldsymbol{H}_{ush}(x_0, y_0, z_0)|}{|\boldsymbol{H}_{sh}(x_0, y_0, z_0)|}\right| \tag{17}$$

A VNA based measurement method was introduced by Feldhues et al. (2014) to investigate the current distribution of shielded HV cables for low frequency magnetic fields. In this context the same method is used to perform measurements of the magnetic SE. Port 1 feeds the cable configurations with impressed current. A current clamp is connected to port 2 and is first put through an unshielded section and second through a shielded part of the HV cable. The ratio between both scattering parameters S_{21} leads to the magnetic SE with:

$$SE_{H,meas} = 20\log_{10}\left|\frac{\underline{S}_{21,ush}}{\underline{S}_{21,sh}}\right| \tag{18}$$

Figure 8 proves that both numerical field simulation and laboratory measurements achieve well matching results in comparison to the models introduced in Sect. 3.1. The shielding behaviour can be described until approximately 1 MHz, since capacitive properties become relevant for even higher

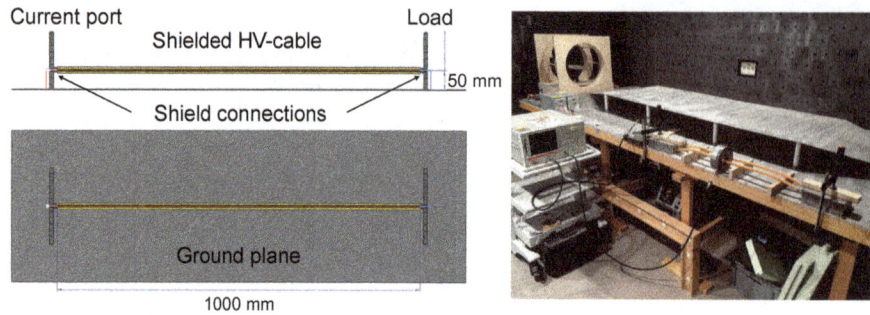

Figure 7. Left: 3-D Simulation Model (CM), right: measurement setup (DM).

frequencies. Having determined typical conditions for automotive HV systems, where $R_{gr} \ll R_{sh}$, magnetic SE values for CM and DM are nearly identical (see Eqs. 13 and 14). However, simulation and measurement data show a constant offset of about 2 dB, when inductive influence becomes dominant. The reason behind are slightly different l.p.u. inductances following from Eqs. (15) and (16), which can be used to calculate the offset with $20\log_{10}\left(L_{\text{CM,in}}/L_{\text{DM,in}}\right)$.

3.4 Reference to automotive applications

In general magnetic fields are attenuated by 20 dB per decade, not before the mutual inductance of the HV cable outweighs the DC-resistance of the shield. Low ohmic screens and screen transitions are therefore decisive to achieve a good shielding performance. At the moment OEMs request 3 mΩ per metre screen and 9 mΩ for each screen connection, based on investigations on the transfer impedance (Jacob, 2013). The method presented in this contribution can be used to review these EMC-requirements for electric vehicles in terms of magnetic fields.

Furthermore the ratio between mutual inductance and self-inductance of the shield plays a significant role. For ideal conditions they are equal and the magnetic field efficiency is continuously increasing with the frequency. In case of additional screen inductances, which are not part of the mutual inductive coupling, it converges to a specific maximum. In real application latter is the usual case, since HV connectors always have small but relevant parasitic inductances that restrict the shielding effect.

When refering to magnetic fields from CM currents, the influence of the ground also needs to be minded (see Eq. 13). For typical conditions, where the ground resistance is far lower than the shield resistance, the CM is basically similar to the DM. But there also might be constellations with high-ohmic ground paths. Even for no or low induced shield voltages, the current is now forced to flow via the shield path, leading to a significant SE even for DC and the very low frequency range. Transferring it to automotive power trains, this condition exemplary will be fulfilled, if HV-components are isolated from vehicle chassis, due to missing or bad protective earth (PE) connections. If there are major issues with

Table 3. Figure 9 legend.

Nr.	Description
1	Dummy load (enclosed by metallic box)
2	Power electronics (enclosed by metallic box)
3	ANs for HV+ and HV− (enclosed by metallic box)
4	Copper plane
5	Phase (AC) lines (u,v,w)
6	Traction (DC) lines (HV+/HV−)
7	Control unit
8	EMI Test Receiver
9	HV DC Voltage Source
10	Active loop antenna

CM currents, this behaviour can be used to improve magnetic shielding.

4 Experimental validations on system level

In the following derived findings of previous sections are validated in terms of magnetic field measurements on system level.

4.1 Test setup

The EMC tests are performed in an anechoic chamber, which enable radiated EMI measurements. In Fig. 9 the basic structure of the setup can be recognized. The centre box includes a prototypal automotive power electronics (EPF) from Continental with a switched-off internal DC/DC converter. Its native enclosure is connected to a surrounding metallic box, that is further grounded to a copper plane. To emulate the EMC behaviour of the electric motor a dummy load is used. It consists of three inductors with an inductance value of $100\,\mu\text{H}$ each. As parasitic capacitive coupling is very low for this configuration, three discrete capacitances (2 nF) are added to obtain realistic conditions in terms of neutral-to-ground and phase-to-phase coupling. The dummy load is also surrounded by a conductive box. Two automotive artificial networks (ANs) are implemented to decouple the

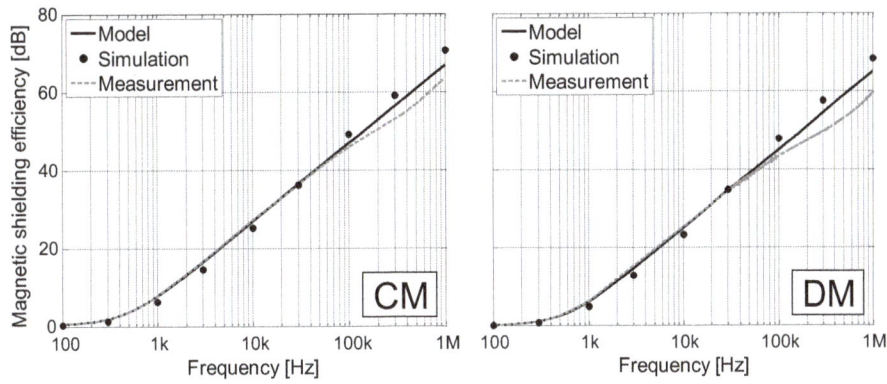

Figure 8. Magnetic shielding efficiency of typical HV cable configurations (left: CM; right: DM).

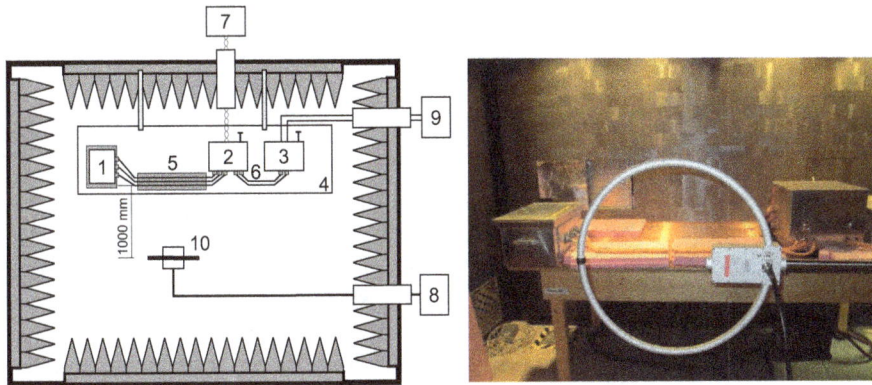

Figure 9. Test setup – H-field measurements of HV system demonstrator.

HV DC power supply from the power electronics. Detailed information about components insight the ANs are not required within the scope of the following investigations, as the primary focus is laid on the AC-side. Coroplast HV cables with Tyco connector systems are used for power transmission, where phase lines are 2 m and traction lines are 1 m long.

There are two operational modes used for experimental investigations below. First one is the so called "Idle Mode", where the operating load current is 0 A and the voltage switching characteristic equals a PWM with a constant duty cycle of 50 % for all three phases. In an electric vehicle this mode correlates to the "Ready to Drive" state. Here, the inverter is already activated, but still causes no torque to run the electric engine. The second one is called the "Load Mode" with sinusoidal pulse-wide modulated phase-to-phase voltages to generate 50 A rotary current at a fundamental frequency of 400 Hz. For both operational conditions the clock frequency is 10 kHz.

4.2 Validation of the current spectrum

Magnetic field measurements are performed with an unshielded cable configuration to correspond them with current spectra introduced in Sect. 2. The screens of the HV cables are still present, but disconnected at the load side. Because of the high-ohmic shield path, the magnetic SE is thus reduced to zero (see Eq. 12). Since currents are proportional to magnetic fields, there is a constant coupling factor over the observed frequency range, which is depend on the geometry of current paths and the distance as well as the polarisation of the loop antenna.

According to Fig. 10 the "Idle Mode" measurement follows the derivated CM current spectrum (see Sect. 2.1). This behaviour is expectable, because the in-phase PWM voltage switching is causing a set of pure CM harmonics. Considering "Load Mode", the spectral characteristic changes and a significant inductive influence from the dummy coils can be observed for lower frequencies. Next to the phase-to-neutral voltages, there are now additional phase-to-phase voltages and therefore respective DM currents (see Sect. 2.2). At approximately 2.5 MHz a resonance point is shaped for both paths, where the current is only restricted by the DC-resistance of the overall path. An EMC requirement for a country-specific type approval is indicated by a threshold according to GB/T 18387 (2008). In reference to these unshielded magnetic field measurements, the threshold is widely exceeded.

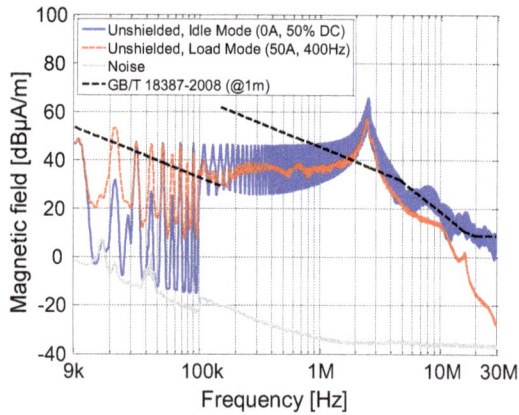

Figure 10. Experimental results to validate current spectra.

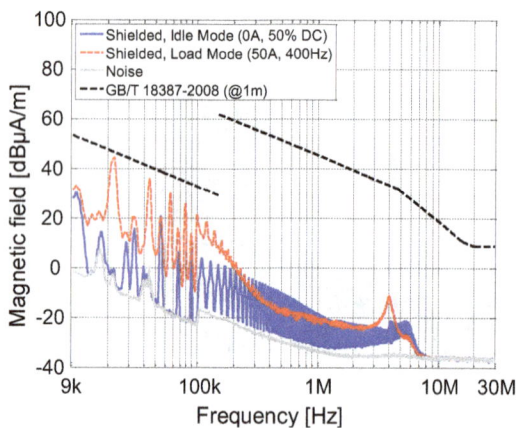

Figure 11. Experimental results to validate magnetic shielding efficiency.

4.3 Validation of magnetic shielding efficiency

In Figure 11 magnetic fields are measured for the shielded situation with cable screens, connected at both sides to the metallic enclosures. Hence, there is a magnetic shielding effect, due to the compensating induced shield currents described in Sect. 3. When comparing unshielded and shielded case, the magnetic SE is approximately 20 dB at 10 kHz (clock frequency). Like expected it increases with 20 dB per decade. Remembering that CM and DM currents are simultaneously present in the "Load Mode", it is verified that respective magnetic fields are attenuated in the same way. Considering again the magnetic field threshold, EMC requirements for the shielded environment are now complied. However, for the loaded operating condition still critical magnetic fields are measured from EMC point of view. They can be directly referred to symmetrical current ripples and thus needs to be focused on, when designing HV topologies.

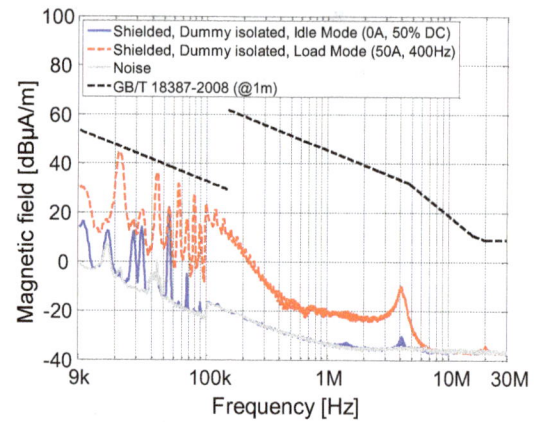

Figure 12. Experimental results to validate CM and DM current paths.

4.4 Validation of current paths

It might be relevant, whether either CM or DM currents are causing dominant magnetic fields. As they are superpositioned in the "Load Mode", the demonstrator must be slightly adapted to enable this kind of investigation. From Sect. 2.1 it is known, that CM currents are generated due to the capacitive coupled reference system, which is represented by the copper plane and cable screens in the test setup. If the copper plane is now isolated from the dummy load box, the return path is restricted to the screens. An ideal condition like in coaxial transmission lines is forced, where the inner wire equals the inverted shield current. Magnetic fields compensate and the shielding efficiency is high, even for DC and the very low frequency range (see Eq. 13). This effect is only related to CM currents, whereas magnetic fields from DM currents are attenuated like before (see Eq. 14).

In Fig. 12 respective measurements with isolated dummy boxes for shielded cables are displayed. As expected nearly no magnetic fields can be measured in the "Idle Mode", because of a set of pure CM current harmonics. Since there are always systemic asymmetries or other coupling paths like the DC-network, some minor current harmonics are left. Observing now the "Load Mode", the magnetic spectrum stays the same, although the CM part is completely removed. Hence, the loaded operating condition is dominated by magnetic fields caused by DM currents over the entire frequency range.

5 Conclusions

This contribution pointed out, that EMC-concepts for HV topologies of electric vehicles require the consideration of parasitic harmonic currents in the scope of magnetic fields. An appropriate countermeasure is to decrease parasitic X- and Y-capacitances distributed in electric motors and power electronics. Since it will never be possible to reduce them to zero and additional currents with inductive origins are present, the routing of HV cables gets a crucial role.

The introduced current analysis revealed the decisive current paths, which must be held geometrically small to minimize inductive coupling. For the CM case HV cables need to be positioned as near as possible to their reference system, which in general is the chassis of the vehicle. In order to reduce the area of DM current loops the single HV lines must lie as close as possible to each other. Especially in the region of the connectors, where cables are often spread, this requirement is difficult to follow. The use of sum shields and consequently less connectors might be advantageous in this context.

The magnetic shielding efficiency can be improved with low ohmic and low inductive screen connections. Particularly for low frequencies from DC to 100 kHz it was demonstrated that attenuation is not sufficient to reduce magnetic fields far below the limit values. Adequate cable routing and elimination of parasitics thus remain an integral component of related EMC-concepts. If unshielded cables are used, an enhancement of magnetic fields up to 60 dB must be expected. In this case a well considered filtering system at the AC- and DC-side of the HV system becomes unavoidable.

Acknowledgements. This contribution was developed within the scope of the project EM4EM (Electromagnetic Reliability of Electronic Systems for Electro Mobility – Subproject: Entwurfs- und EMZ-Messmethodik für EMZ-Analysen auf EV-Gesamtsystemebene) which is funded by the BMBF (Bundesministerium für Bildung und Forschung) under the grant no. 16M3092A. The responsibility for this publication is held by the authors only.

Edited by: M. Chandra

References

Cacciato, M., Consoli, A., Scarcella, G., and Testa, A.: Reduction of Common-Mode Currents in PWM Inverter Motor Drives, IEEE T. Ind. Appl., 35, 469–476, 1999.

CST EM Studio 2014: Workflow & Solver Overview, CST – Computer Simulation Technology AG, Darmstadt, Germany, 2014.

Feldhues, K., Diebig, M., and Frei, S.: Analysis of the Low Frequency Shielding Behavior of High Voltage Cables in Electric Vehicles, International Symposium on Electromagnetic Compatibility (EMC Europe 2014), Gothenburg, Sweden, 1–4 September 2014, 408–413, 2014.

GB/T 18387-2008: Limits and test methods of magnetic and electric field strength from electric vehicles Broadband 9 kHz to 30 MHz, EMC norm, 2008.

Grandi, G., Casadei, D., and Reggiani, U.: Common- and Differential-Mode HF Current Components in AC Motors Supplied by Voltage Source Inverters, IEEE T. Power Electr., 19, 16–24, 2004.

Hohloch, J., Tenbohlen, S., Köhler, W., Aidam, M., and Ludwig, A.: Measurement of Transfer Impedances of Components for Automotive High-Voltage Power Networks, 2012 International Symposium on Electromagnetic Compatibility (EMC EUROPE), Rome, Italy, 17–21 September 2012 , 1–6, doi:10.1109/EMCEurope.2012.6396708, 2012.

Jacob, F., Heyen J., Rinkleff T., and Golisch F.: EMV-Anforderungen an das Hochvoltbordnetz, GMM-Fachbericht 77: EMV in der Kfz-Technik, VDE Verlag GmbH, Stuttgart, 2013 (in German).

Jettanasen C.: Influence of Power Shielded Cable and Ground on Distribution of Common Mode Currents Flowing in Variable-Speed AC Motor Drive Systems, 2010 Asia-Pacific International Symposium on Electromagnetic Compatibility, Beijing, China, 12–16 April 2010, 953–956, 2010.

Kempski, A., Smolenski, R., and Strzelecki, R.: Common Mode Current Paths and Their Modeling in PWM Inverter-Fed Drives, Power Electronics Specialists Annual IEEE Conference – PESC, Cairns, Australia, 23–27 June 2002, 1551–1556, 2002.

Kempski, A. and Smolenski, R.: Decomposition of EMI Noise into Common and Differential Modes in PWM Inverter Drive System, Electrical Power Quality and Utilisation, XII, 53–58, 2006.

Paul, C. R.: Introduction to electromagnetic compatibility, 2nd Edn., Hoboken, NJ, Wiley, 2006.

5

Improving the range of UHF RFID transponders using solar energy harvesting under low light conditions

A. Ascher, M. Lehner, M. Eberhardt, and E. Biebl

Fachgebiet Höchstfrequenztechnik, Technische Universität München, Munich, Germany

Correspondence to: A. Ascher (alois.ascher@tum.de)

Abstract. The sensitivity of passive UHF RFID transponders (Radio Frequency Identification) is the key issue, which determines the maximum read range of an UHF RFID system. During this work the ability of improving the sensitivity using solar energy harvesting, especially for low light conditions, is shown. To use the additional energy harvested from the examined silicon and organic solar cells, the passive RFID system is changed into a semi-active one. This needs no changes on the reader hardware itself, only the used RFIC (Radio Frequency Integrated Circuit) of the transponder has to possess an additional input pin for an external supply voltage. The silicon and organic cells are evaluated and compared to each other regarding their low light performance. The different cells are examined in a shielded box, which is protected from the environmental lighting. Additionally, a demonstrator is shown, which makes the measurement of the extended read range with respect to the lighting conditions possible. If the cells are completely darkened, the sensitivity gain is ascertained using high capacity super caps. Due to the measurements an enhancement in range up to 70 % could be guaranteed even under low light conditions.

1 Introduction

RFID denotes a system that allows an automatically detection and localizing process. The communication between an UHF RFID transponder and the corresponding reader comprises of an UHF reader sending a RF signal, which is detected by the transponder. The transponder, which is consisting of an antenna and an RFIC chip, is operated with the received power of the reader. The reader signal is used for supplying the internal logic of the IC and the backscatter to the reader itself. The maximum detection range is limited by law because of sending-power restrictions and the given environmental conditions for a specific case of application. Depending on the environmental requirements the detection range of an RFID system can be improved. Nowadays RFID systems are utilized in different applications, which exhibit various requirements in case of detection range and reliability. For monitoring of issued goods in a warehouse the dominating factor is reliability. The detection range is not that important, because the goods are very close to the reader. In contrast a system for gathering the freeway toll must be both, very reliable and with a large detection range. To guarantee high ranges with RFID systems without using an additional battery a solution using solar energy harvesting is proposed. Based on the concept of a passive UHF RFID system a semi-active localizing system is developed. Thereby no changes on the reader itself are needed. Only the transponder is changed with the goal to get the internal supply of the used RFIC independent of the received reader signal power. For that issue solar energy harvesting should be applied (Spies et al., 2007). Thereby amorphous silicon solar cells from Vimun and Powerfilm, which are commonly used in energy harvesting solutions (Georgiadis and Collado, 2011; Sample et al., 2011), and organic solar cells from the Fraunhofer ISE, which have some advantages in low light performance (Dennler and Sariciftci, 2005), are evaluated and compared to each other. In many cases solar energy harvesting in RFID applications is described using optimal illumination scenarios or no further information about the needed illuminance is given. So the focus shall be onto the low light performance of a possible transponder for indoor applications, using different types of solar cells linked to the maximum read range of the RFID system. Several research activities for performant silicon solar cells with high efficiencies at low light conditions (Glunz et al., 2002) show,

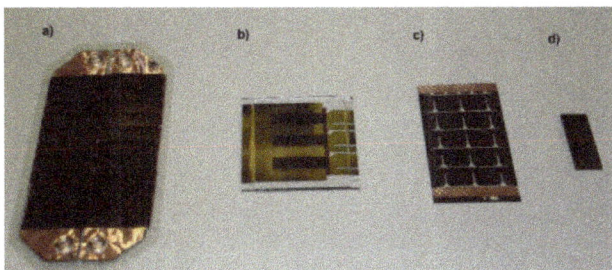

Figure 1. Used solar cells: (**a**) organic panel; (**b**) organic single cells; (**c**) Powerfilm SP 3-37; (**d**) Vimun Sc-3012.

that the operation of autonomous solar powered devices for indoor applications is possible. The case of completely darkened solar cells should be intercepted in our case and a potential transponder should work with its higher sensitivity for a few hours using an additional capacitor. Finally, based on the previous results, a variable demonstrator is build with the possibility to examine different types of antennas, capacitors and solar cells regarding the illuminance and the achievable reading range.

The following demonstrator-transponder uses the Impinij Monza X-2K RFIC, which has a reading sensitivity of -17 dBm without- and a reading sensitivity of -24 dBm with an additional external voltage supply. Therefore, the minimum voltage supply is 1.6 V at a minimum current of $-17\,\mu$A. Using the demonstrator read sensitivity of -17 dBm and the Friis equation (Finkenzeller, 2008) the maximum detection range is 11.15 m with a dipol as transponder antenna and the maximum EIRP sending power of 33 dBm at the reader. The maximum detection range for the same setting, excepting a read sensitivity of -24 dBm, is 24.96 m. The difference in distance is the maximum achievable gain for the detection range limited by the sensitivity of the demonstrator-transponder.

2 Evaluating solar cells

For evaluating the amorphous silicon solar cells Powerfilm SP 3-37 and Vimun Sc-3012, the organic solar cells (Fig. 1) and the comparison of these types, a reproducible lighting environment is needed. For that reason a lighting box is build and verified (Fig. 2). Three different lighting concepts are integrated in the reference box which can be regulated using analog dimming. The measurement can be performed using a controllable dc power supply. The three different lighting types are two modes using LEDs for 1200 lx illuminance (winter in the shade) and 9000 lx illuminance (summer in the shade) and one mode using halogen lamps for an illuminance of 110 000 lx (summer in the sun). Thereby, first attempts have shown that the high illuminance mode using the halogen lamps is uncritical for all solar cells evaluated regarding the power provided. Following the dc power sup-

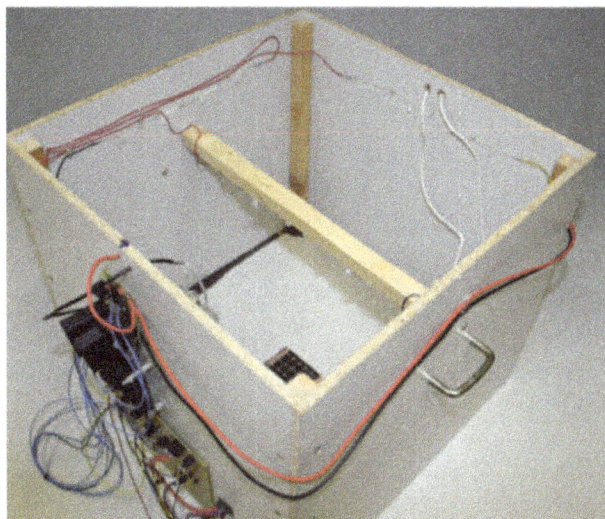

Figure 2. Illuminance box for reference lighting.

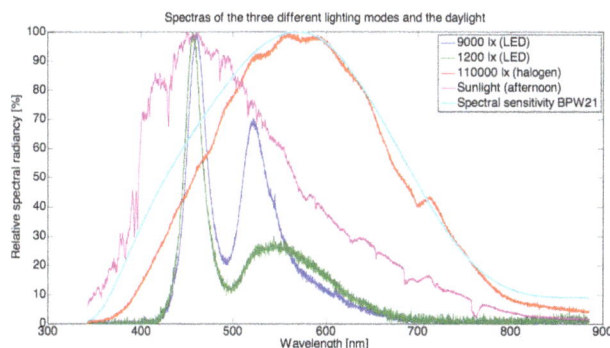

Figure 3. Different spectra of the reference box.

ply for the LED modes starts with the corresponding voltage for 1200 or 9000 lx illuminance. Then the supply voltage is stepping down until no light is emitting anymore. That implies in case of the 1200 lx- or the 9000 lx LED mode that because of the circuitry-wise realization a total number of 280 dimming steps with an increment of 0.01 V are possible. The illuminance of the different modes is investigated with using a luxmeter with an integrated BPW21 photodiode. This photodiode has approximately the same weighting function as the human eye, so the illuminance measurement in lux is possible. The different spectra of the three different modes, which were recorded using a spectrometer from Aseq, and the spectral sensitivity of the BPW21 photodiode can be seen in Fig. 3.

For a better comparison of the different lighting conditions it would be possible to transform the illuminance into the corresponding irradiance using the photometric radiation equivalent, because the spectrum is known. But for the considered kinds of solar cells there is no need for this transformation, because their spectral sensitivity is approximately equal.

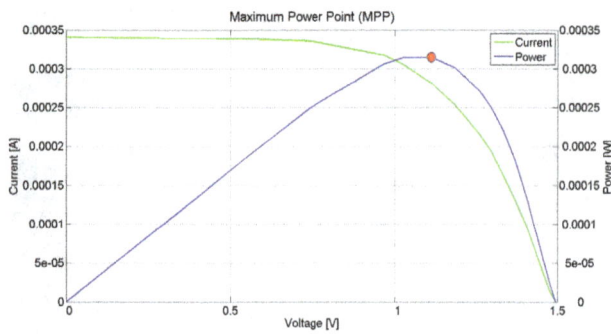

Figure 4. MPP of the Powerfilm SP 3-37 cell with an illuminance of 1200 lx.

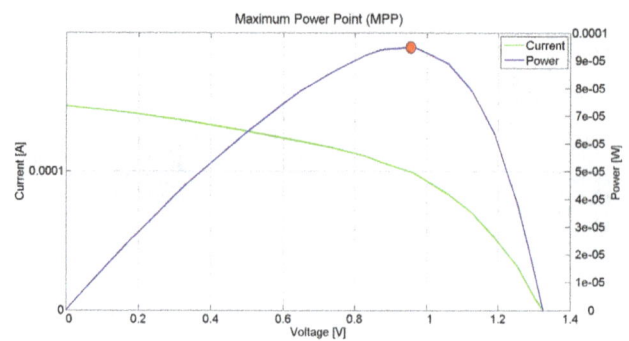

Figure 5. MPP of the organic solar cell with an illuminance 1200 lx.

To compare the different cells with each other maximum power point (mpp) measurements are used (McEvoy et al., 2011). The maximum power point defines the point of maximum available power in the voltage-current diagram of a solar cell. This point depends on the temperature and the lighting. In this case both, the Vimun Sc-3012 and the organic solar cell are compared under the same illumination conditions at an illuminance of 1200 lx. With less than 1.6 V maximum voltage (Fig. 4) the Powerfilm SP 3-37 is not guaranteeing the needed voltage at an illuminance of 1200 lx and is for that reason not suitable for powering the IC. Further on the SP 3-37 would require a lot of space on a possible transponder because of its greater size and is therefore inappropriate for a realization. In Figs. 4–6, the mpp is highlited red.

As shown in Figs. 4–6 such comparatively bad illumination conditions already ensure a power of about 90 µW. To use the Monza X-2K a voltage of 1.6 V and a current of 15 µA must be guaranteed. Furthermore it is important that the examined organic solar cells are strongly degenerated. That implies that a fully operative organic solar cell can provide five times more power than the evaluated one. The used degenerated organic cell can already ensure a voltage of 1.4 V and a current of around 65 µA in its maximum power point (Fig. 5). If a non-degenerated panel would be used a current of about 300 µA would appear at the same voltage.

For the amorphous silicon solar cell Vimun Sc-3012 the mpp is around 0.95 V at an illuminance of 1200 lx (Fig. 6).

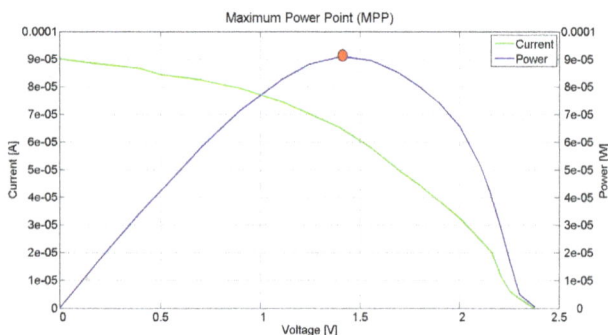

Figure 6. MPP of the Vimun Sc-3012 with an illuminance of 1200 lx.

As a voltage of 1.6 V is needed to get the IC operational and the maximum voltage is only 1.3 V a enhanced reading range is not possible. However it is possible to build a series circuit of two Vimun Sc-3012 solar cells. These series connection is ensuring enough current for operating the Monza X-2K consequently. Because of the tiny form factor of the Vimun Sc-3012 a second cell can easily be attached at the surface of a possible transponder. The different cells have furthermore been evaluated under higher illuminance, which indicated as uncritical for the service of the RFIC. Further measurement results concerning different types of solar cells are mentioned in Mathews et al. (2014). Thereby GaAs – and dye sensitized solar cells were examined for indoor applications by varying the illuminance.

3   Used capacitor

Different capacitors of four electronic contributors got investigated regarding their loading characteristic, leakage current and operating time (Brunelli et al., 2009). To evaluate the loading characteristic of the different capacitors with typical values from 50 to 220 mF, it is directly linked to the used solar cells under certain lighting conditions. The time to load a 220 mF capacitor until the minimum operating voltage of the Monza X-2K is reached with respect to an illuminance of 1200 lx is for the organic non-degraded cell one hour, for the degraded organic cell five hours and for the Vimun Sc-3012 five hours too. If an additionally schottky diode with a typical dropout voltage of 0.2 V in forward direction is used for discharge protection, the loading time for the non-degraded organic cell increases up to two hours and for the degraded organic cell and the Vimun Sc-3012 up to eight hours assuming an illuminance of 1200 lx. For higher illuminance the required loading time is decreasing very fast. The typical leakage current is around 2 µA which is the reason for the voltage drop over time. If a 220 mF capacitor has a voltage of 2.5 V and exhibits a leakage current of 2 µA it takes 35 h until the voltage drops below 1.6 V. Comparing two different fully loaded capacitors with a capacity of 50 and 220 mF

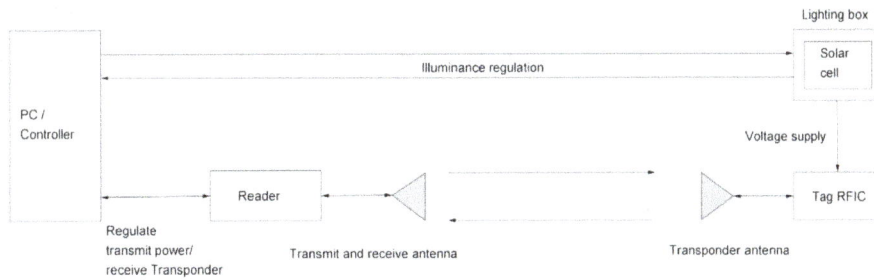

Figure 7. Measurement setup.

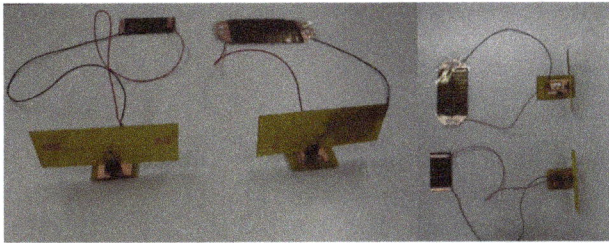

Figure 8. Demonstrators using meander dipole antennas.

concerning their maximum operating time shows, that a capacity of 50 mF allows an operating time of one hour and a 220 mF capacitor an operating time of 7 h. Finally, it appears that the Panasonic Goldcaps are the most common for the intended application. They combine a tiny form factor with a small leakage current and a cheap price.

4   Range improvement

Initially different modular demonstrators are build, which allows the evaluation of the possible range improvement using a combination of the examined solar cells, the chosen capacitors and different antenna types. These demonstrators are examined concerning the reachable sensitivity improvement. Thereby, the different solar cells are in a series connection with a schottky diode for discharge protection followed by a supercapacitor in a parallel connection between power and ground. This circuit is connected to the VDD pin of the RFIC. The demonstrators are shown in Fig. 8.

The modularized demonstrators are developed to use different antenna types. One possibility is to use meander dipole antennas, which allows a decrease in required space of the transponder. The effective length remains equal and the directivity is approximately omnidirectional. A second possibility is the usage of planar patch antennas on different substrates. Due to the small bandwidth of about one percent, the high substrate losses using FR4 and the large amount of space needed for a patch antenna lead to the conclusion that this antenna type is impractical for the planned application (Balanis, 2005).

Figure 9. The mentioned setup from Fig. 6 in the anechoic chamber.

The chosen meander dipole antennas can be plugged into a basic module that consists of a capacitor for buffering and a diode as discharge protection. For measurements it is possible to change each of the components, either the antenna, the capacitor or the solar cell.

The read sensitivity depending on the illuminance should be evaluated for the different demonstrator setups. For the following measurements, using the measurement setup shown in Fig. 6, a 3 V reference, the Powerfilm SP 3-37, the Vimun Sc-3012 and the degraded organic cells are compared to each other. The measurements were realized in an anechoic chamber (Fig. 9), with a fixed distance between the reader antenna and the demonstrator of one meter at a height of 91 cm. Two different measurements were made.

First of all the transponder sensitivity for a used dipole antenna with vertical meander structure and a dipole antenna with horizontal meander structure are measured as a function of the used antenna type and the applied solar cell. The illuminance remains constant during one measurement cycle. The read rate (transponder readings per second) was recorded while varying the transmitted power of the reader. It can be seen in Fig. 10 that for the 3 V reference as well as for the different voltage sources the vertical meander dipole needs 2 dB less reader sending power then the horizontal dipole. To determine the maximal sensitivity enhancement of the

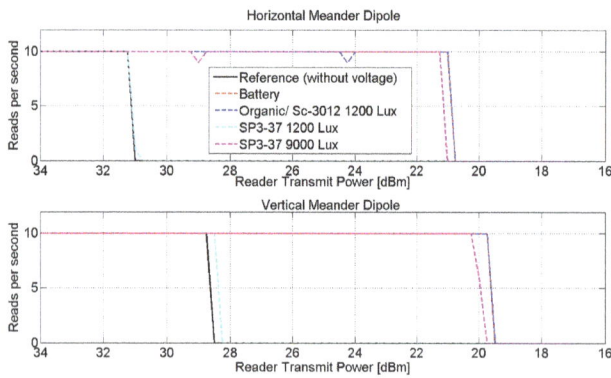

Figure 10. Reads per second using different solar cells and varying the reader sending power.

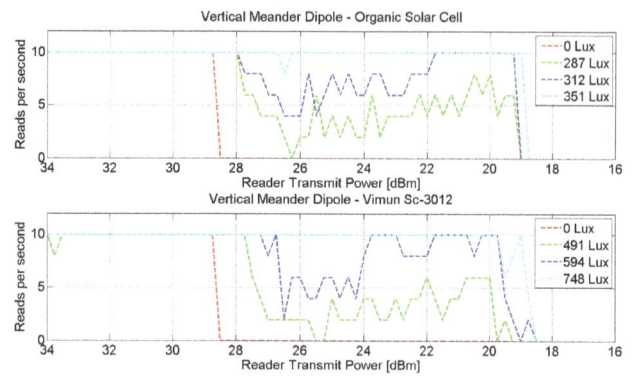

Figure 11. Sensitivity improvement depending on the illuminance for the organic solar cell and the Vimun Sc-3012 using a vertical meander dipole antenna.

transponder the minimal necessary reader sending power was detected using a 3 V reference voltage independent from the illuminance.

Measurements with the Powerfilm SP 3-37, the Vimun Sc-3012 and the organic degraded cells at an illuminance of 1200 and 9000 lx were done additionally. It showed, that the Vimun Sc-3012 and the organic cells could provide enough power to the RFIC at an illuminance of 1200 lx. The measurement graphs of the Vimun and the organic cells in Fig. 10 are congruent with the 3 V reference. Furthermore it can be seen, that the additional voltage supply of the RFIC leads to a 9 dB lower sending power needed at the reader. That 9 dB less sending power equals a 70 % higher detection range. The listed SP 3-37 did not provide enough supply voltage at 1200 lx. Even if the illuminance is 9000 lx, there is a narrow difference in the needed reader sending power compared to the 3 V reference.

A second performance evaluation is realized to examine the necessary illuminance to guarantee the maximal sensitivity improvement of the transponder. For this evaluation only the vertical meander dipole antenna is used as the same result can be seen for the horizontal meander dipole.

The measurements in Fig. 11 have shown, that the sensitivity improvement for the organic solar cells and a series circuit of two Vimun Sc-3012 start at different illuminances. With an illuminance of 287 lx the sensitivity improvement initiates for the organic solar cells and reaches its maximum at a value of 351 lx. This effect appears considerably later for the series circuit of the two Vimun cells. With an illuminance of 491 lx the sensitivity improvement initiates and reaches its maximum value at 748 lx.

Regarding the mentioned measurement setup and the performed evaluations it is possible to decide whether the solar cells of an energy harvesting transponder are suitable for the envisaged application. In summary, the organic solar cells are more suitably at bad lighting conditions. However, the organic cells require more space than the conventional amor-

phous silicon solar cells and are more expensive disposed of their development stage.

5 Conclusions

Organic solar cells and amorphous silicon solar cells are proposed and evaluated to power RFICs that contain a pin for an external voltage supply in order to increase the read range of a transponder under bad lighting conditions. The cells are evaluated by maximum power point measurements using a reproducible measurement environment. To guarantee the higher sensitivity of potential semi-active transponder, even if the cells are darkened for a short period of time, an adequate capacitor is chosen. Finally a modular demonstrator is build with the possibility to use different antenna types. This demonstrator is used to examine the transponder sensitivity improvement depending on the illuminance regarding the different types of solar cells. As a result of the examinations a 70 % higher detection range could be ensured using organic cells starting at an illuminance of 351 and 748 lx in case of the Vimun silicon solar cells. With the measurement setup shown in Fig. 7 it is possible to make a decision whether the solar cells of an energy harvesting transponder are suitable for the desired application. Future work consists of developing a completely integrated transponder using organic solar cells for indoor applications.

Acknowledgements. The authors acknowledge the Fraunhofer ISE for providing the organic solar cells used in this work.

References

Balanis, C. A.: Antenna Theory, 3rd Edn., John Wiley & Sons, Hoboken, USA, 2005.

Brunelli, D., Moser, C., Thiele, L., and Benini, L.: Design of a Solar-Harvesting Circuit for Batteryless Embedded Systems, IEEE T. Circ. Syst., 1, 2519–2528, 2009.

Dennler, G. and Sariciftci, N. S.: Flexible Conjugated Polymer-Based Plastic Solar Cells: From Basics to Applications, Proc. IEEE, 93, 1429–1439, 2005.

Finkenzeller, K.: RFID Handbuch, 5. Aktualisierte und erweiterte Auflage, Carl Hanser Verlag GmbH, Munich,, 2008.

Georgiadis, A. and Collado, A.: Improving Range of Passive RFID Tag Utitilizing Energy Harvesting and High Efficiency Class-E Oscillators, Centre Tecnologic de Tele-comunications de Catalunya (CTTC), Prague, 2011.

Glunz, S. W., Dicker, J., Esterle, M., Hermle, M., Isenberg, J., Kamerewerd, F. J., Knobloch, J., Kray, D., Leimenstoll, A., Lutz, F., Osswald, D., Preu, R., Rein, S., Schaffer, E., Schetter, C., Schmidhuber, H., Schmidt, H., Steuder, M., Vorgrimler, C., and Willeke, G.: High-Efficiency Silicon Solar Cells for Low-Illumination Applications, IEEE Photovoltaic Specialist Conference (PVSC), New Orleans, USA, 2002.

Mathews, I., Kelly, G., King, P. J., and Frizzell, R.: GaAs solar cells for Indoor Light Harvesting, 40th IEEE Photovoltaic Specialist Conference (PVSC), Denver, USA, 2014.

McEvoy, A., Markvart, T., and Castaner, L.: Practical Handbook of Photovoltaics: Fundamentals and Applications, 2nd Edn., Academic Press, Waltham, 2011.

Sample, A. P., Braun, J., Parks, A., and Smith, J. R.: Photovoltaic Enhanced UHF RFID Tag Antennas for Dual Purpose Energy Harvesting, IEEE International Conference on RFID 2011, Orlando, USA, 2011.

Spies, P., Pollak, M., and Rohmer, G.: Energy Harvesting for Mobile Communication Devices, 29. Int. Telecommunications Energy Conference, Fraunhofer ILS, Nürnberg, 2007.

6

Efficient determination of the left-eigenvectors for the Method of Lines

S. F. Helfert

FernUniversität in Hagen, Hagen, Germany

Correspondence to: S. Helfert (stefan.helfert@fernuni-hagen.de)

Abstract. The efficient determination of left eigenvectors in the method of lines (MoL) is described in this paper. The electromagnetic fields are expanded into eigenmodes and the eigenmodes are determined from an explicit matrix eigenvector problem. To study complicated structures with a moderate numerical effort, the analysis is done with a reduced set of these eigenmodes. The enforcements of the continuity of the transverse electric and magnetic fields at interfaces leads to expressions with rectangular matrices. Now left eigenvectors can be considered as inverse of these rectangular matrices. Until now, the left eigenvectors were determined from a second explicit eigenvalue problem. Here, it is shown how they can be determined with simple matrix products from previously determined right eigenvectors. This is done by utilizing the relation between the transverse electric and magnetic fields. The derived formulas hold for structures with Dirichlet, Neumann or periodic boundary conditions and the materials may be lossy. Open structures are modeled with perfectly matched layers (PML). To verify the expressions, various devices that contain such PMLs and lossy metals were studied. In all cases, error measures show that the algorithm derived in this paper works very well.

1 Introduction

Complicated waveguide structures can be analyzed analytically only in exceptional cases. Therefore, usually numerical methods are used for this purpose. There are various possibilities to classify these methods, e.g. according to their analytical/numerical part. In the well known finite difference time domain method (FDTD, see e.g. Taflove, 1995) all derivatives that occur in Maxwell's equations are discretized with finite differences. Since all points in space and time are dis-

cretized, the simulation of complicated structures can be numerically demanding.

On the other side there are eigenmode algorithms (see e.g. Sudbø, 1993; Sztefka, 1992). Here, analytic expressions are used at least in the direction of wave propagation. Continuously varying structures are typically modeled with a stair-case approximation. Besides, usually the analytic expressions exist of infinite series, which must be truncated in practical applications. Therefore, in spite of its analytic formulation, approximations are also introduced in case of eigenmode methods. There exist various ways in computing the eigenmodes and it is not always easy to determine all important ones with analytic expressions in case of complex structures.

Here, we use the Method of Lines (MoL) where the eigenmodes are determined after discretization with finite differences from an explicit eigenvalue problem.

Since the MoL is well documented in the literature we just mention some book chapters (Pregla and Pascher, 1989) (Pregla, 1995) and the book (Pregla, 2008), here. For three-dimensional structures a two-dimensional discretization is required and the occurring matrices become very large. To keep the numerical effort moderate, it was proposed in the past to use a reduced number of eigenmodes (Gerdes, 1994). At interfaces between different sections the transverse field components have to be continuous. When a reduced set of eigenmodes is used one has to determine inverses of rectangular (not square) matrices. For this reason, (Schneider, 1999) proposed the use of a pseudo inverse (Strang, 1986). In contrast, the utilization of left-eigenvectors was suggested in (Helfert et al., 2003). The given methods work quite well, but the numerical cost for determining the pseudo inverse or the left eigenvectors is quite high. Here, we will also apply left-eigenvectors but show how they can be determined with sim-

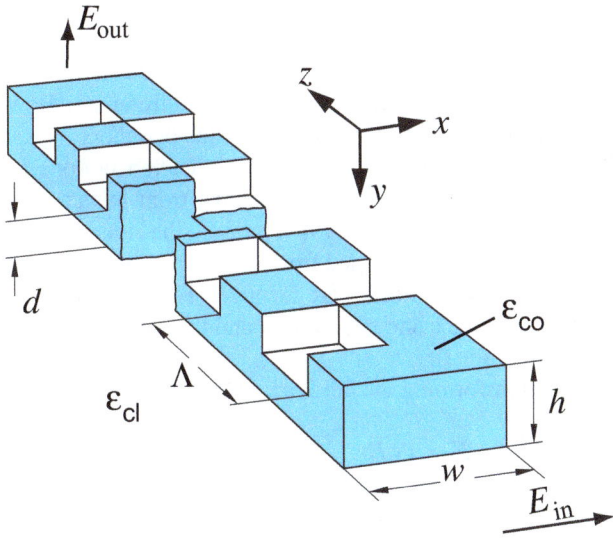

Figure 1. Polarization converter as example of a complex three-dimensional structure.

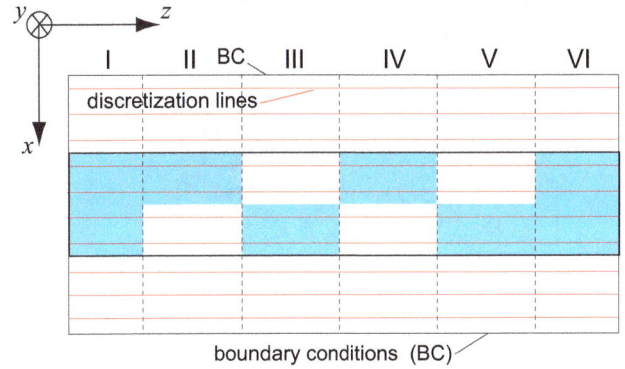

Figure 2. Analysis with the MoL: dividing the structure into homogeneous sections and discretization.

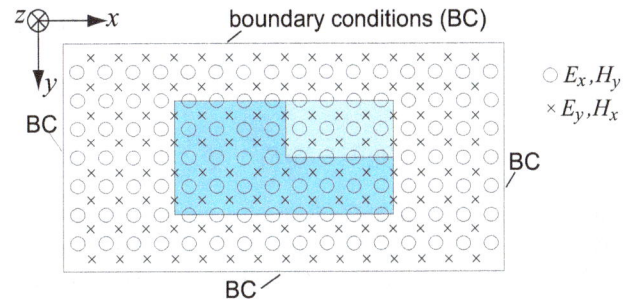

Figure 3. Discretization points in the cross-section.

ple matrix products from previously determined right eigenvectors. The derivations can also be understood as proof of the orthogonality of eigenmodes in closed structures.

The paper is organized as follows: we start with a repetition of the principles of the MoL. Following is a description of the left-eigenvectors, and we will particularly show how they can be determined in an efficient way. After that, numerical results are shown where we demonstrate that the developed expressions also work in case of perfectly matched layers and for lossy materials. The paper ends with a summary.

2 Theory

2.1 Method of Lines

In the Method of Lines (MoL) the wave propagation is described with eigenmodes. It allows the analysis of complicated devices. As example the polarization converter taken from (Mustieles et al., 1993) is shown in Fig. 1. It consist of a dielectric waveguide at the input followed by a region with a periodic modification of the core. With a suitable choice of the length Λ, a polarization rotation is possible (see Mustieles et al., 1993).

The eigenmodes in the MoL are determined after a discretization with finite differences. In this section, we describe the basic algorithm. To model waveguide structures with the MoL, we first divide it into homogeneous sections with respect to the direction of propagation as shown in Fig. 2. Then, solutions for the homogeneous sections and the interfaces are determined.

Solution in homogeneous sections

For the numerical analysis, we start with generalized transmission line (GTL) equations, which relate the transverse electric and magnetic fields. In what follows, the coordinates are normalized with the free space wave number $k_0 = \omega\sqrt{\varepsilon_0\mu_0} = 2\pi/\lambda_0$: $\bar{u} = k_0 u$ ($u = x, y, z$). Further, the magnetic field is normalized with the free space wave impedance $Z_0 = \sqrt{\mu_0/\varepsilon_0} = 120\pi\,\Omega$: $\tilde{H}_u = Z_0 H_u$. Then, the GTL-equations read (see e.g.Pregla, 1999; Pregla, 2008):

$$\frac{\partial [H]_t}{\partial \bar{z}} = -j R_E [E]_t \qquad \frac{\partial [E]_t}{\partial \bar{z}} = -j R_H [H]_t \qquad (1)$$

with

$$[E]_t = \begin{pmatrix} E_y \\ E_x \end{pmatrix} \qquad [H]_t = \begin{pmatrix} -\tilde{H}_x \\ \tilde{H}_y \end{pmatrix} \qquad (2)$$

Note: the continuous physical vectors were put in brackets here, to distinguish them from mathematical vectors that originate from discretization. These mathematical vectors are written in bold.

In the analysis, anisotropic materials of the following form are considered:

$$\overset{\leftrightarrow}{v}_r = \begin{bmatrix} v_x & 0 & 0 \\ 0 & v_y & 0 \\ 0 & 0 & v_z \end{bmatrix} \qquad \text{with} \qquad \overset{\leftrightarrow}{v}_r = \overset{\leftrightarrow}{\varepsilon}_r, \overset{\leftrightarrow}{\mu}_r \qquad (3)$$

With these tensors for the material parameters the operators R_H und R_E are given as:

$$R_E = \begin{bmatrix} \varepsilon_y + D_{\overline{x}}\mu_z^{-1}D_{\overline{x}} & -D_{\overline{x}}\mu_z^{-1}D_{\overline{y}} \\ -D_{\overline{y}}\mu_z^{-1}D_{\overline{x}} & \varepsilon_x + D_{\overline{y}}\mu_z^{-1}D_{\overline{y}} \end{bmatrix}$$

$$R_H = \begin{bmatrix} \mu_x + D_{\overline{y}}\varepsilon_z^{-1}D_{\overline{y}} & D_{\overline{y}}\varepsilon_z^{-1}D_{\overline{x}} \\ D_{\overline{x}}\varepsilon_z^{-1}D_{\overline{y}} & \mu_y + D_{\overline{x}}\varepsilon_z^{-1}D_{\overline{x}} \end{bmatrix}$$

Note: in the following the derivatives in x, y direction will be treated differently than the derivative with respect to z. To indicate this, the abbreviation $D_{\overline{v}} = \partial/\partial\overline{v}$ (with $v = x, y$) was introduced for these derivatives.

By combining the GTL-equations (1), the following coupled wave equations are obtained:

$$\frac{\partial^2[H]_t}{\partial\overline{z}^2} - Q_H[H]_t = [0] \qquad \frac{\partial^2[E]_t}{\partial\overline{z}^2} - Q_E[E]_t = [0] \quad (4)$$

with

$$Q_H = -R_E R_H \qquad Q_E = -R_H R_E \qquad (5)$$

The next step of the analysis is the discretization of Eq. (4) with finite differences in the direction of the cross–section, while the z-dependency remains. Figures 2–3 show this discretization process from the top and in the cross-section.

As indicated, the components of the electric or magnetic fields (i.e. E_x, E_y resp. H_x, H_y) are determined at different positions. With this shift, the coupling between the components that is described by the product of first derivatives in x and y direction (see off-diagonal elements of $R_{E,H}$) can be modeled in an optimal way. [More details can be found e.g. in Pregla, 2008].

Now, the fields are combined in vectors, and the operators Q, R becomes operator matrices that contain the approximations of the derivatives with finite differences and the material parameters.

$$[H]_t \Rightarrow H \quad [E]_t \Rightarrow E \qquad Q_{H,E} \Rightarrow \mathbf{Q}_{H,E} \quad R_{H,E} \Rightarrow \mathbf{R}_{H,E}$$

In what follows, we describe the solution of the wave equation for the electric and magnetic field in parallel. However, these fields are coupled by Eq. (1). Therefore, in practice, we have to solve the wave equation for only one of the fields and determine the remaining one with the help of Eq. (1).

After discretization, Eq. (4) becomes:

$$\frac{\partial^2 H}{\partial\overline{z}^2} - \mathbf{Q}_H H = 0 \qquad \frac{\partial^2 E}{\partial\overline{z}^2} - \mathbf{Q}_E E = 0 \qquad (6)$$

The expressions in Eq. (6) are coupled linear differential equation systems. As known from mathematics, such equation systems can be solved by transformation to principal axes

$$\mathbf{Q}_{E,H} = \mathbf{T}_{E,H}\mathbf{\Gamma}^2\mathbf{T}_{E,H}^{-1} \qquad (7)$$

with $\mathbf{T}_{E,H}$ and $\mathbf{\Gamma}^2$ being the eigenvectors and eigenvalues of $\mathbf{Q}_{E,H}$.

Here, the matrices $\mathbf{Q}_E$ and $\mathbf{Q}_H$ were determined from Eq. (5) and then discretized with finite differences. However, identical matrices are obtained if the operators R_E and R_H are discretized first and the multiplication is done for these discrete matrices. Now, as known from mathematics (see e.g. Zurmühl and Falk, 1984, pp. 163) the eigenvalues of matrices that are determined from the product of two square matrices in reversed order are identical. For this reason $\mathbf{Q}_E$ and $\mathbf{Q}_H$ have the same eigenvalues and the subscript E or H was omitted for $\mathbf{\Gamma}^2$.

After transforming the fields according to:

$$H = \mathbf{T}_H\overline{H} \qquad E = \mathbf{T}_E\overline{E} \qquad (8)$$

a system of decoupled equations is developed:

$$\frac{\partial^2\overline{H}}{\partial\overline{z}^2} - \mathbf{\Gamma}^2\overline{H} = 0 \qquad \frac{\partial^2\overline{E}}{\partial\overline{z}^2} - \mathbf{\Gamma}^2\overline{E} = 0 \qquad (9)$$

whose solution can be given immediately as:

$$\overline{E} = e^{-\mathbf{\Gamma}\overline{z}}\overline{E}_f + e^{\mathbf{\Gamma}\overline{z}}\overline{E}_b \qquad (10)$$

$$\overline{H} = e^{-\mathbf{\Gamma}\overline{z}}\overline{H}_f + e^{\mathbf{\Gamma}\overline{z}}\overline{H}_b \qquad (11)$$

Equations (7)–(11) show that the solution of the wave equation is given in terms of eigenmodes. Each column of the eigenvector matrices $\mathbf{T}_{E,H}$ represents the electric resp. magnetic fields of these eigenmodes, $\mathbf{\Gamma}$ is a diagonal matrix with the propagation constants and the overlined quantities $\overline{E}_{f,b}$ $\overline{H}_{f,b}$ represent the amplitudes of the forward and backward propagating modes. By introducing the solution for the electric or magnetic field Eqs. (10), (11) into the GTL-expressions in discretized domain Eq. (1) one can obtain following relation between the amplitudes:

$$\overline{E}_f = \overline{H}_f \qquad \overline{E}_b = -\overline{H}_b \qquad (12)$$

In this case, the transformation matrices are related according to:

$$\mathbf{T}_E = j\mathbf{R}_H\mathbf{T}_H\mathbf{\Gamma}^{-1} \qquad \mathbf{T}_H = j\mathbf{R}_E\mathbf{T}_E\mathbf{\Gamma}^{-1} \qquad (13)$$

Interfaces

After solving the wave equation in homogeneous sections, interfaces must be considered. Here, the transverse electric and magnetic field components have to be continuous. Figure 4 shows an interface between two sections (I and II) with individual eigenmodes. From the continuity condition for the

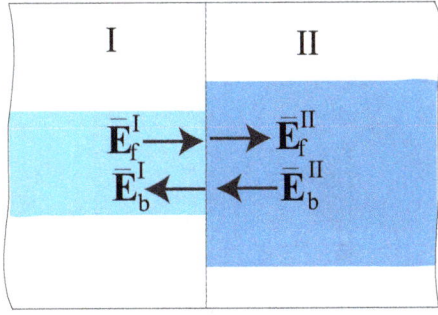

Figure 4. Interface between two homogeneous sections.

transverse fields, a relation for the amplitudes of the eigenmodes is obtained, where Eqs. (8)–(12) were considered

$$\begin{bmatrix} \mathbf{T}_E^I & \mathbf{T}_E^I \\ \mathbf{T}_H^I & -\mathbf{T}_H^I \end{bmatrix} \begin{pmatrix} \overline{E}_f^I \\ \overline{E}_b^I \end{pmatrix} = \begin{bmatrix} \mathbf{T}_E^{II} & \mathbf{T}_E^{II} \\ \mathbf{T}_H^{II} & -\mathbf{T}_H^{II} \end{bmatrix} \begin{pmatrix} \overline{E}_f^{II} \\ \overline{E}_b^{II} \end{pmatrix} \qquad (14)$$

From Eq. (14) we can e.g. compute the amplitudes of the eigenmodes in region II from the ones in region I in the following way:

$$\overline{E}_f^{II} + \overline{E}_b^{II} = \mathbf{a}_{II,I}(\overline{E}_f^I + \overline{E}_b^I) \qquad (15)$$

$$\overline{E}_f^{II} - \overline{E}_b^{II} = \mathbf{b}_{II,I}(\overline{E}_f^I - \overline{E}_b^I) \qquad (16)$$

with

$$\mathbf{a}_{II,I} = (\mathbf{T}_E^{II})^{-1} \mathbf{T}_E^I \quad \text{and} \quad \mathbf{b}_{II,I} = (\mathbf{T}_H^{II})^{-1} \mathbf{T}_H^I \qquad (17)$$

In principle, we are now in position to analyze whole devices like the polarization converter shown in Fig. 1. For this purpose, we must introduce conditions at the input and output of the device and apply the expressions derived for the homogeneous sections and the interfaces. However, the exponential increasing terms in Eqs. (10)–(11) can cause numerical problems if e.g. transfer matrices are used. Therefore, numerical stable expressions have been developed in the past, to avoid the numerical computation of these exponential increasing terms. These are algorithms with impedances/admittances (see e.g. Rogge and Pregla, 1993; Pregla, 2008) or with scattering parameters (e.g. Helfert and Pregla, 2002). At this point, we do not go into details, but just refer to the relevant literature.

2.2 Reduction of the eigenmode system

The presented algorithm works well, if the number of discretization lines is not too high. However, particularly if vectorial, 3-D problems are considered, the occurring matrices for the electric/magnetic field distribution become very large. The determination of all eigenmodes in Eq. (7) is very time consuming and the memory requirement becomes very high.

For this reason, the analysis with a reduced number of eigenmodes was proposed in Gerdes (1994). However, in that paper all eigenmodes were determined first. Only after that, some of them were omitted for the further simulations.

In Schneider (1999) was described how this computation of all eigenmodes can be avoided. For this purpose the Arnoldi algorithm that is implemented in the MATLAB program package (MATLAB, 2014) was used. This was done here as well. Information about this method can be found in the literature (e.g. Arnoldi, 1951; Golub and van Loan, 1989) so that we do not go into details here.

With a reduced set of eigenmodes the continuity of the transverse fields at interfaces leads to expressions where the matrices have the following shape:

$$\begin{pmatrix} \mathbf{T}_E^I & \mathbf{T}_E^I \\ \mathbf{T}_H^I & -\mathbf{T}_H^I \end{pmatrix} \begin{pmatrix} \overline{E}_f^I \\ \overline{E}_b^I \end{pmatrix} = \begin{pmatrix} \mathbf{T}_E^{II} & \mathbf{T}_E^{II} \\ \mathbf{T}_H^{II} & -\mathbf{T}_H^{II} \end{pmatrix} \begin{pmatrix} \overline{E}_f^{II} \\ \overline{E}_b^{II} \end{pmatrix}$$

Hence, the expressions in Eqs. (15)–(17) require the inversion of non-square matrices. The use of a pseudo-inverse (see Strang, 1986) was proposed in Schneider (1999) for this purpose. Here, we follow a different path and apply left eigenvectors as described in Helfert et al. (2003).

In principle the left eigenvectors $\mathbf{Y}$ of a matrix $\mathbf{A}$ are solutions of

$$\mathbf{Y}\mathbf{A} = \lambda \mathbf{Y} \qquad (18)$$

in contrast to the more known problem for the right eigenvectors $\mathbf{X}$

$$\mathbf{A}\mathbf{X} = \mathbf{X}\lambda \qquad (19)$$

The eigenvalues λ for both problems are identical. After suitable sorting and normalization of the left and right eigenvectors we may write:

$$\mathbf{Y}\mathbf{X} = \mathbf{I} \qquad (20)$$

where $\mathbf{I}$ is the identity matrix. For our purpose, it is important that this relation is also true when a reduced set of eigenmodes is used. This is indicated in Fig. 5.

Note: Eq. (20) is true as long as the eigenvalues are different. For degenerated modes with identical eigenvalues this condition does not have to be fulfilled. However as shown e.g. in Zurmühl and Falk (1984), pp. 155 it is possible to transform the eigenvectors in such a way that condition Eq. (20) is enforced. A summary of this transformation is given in the Appendix.

Once the left eigenvectors (indicated by the superscript "L") have been computed, we determine the expressions given in Eq. (17) as:

$$\mathbf{a}_{II,I}^r = \mathbf{T}_E^{L,II} \mathbf{T}_E^I \quad \text{and} \quad \mathbf{b}_{II,I}^r = \mathbf{T}_H^{L,II} \mathbf{T}_H^I \qquad (21)$$

The superscript "r" indicates that a reduced set of eigenmodes was used. It is now worth noting that the matrices in Eq. (21) are sub-matrices of the ones given in Eq. (17). The illustration shown in Fig. 5 can be interpreted as a special case of this feature.

Unfortunately, the author is not aware of a standard algorithm that determines right- and left-eigenvectors at the same time. As consequence, in the past, the right and left eigenvectors were determined independently of each other (Helfert et al., 2003). This was done with $\mathbf{Q}_{E,H}$ and its transposed. Obviously, the numerical effort is quite high. As second problem the numerical orthogonality of the left and right eigenvectors Eq. (20) can be quite bad when they are computed independently.

2.3 Efficient determination of the left eigenvectors

In this section we describe how we can avoid solving the eigenvalue problem twice. To develop suitable expressions, we start with the operators $R_{E,H}$. For Dirichlet, Neumann, or periodic boundary conditions, we may write the discretized operators with matrix products of the following form:

$$\mathbf{R}_E = \begin{bmatrix} \boldsymbol{\varepsilon}_y - \mathbf{D}_{\overline{x}}^{T}\boldsymbol{\mu}_z^{-1}\mathbf{D}_{\overline{x}} & -\mathbf{D}_{\overline{x}}^{T}\boldsymbol{\mu}_z^{-1}\mathbf{D}_{\overline{y}}^{T} \\ -\mathbf{D}_{\overline{y}}\boldsymbol{\mu}_z^{-1}\mathbf{D}_{\overline{x}} & \boldsymbol{\varepsilon}_x - \mathbf{D}_{\overline{y}}\boldsymbol{\mu}_z^{-1}\mathbf{D}_{\overline{y}}^{T} \end{bmatrix} \quad (22)$$

$$\mathbf{R}_H = \begin{bmatrix} \boldsymbol{\mu}_x - \mathbf{D}_{\overline{y}}\boldsymbol{\varepsilon}_z^{-1}\mathbf{D}_{\overline{y}}^{T} & -\mathbf{D}_{\overline{y}}\boldsymbol{\varepsilon}_z^{-1}\mathbf{D}_{\overline{x}}^{T} \\ -\mathbf{D}_{\overline{x}}\boldsymbol{\varepsilon}_z^{-1}\mathbf{D}_{\overline{y}}^{T} & \boldsymbol{\mu}_y - \mathbf{D}_{\overline{x}}\boldsymbol{\varepsilon}_z^{-1}\mathbf{D}_{\overline{x}}^{T} \end{bmatrix} \quad (23)$$

The matrices $\mathbf{D}_{\overline{x}}$, $\mathbf{D}_{\overline{y}}$ and their transposed (indicated by the superscript "T") contain the finite difference approximation of the derivatives with respect to $\overline{x}$ and $\overline{y}$. $\boldsymbol{\varepsilon}_{x,y,z}$ and $\boldsymbol{\mu}_{x,y,z}$ are diagonal matrices containing the permittivities and permeabilities on the discretization points. As can be seen, the matrices are symmetric (not Hermitian)

$$\mathbf{R}_{E,H}^{T} = \mathbf{R}_{E,H} \quad (24)$$

Next we look at the matrices that occur in the wave equation Eq. (6). As mentioned before, we may write

$$\mathbf{Q}_H = -\mathbf{R}_E\mathbf{R}_H \qquad \mathbf{Q}_E = -\mathbf{R}_H\mathbf{R}_E$$

If we now transpose e.g. $\mathbf{Q}_H$, we find, due to Eq. (24), the following relation:

$$\mathbf{Q}_H^{T} = -(\mathbf{R}_E\mathbf{R}_H)^{T} = -\mathbf{R}_H^{T}\mathbf{R}_E^{T} = -\mathbf{R}_H\mathbf{R}_E = \mathbf{Q}_E \quad (25)$$

Next, this transposed matrix is diagonalized. We use the left eigenvector matrix (indicated by the superscript "L", as before) as inverse of $\mathbf{T}_H$ and obtain

$$\mathbf{Q}_H^{T} = (\mathbf{T}_H\boldsymbol{\Gamma}^2\mathbf{T}_H^{L})^{T} = (\mathbf{T}_H^{L})^{T}\boldsymbol{\Gamma}^2\mathbf{T}_H^{T} \quad (26)$$

Similar, the diagonalization of $\mathbf{Q}_E$ reads:

$$\mathbf{Q}_E = \mathbf{T}_E\boldsymbol{\Gamma}^2\mathbf{T}_E^{L} \quad (27)$$

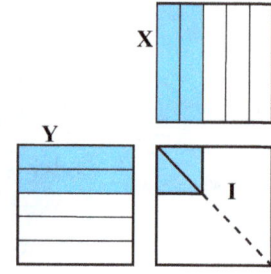

Figure 5. Multiplication of left- and right-eigenvectors resulting in an identity matrix.

A comparison of Eq. (26) and Eq. (27) with consideration of Eq. (25) shows that $\mathbf{T}_E$, $\mathbf{T}_H$ and the corresponding left eigenvectors (i.e. their inverse matrices) are related according to:

$$\mathbf{T}_E^{T} = \mathbf{T}_H^{L} \qquad \mathbf{T}_H^{T} = \mathbf{T}_E^{L} \quad (28)$$

Note: since eigenvectors can be determined up to an arbitrary scaling factor, Eq. (28) is only true, if the eigenvectors have been normalized accordingly. This is not a principle problem, but has to be kept in mind.

As final step we introduce Eq. (28) into Eq. (13) and obtain:

$$(\mathbf{T}_H^{L})^{T} = j\mathbf{R}_H\mathbf{T}_H\boldsymbol{\Gamma}^{-1} \quad (\mathbf{T}_E^{L})^{T} = j\mathbf{R}_E\mathbf{T}_E\boldsymbol{\Gamma}^{-1} \quad (29)$$

The products in Eq. (29) can be computed with verly low numerical effort, because $\mathbf{R}_H$ and $\mathbf{R}_E$ are sparse–matrices. Further, the eigenvalues are combined in the diagonal matrix $\boldsymbol{\Gamma}$ so that its inverse can be computed very easily. Hence, Eq. (29) shows that the left eigenvectors can be determined with simple matrix products from the previously determined right eigenvectors. The numerical effort for this procedure is clearly lower than that for solving the eigenvalue/eigenvector problem twice or as determining a pseudo inverse.

2.4 Open structures

As mentioned before, the derived expression Eq. (29) only holds for special boundary conditions, for which the operator matrices $\mathbf{R}_{E,H}$ are symmetric. Now outgoing waves are completely reflected at Dirichlet or Neumann boundaries ("hard boundaries") leading to modeling errors. There exist various ways to reduce such reflections, e.g. absorbing boundary conditions (ABC) as described in Pregla (1995) or Vassallo and Collino (1997). Unfortunately, with the ABCs mentioned above, the symmetry condition in Eq. (24) does not hold any longer. Hence, we need to apply methods that permit the inclusion of hard BCs.

Particularly, we decided to use perfectly matched layers (PMLs). These are lossy layers positioned at the boundary of the computational window. Waves that enter this region are damped and after being reflected at the hard wall, they are damped further to negligible amplitudes before re-entering the original computational window.

Generally, PMLs can be interpreted in various ways (see e.g. Berenger, 1994; Mittra and Pekel, 1995; Rappaport, 1995; Al-Bader and Jamid, 1998). Here, they are considered as lossy anisotropic layers, which are described by tensors that only contain non-zero elements on the main diagonal (see e.g. Sacks et al., 1995; Werner and Mittra, 1997; Cucinotta et al., 1999), i.e. tensors of the form given by Eq. (3), for which the formulas were derived. Hence, Eq. (29) is also true for PMLs.

2.4.1 Orthogonality of the eigenmodes

By considering the conservation of energy and from the reciprocity theorem one can derive orthogonality relations for the eigenmodes in waveguide sections (see e.g. Syms and Cozens, 1992; Snyder and Love, 1983). For the fields of two different eigenmodes (labels i, k) that propagate in z direction the orthogonality condition reads

$$\int\int_A ([E]_i \times [H]_k) \cdot [e]_z dA = 0 \tag{30}$$

Note: as before, the physical vectors ($[E], [H], [e]$) were written in brackets.

In the MoL, the discretized field distributions of the eigenmodes are combined in the columns of the matrices $\mathbf{T}_E$, $\mathbf{T}_H$. Therefore, the discretized integral in Eq. (30) contains the product

$$(\mathbf{T}_H)^T \cdot (\mathbf{T}_E)$$

which results in a diagonal matrix, as derived. Hence, the orthogonality of the discretized eigenmodes in a closed structure can be seen directly from Eq. (28). We should point out that this orthogonality relation does also hold for lossy structures.

3 Numerical results

In this section we show some numerical examples to evaluate the derived expression. For comparison the left eigenvectors were (a) determined with Eq. (29) and (b) by solving a second eigenvalue/eigenvector problem as described in (Helfert et al., 2003). For a quantitative assessment, an error measure was introduced. Ideally, the product between left- and right-eigenvectors results in an identity matrix. The deviation from this ideal result can be written as:

$$\mathbf{T}^L \mathbf{T}^R = \mathbf{I} + \mathbf{M}^{err} \tag{31}$$

In the following we will use $||\mathbf{M}^{err}||_{max}$ i.e. the maximum element of this matrix as error measure.

There are a few factors that contribute to this error. The first one is caused by the finite machine precision so that the

eigenvectors can be determined only up to a certain accuracy. A second contribution to the error comes from degenerated eigenmodes, as mentioned in Sect. 2.2. As will be shown numerically in this section this error can be reduced by a transformation of these modes.

To see if the developed expressions work in principle and to study the behavior of the PMLs, we started with two simple 2-D structures (TE-polarization with the components E_y, H_x, H_z). For both cases Dirichlet conditions were introduced at the lateral boundaries and results obtained with and without PMLs were compared. As mentioned before, the PMLs were introduced as anisotropic materials (18 lines close to the boundaries) where the following values for the materials were chosen: $\mu_z = \varepsilon_y = 1 - 0.36j$, $\mu_x = 1/\mu_z$. For details about the relation between these values see e.g. Sacks et al. (1995), Werner and Mittra (1997), and Cucinotta et al. (1999).

We should point out that the z direction is treated with analytic expressions. Hence, to model the infinite long structures no PMLs are needed, but we assume that only forward propagating modes occur for these regions in Eqs. (10)–(11), thus: $\overline{E}_b = \overline{H}_b = 0$. Since we are dealing with 2-D problems, a reduction of the eigenmode system is not necessary, and all eigenmodes were used for the computations. Particularly, the number of discretization points ($=$ number of eigenmodes) was 107.

First of all, a tilted Gaussian beam was injected into a homogeneous air region. The input field is given as:

$$E_{yin}(x, z = 0) = E_0 e^{-(x/w)^2} e^{-jk_0 x \cos(\varphi)} \tag{32}$$

with $\lambda_0 = 1.55\,\mu m$, $w = 2\,\mu m$ and $\varphi = 45°$. To compute the wave propagation of the Gaussian beam, we discretize the input field in x direction and must invert Eq. (8) after that. As mentioned before, an infinite long region with only forward propagating modes is assumed. Hence, the inversion (written with left eigenvectors) reads:

$$\overline{E}_{f\,in} = \mathbf{T}_E^L E_{in} \tag{33}$$

In the following, the propagation in the region is computed with Eq. (10) and the original fields are obtained from Eq. (8). As can be seen, even in this simple case left eigenvectors are used.

As second example the abrupt ending of a waveguide was studied. As can be seen in Fig. 6, we have the concatenation of a waveguide section and an infinite long air region. In the simulations the fundamental waveguide mode was injected on the left. Then, the wave propagation was simulated with the expressions given in Sect. 2. For this structure the left eigenvectors were needed to enforce the continuity of the fields at the interface between the waveguide and the air region.

Before looking at the fields, let us examine the error. The structure in Fig. 6 consists of two parts (waveguide, homogeneous air region). Therefore, for each of these regions a different error value is obtained. For the further considerations

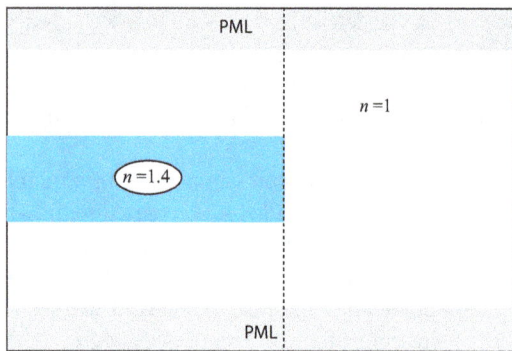

Figure 6. Abrupt ending of a waveguide.

Table 1. Error determined with Eq. (31) for the structures shown in Fig. 6; in case "PML symmetric a)" the eigenmodes were taken as obtained by the eigenvalue procedure; in "PML symmetric b)" the degenerated eigenmodes were transformed.

n_{core}	1.4	$1.4 - 0.01j$
electric wall	10^{-12}	10^{-11}
PML		
symmetric a)	0.15	0.7
symmetric b)	10^{-9}	10^{-9}
PML		
asymmetric	10^{-11}	10^{-11}

the worse one was taken. Since the homogeneous air region is already included in these studies, there are no additional error measures required for the Gaussian beam problem.

In the simulations, several parameters were varied. We should repeat that the expression in Eq. (29) was derived for a general case incl. losses. These losses can be caused by dielectric media (i.e. with complex permittivity) or by PMLs (anisotropic media with complex permittivity and permeability). Hence, the influence of the losses on the error is of particular interest.

The results for the error are summarized in Table 1. As can be seen, two cases were compared; for a lossless core ($n_{core} = 1.4$) and a lossy one ($n_{core} = 1.4 - 0.01j$). The lossless case (row "electric wall") leads to the small error $||\mathbf{M}^{err}||_{max} = 10^{-12}$, which deteriorates only moderately for a lossy core.

The next rows show the influence of the PMLs on the error. In row "PML symmetric a)" the determined error-value was computed directly after determining the right and left eigenvectors with Eqs. (7) and (29). Here, degenerated eigenmodes (i.e. pairs with identical propagation constant) occur, resulting in very high error values. This error can be reduced drastically by a transformation of the eigenmodes leading to the results "symmetric b)".

Since the error in case of PMLs is caused by degenerated modes, it was examined next, what happens if a small

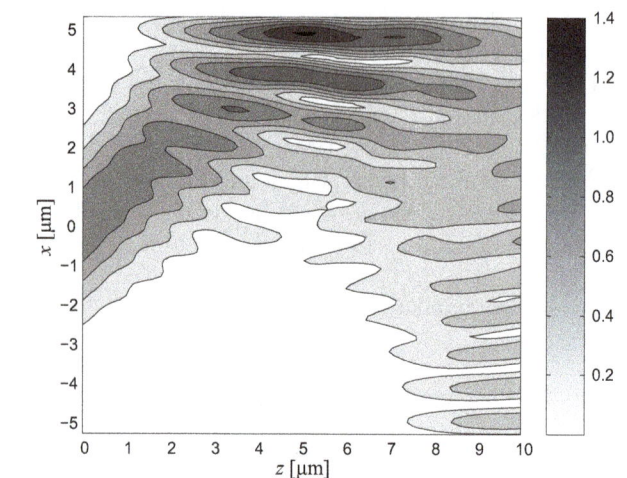

a)

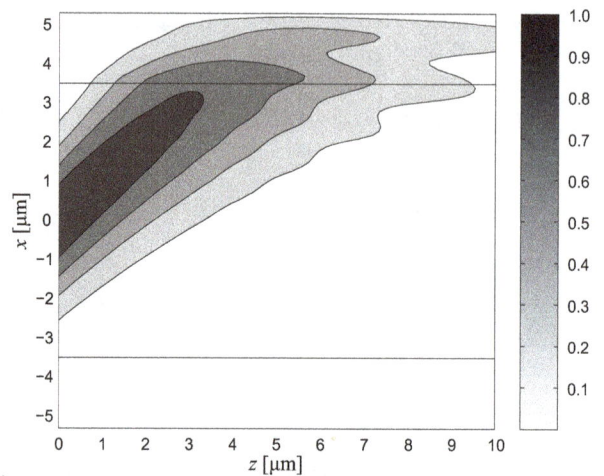

b)

Figure 7. Injection of a tilted Gaussian beam into a homogeneous section, computed electric field, **(a)** without PMLs, **(b)** with PMLs, the horizontal lines indicate the boundaries of the PMLs.

asymmetry in the PMLs is introduced. In particular, on the lower boundary, the imaginary part of μ_z was increased by the factor 1.0001 and the other values (ε_y, μ_x) were modified accordingly. Then, the degeneration of the modes vanishes, leading to the error presented in row: "PML asymmetric". These errors are close to the lossless case.

Keeping in mind that the values in row "PML symmetric a)" can be improved to "PML symmetric b)", it is found that all errors are quite small. Hence the formulas Eq. (31) work without problems for the 2-D-case incl. losses due to lossy materials or PMLs.

For comparison, the left eigenvectors were also determined from solving a second eigenvalue problem. Here, the error for the lossless case is reduced to 10^{-15}. For all other cases, however, the error is similar or slightly higher than given in Table 1.

a)

b)

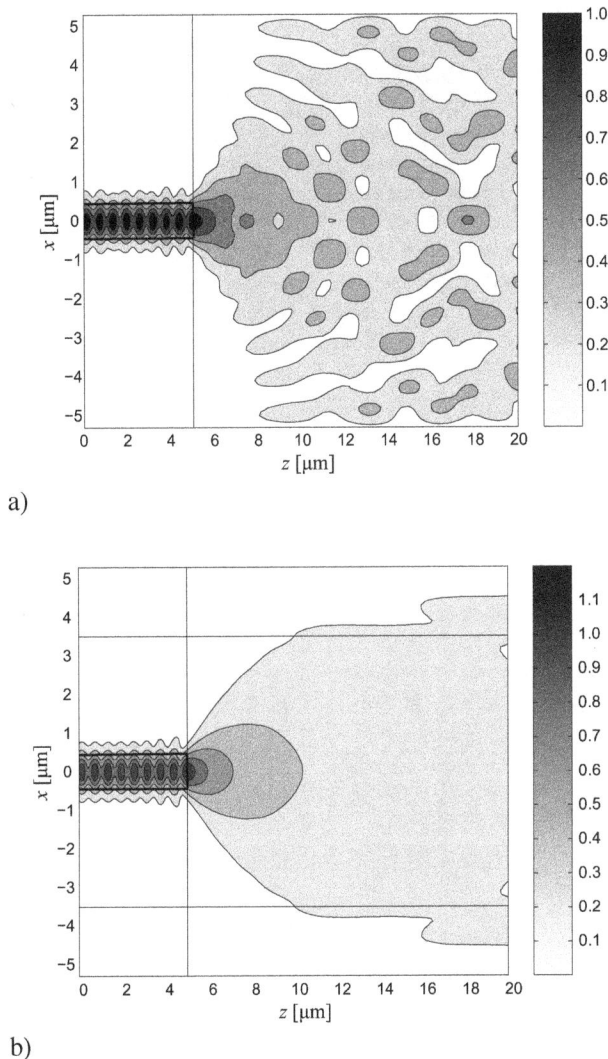

Figure 8. Electric field in the structure shown in Fig. 6; (**a**) without PMLs, (**b**) PMLs included, the horizontal lines indicate the boundaries of the PMLs.

The computed electric field distributions are presented in Fig. 7 for the Gaussian beam and in Fig. 8 for the abrupt ending of the waveguide. The PML-graphs were obtained with symmetric PMLs. However, there are no visual differences recognizable when a small asymmetry is introduced.

For both structures it can be seen, how the reflections caused by the Dirichlet boundaries Fig. 7a (resp. Fig. 8a) can be suppressed with the PMLs Fig. 7b (resp. Fig. 8b).

The results presented so far show that the algorithm works in principle in 2-D including losses. However, the analysis of such a 2-D-structures is numerical not very demanding and (as was done) the full eigenmode system can be used.

Therefore, the real reduction of the numerical effort is expected for 3-D-structures. Two examples were considered here.

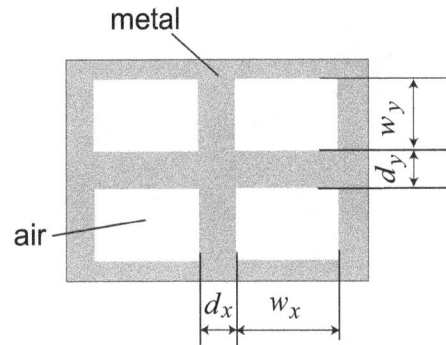

Figure 9. Array of hollow waveguides; dimensions: $w_x = 0.8\lambda_0$ $w_y = 0.6\lambda_0, d_x = d_y = 0.04\lambda_0$, with the free space wavelength λ_0.

The polarization converter shown in Fig. 2 had already been examined with the MoL (Helfert et al., 2003). In that study the left eigenvectors were determined from an independent eigenvalue/eigenvector computation. The structure was analyzed with $N_x \times N_y \times 2 = 100 \times 75 \times 2 = 15\,000$ discretization points (factor "2" because of the vectorial analysis) and 300 eigenmodes. When the left eigenvectors were determined with the algorithm described here, the value 10^{-8} was obtained for the error Eq. (31). The separate left eigenvector determination (Helfert et al., 2003) results in the error value 10^{-7}. Both values are satisfactory. We should point out here, that most of the CPU-time for analyzing the polarization converter is required for the determination of the eigenmodes. In particular, computing the left eigenvectors separately, required 3 min, which could be reduced to 4 s with the algorithm presented here (computations on a PC). This shows the significant reduction of the numerical effort.

The features of this polarization converter had been studied with the MoL in detail in (Helfert et al., 2003). Since these results do not change, when the eigenvectors are determined differently, they are not repeated here. Instead we refer to Helfert et al. (2003).

As second example for a 3-D-device, an array of hollow waveguides as shown in Fig. 9 was examined. The idea is to use such hollow waveguide arrays as polarization converting elements. For this purpose, the difference of the propagation constants of the vertically and horizontally polarized eigenmodes is utilized ("half–wave plate") and the length L_z of the device was chosen in such a way that these wave experience a phase difference of π. Hence:

$$k_0 \Delta n_{\text{eff}} L_z = \pi \qquad (34)$$

A detailed description of the structure is given in Helfert et al. (2014) and Helfert et al. (2015). Therefore, we concentrate here on the results as far as the numerical algorithm is concerned and do not repeat all the features of this device.

Since the permittivity of the metal is very different in optics and for THz-frequencies, studies for these two frequency regimes were performed. Particularly, silver was taken as

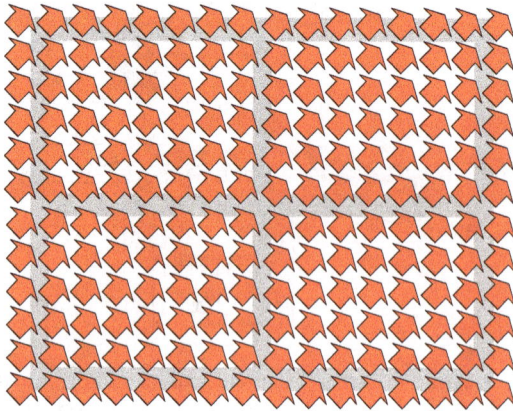

Figure 10. Injection of a plane wave into an array of hollow waveguides, magnetic field in front of the waveguides.

metal. Its relative permittivity was determined with a Dude-model, resulting in:

- THz ($\lambda_0 = 0.48$ mm, $f = 0.625$ THz)

    $\varepsilon_{rAg} = -6.2 \times 10^5 - 2.25 \times 10^6 j$

- Optics ($\lambda_0 = 1000$ nm, $f = 3 \times 10^{14}$ Hz)

    $\varepsilon_{rAg} = -47 - 1.89 j$

As can be seen, losses in the metal are considered.

With the values for the permittivity given above, the eigenmodes were computed. Then, the following lengths were determined from Eq. (34):

$$L_{z\text{THz}} = 2.2\lambda_0 \qquad L_{z\text{optics}} = 3\lambda_0 \qquad (35)$$

Now, plane waves were injected into the structures as shown in Fig. 10. The magnetic field at the output is sketched in Fig. 11. The rotation of the fields is clearly seen. In the THz case (Fig. 11a) we recognize that the whole field is inside the air-region, whereas a non–negligible part of the field is inside the metal in optics (Fig. 11b).

The computations were done with $N_x \times N_y \times 2 = 170 \times 130 \times 2 = 44\,200$ discretization points. The number of eigenmodes was 20 i.e. much lower. With these parameters, the following error values occur: $||\mathbf{M}^{\text{err}}||_{\max} = 10^{-9}$ (THz), $||\mathbf{M}^{\text{err}}||_{\max} = 10^{-8}$ (optics). When the left eigenvectors were determined separately (see (Helfert et al., 2003)), errors of the same order were obtained. So, Eq. (29) does also work very well for the considered full vectorial 3-D-structures.

4   Summary and conclusions

In this paper was shown how left eigenvectors can be determined very efficiently. For this purpose the relation between the transverse electric and magnetic fields of the eigenmodes was utilized. It was found that the matrix of the left eigenvectors can be computed by simple matrix products from previously determined right eigenvalues in case of hard walls

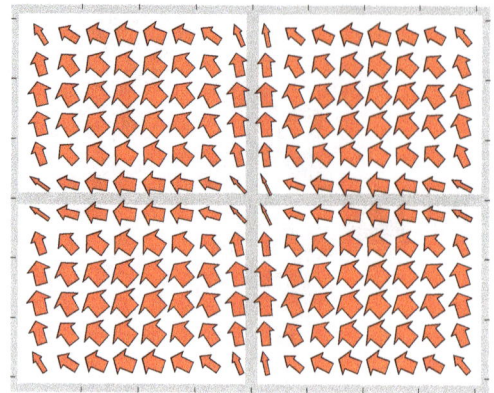

a)

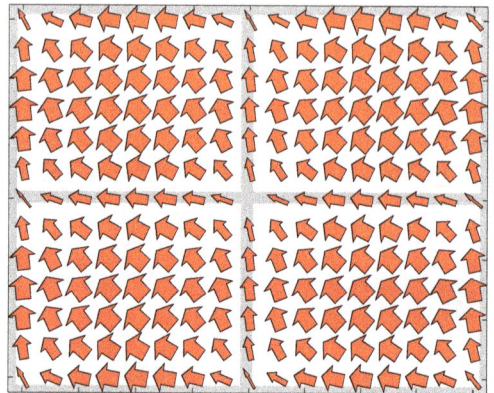

b)

Figure 11. Magnetic field at the output of the hollow waveguides, (**a**) THz-frequencies, (**b**) optics.

(Dirichlet, Neumann) or periodic boundaries. Compared to the erlier determination from a second eigenvalue problem, the numerical effort could be reduduced significantly.

Absorbing boundary condition that are often employed in the MoL, cannot be used here, because of symmetry requirements. Therefore, open structures were modeled with perfectly matched layers where the PMLs were introduced as lossy anisotropic materials. The numerical results (incl. error measures) for various structures in 2-D and for 3-D show that the method works very well also when losses are considered.

Appendix A

If multiple eigenmodes with identical propagation constant are computed, Eq. (20) does not necessarily hold. However, as described e.g. in (Zurmühl and Falk, 1984), it is always possible to transform the corresponding eigenvectors to enforce this conditions. Here, we will briefly repeat the procedure given in (Zurmühl and Falk, 1984). In what follows, X_m and Y_m are the submatrices of the right and left eigenvector matrix that correspond to the multiple eigenvalues. Generally, their product results in a full matrix M:

$$Y_m X_m = M \tag{A1}$$

Note: in this paper the left eigenvectors were introduced as row-vectors. This is in contrast to (Zurmühl and Falk, 1984) where they were written as column vectors. So, some of the expressions in (Zurmühl and Falk, 1984) are transposed compared to the formulas given here.

Now, the eigenvectors are transformed in the following way:

$$Y_m = c_y \widetilde{Y}_m \qquad X_m = \widetilde{X}_m c_x \tag{A2}$$

where the new eigenvectors are orthogonal:

$$\widetilde{Y}_m \widetilde{X}_m = I \tag{A3}$$

Then one can write

$$M = Y_m X_m = c_y \widetilde{Y}_m \widetilde{X}_m c_x = c_y c_x \tag{A4}$$

Hence, the matrices c_y and c_x must be chosen such that their product gives M. This choice is not unique. The use of triangular matrices was proposed in (Zurmühl and Falk, 1984) and as second possibility chosing $c_y = I$ (or $c_x = I$) was mentioned. Due to Eq. (29) we chose here:

$$c_y = c_x \tag{A5}$$

resulting in the following expression

$$c_x^2 = M \tag{A6}$$

to compute c_x. We should point out that the size of M (i.e. the number of eigenmodes with identical eigenvalues) is usually very small. (For the structures examined in this paper the maximum size was 4). Therefore, the numerical effort for solving Eq. (A6) is negligible.

References

Al-Bader, S. and Jamid, H. A.: Perfectly matched layer absorbing boundary conditions for the method of lines modeling scheme, IEEE Microwave Guided Wave Lett., 8, 357–359, 1998.

Arnoldi, W.: The Principle of minimized iterations in the solution of the matrix eigenvalue problem, Quarterly Appl. Mathem., 9, 17–29, 1951.

Berenger, J.-P.: A perfectly matched layer for the absorption of electromagnetic waves, J. Computat. Phys., 114, 185–200, 1994.

Cucinotta, A., Pelosi, G., Selleri, S., Vincetti, L., and Zoboli, M.: Perfectly matched anisotropic layers for optical waveguide analysis through the finite element beam propagation method, Microw. Opt, Tech. Lett., 23, 67–69, 1999.

Gerdes, J.: Bidirectional eigenmode propagation analysis of optical waveguides based on method of lines, Electron. Lett., 30, 550–551, 1994.

Golub, G. H. and van Loan, C. F.: Matrix computations, Johns Hopkins University Press, Baltimore, USA, 501–502, 1989.

Helfert, S. F. and Pregla, R.: The method of lines: a versatile tool for the analysis of waveguide structures, Electromagnetics, 22, 615–637, invited paper for the special issue on "Optical wave propagation in guiding structures", 2002.

Helfert, S. F., Barcz, A., and Pregla, R.: Three-dimensional vectorial analysis of waveguide structures with the method of lines, Opt. Quantum Electron., 35, 381–394, 2003.

Helfert, S. F., Edelmann, A., and Jahns, J.: Hollow waveguides as polarization converting elements, in: Europ. Opt. Soc. ann. meet. (EOSAM), p. TOM5 S02: Subwavelength device, Berlin, Germany, 2014.

Helfert, S. F., Edelmann, A., and Jahns, J.: Hollow waveguides as polarization converting elements: a theoretical study, J. Europ. Opt. Soc.: Rap. Publ., 10, 15006, 1–7, 2015.

MATLAB: version 8.4 (R2014b), The MathWorks Inc., Natick, Massachusetts, 2014.

Mittra, R. and Pekel, Ü.: A new look at the perfectly matched layer (PML) concept for the reflectionless absorption of electromagnetic waves, IEEE Microwave Guided Wave Lett., 5, 84–86, 1995.

Mustieles, F. J., Ballesteros, E., and Hernández-Gil, F.: Multimodal analysis method for the design of passive TE/TM converters in integrated waveguides, IEEE Photonics Technol. Lett., 5, 809–811, 1993.

Pregla, R.: MoL-BPM Method of lines based beam propagation method, in: Methods for modeling and simulation of guided-wave optoelectronic devices (PIER 11), edited by: Huang, W. P., Progress in Electromagnetic Research, 51–102, EMW Publishing, Cambridge, Massachusetts, USA, 1995.

Pregla, R.: Novel FD-BPM for optical waveguide structures with isotropic or anisotropic material, in: Europ. Conf. Int. Opt. (ECIO), 55–58, Torino, Italy, 1999.

Pregla, R.: Analysis of electromagnetic fields and waves - The method of lines, Wiley & Sons, Chichester, UK, 2008.

Pregla, R. and Pascher, W.: The method of lines, in: Numerical techniques for microwave and millimeter wave passive structures, edited by: Itoh, T., 381–446, J. Wiley Publ., New York, USA, 1989.

Rappaport, C. M.: Perfectly matched absorbing boundary conditions based on anisotropic lossy mapping of space, IEEE Microwave Guided Wave Lett., 5, 90–92, 1995.

Rogge, U. and Pregla, R.: Method of lines for the analysis of dielectric waveguides, J. Lightwave Technol., 11, 2015–2020, 1993.

Sacks, Z. S., Kingsland, D. M., Lee, R., and Lee, J.-F.: A perfectly matched anisotropic absorber for use as an absorbing boundary, IEEE Trans. Antennas. Propagation, 43, 1460–1463, 1995.

Schneider, V. M.: Analysis of passive optical structures with an adaptive set of radiation modes, Opt. Comm., 160, 230–234, 1999.

Snyder, A. W. and Love, J. D.: Optical waveguide theory, Chapman and Hall, London, New York, 604–605, 1983.

Strang, G.: Linear algebra and its applications, Saunders HBJ College Publishers, Orlando, Fl., USA, 3rd edn., 130–138, 1986.

Sudbø, A. S.: Film mode matching: A versatile method for mode field calculations in dielectric waveguides, Pure Appl. Opt., 2, 211–233, 1993.

Syms, R. and Cozens, J.: Optical guided waves and devices, McGraw-Hill Publishing Co., New York, chap. 6.5, 1992.

Sztefka, G.: A bidirectional propagation algorithm for large refractive index steps and systems of waveguides based on the mode matching method, in: OSA Integr. Photo. Resear. Tech. Dig., 134–135, New Orleans, Louisiana, USA, 1992.

Taflove, A.: The finite-difference time–domain method, Computational electrodynamics, Artech house, inc, Norwood, MA, 1995.

Vassallo, C. and Collino, F.: Comparison of a few transparent boundary bonditions for finite-difference optical mode-solvers, J. Lightwave Technol., 15, 397–402, 1997.

Werner, D. H. and Mittra, R.: A new field scaling interpretation of Berenger's PML and its comparison to other PML formulations, Microw. Opt. Tech. Lett., 16, 103–106, 1997.

Zurmühl, R. and Falk, S.: Matrizen und ihre Anwendungen, Teil 1, Springer Verlag, Berlin, Germany, 5 edn., 1984.

7

Coexistence issues for a 2.4 GHz wireless audio streaming in presence of bluetooth paging and WLAN

F. Pfeiffer[1]**, M. Rashwan**[a]**, E. Biebl**[2]**, and B. Napholz**[3]

[1]perisens GmbH, Munich, Germany
[2]Fachgebiet Höchstfrequenztechnik, Technische Universität München, Munich, Germany
[3]Daimler AG, Sindelfingen, Germany
[a]formerly at: Daimler AG, Sindelfingen, Germany

Correspondence to: F. Pfeiffer (pfeiffer@perisens.de)

Abstract. Nowadays, customers expect to integrate their mobile electronic devices (smartphones and laptops) in a vehicle to form a wireless network. Typically, IEEE 802.11 is used to provide a high-speed wireless local area network (WLAN) and Bluetooth is used for cable replacement applications in a wireless personal area network (PAN). In addition, Daimler uses KLEER as third wireless technology in the unlicensed (UL) 2.4 GHz-ISM-band to transmit full CD-quality digital audio. As Bluetooth, IEEE 802.11 and KLEER are operating in the same frequency band, it has to be ensured that all three technologies can be used simultaneously without interference. In this paper, we focus on the impact of Bluetooth and IEEE 802.11 as interferer in presence of a KLEER audio transmission.

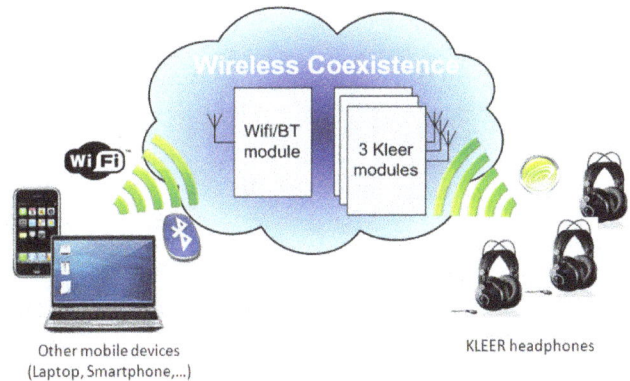

Figure 1. Considered wireless architecture.

1 Introduction

This paper addresses the impact of an IEEE 802.11b/g and a Bluetooth system on a KLEER audio transmission. The considered wireless architecture includes a WiFi (IEEE 802.11b/g) combo module and three independent KLEER sources which allow three independent audio streams (see Fig. 1).

KLEER from SMSC is a short-range radio interface in the 2.4 GHz frequency band that is designed for lossless 44.1 kHz-sampled 16 bit stereo audio transmission in full CD quality (see SMSC, 2007a, b; Devries et al., 2009; Mason et al., 2009). A lossless audio compression is used to reduce the required net data rate of $1.4112 \, \mathrm{Mb \, s^{-1}}$ to approximately $1 \, \mathrm{Mb \, s^{-1}}$ on average. But the short term compression ratio

depends on characteristics of the audio signal. Therefore the full data rate of $1.4112 \, \mathrm{Mb \, s^{-1}}$ has to be supported to achieve lossless full CD quality audio streaming. KLEER is using 16 equally spaced RF channels from 2.403 to 2.478 GHz (having a frequency spacing of 5 MHz) and each channel occupies an RF bandwidth of 3 MHz. For independent audio streams, each stream requires its own RF channel. Thereby KLEER is offering a gross data rate of $2.37 \, \mathrm{Mb \, s^{-1}}$. The resulting excess net data rate is required for retransmission of corrupted packets in case of interference. An audio buffer ensures a continuous audio stream in case of packet loss. KLEER uses a configurable buffer size of up to 100 ms. The buffer size has to be high enough to assure a sufficient number of hops to find a channel without interference. As additional coexistence mechanism for interference mitigation, KLEER uses a dynamic frequency diversity to change the

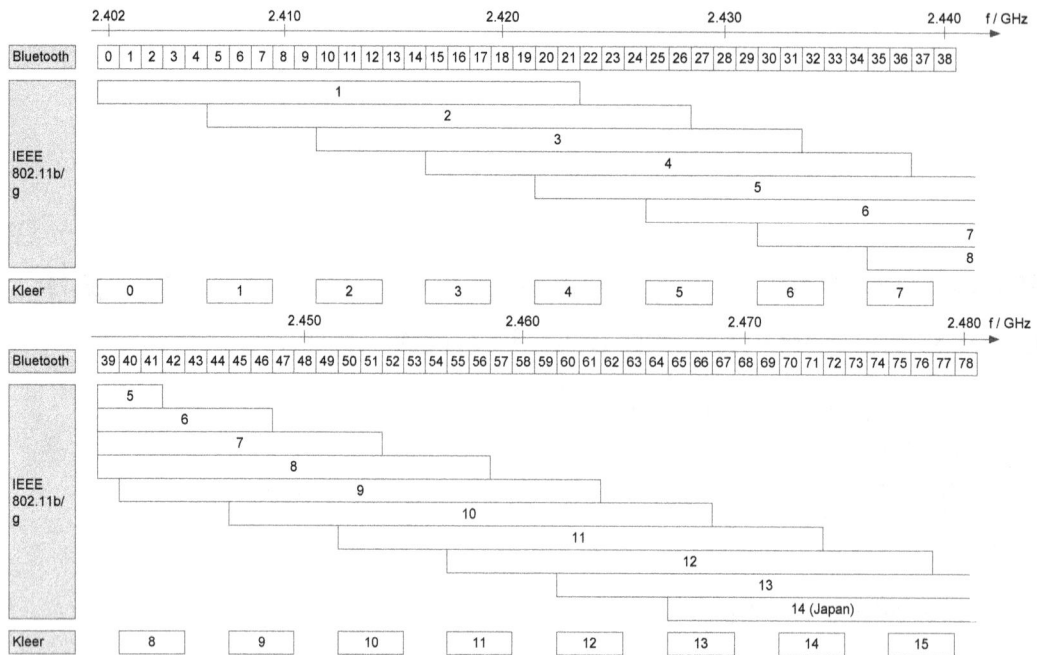

Figure 2. Bluetooth, IEEE 802.11b/g and KLEER channels in the 2.4 GHz-ISM-band.

Table 1. Comparison of wireless technologies.

	KLEER	Bluetooth	IEEE 802.11b	IEEE 802.11g
Gross data rate [Mbit s^{-1}]	2.37	1/2/3	1/2/5.5/11	6/9/12/18/24/ 36/48/54
Max. transmit power [dBm]	1.5	− 20 (class 1) − 4 (class 2) − 0 (class 3)	17	17
20 dB-bandwidth [MHz]	3	1	16	17
Channels	16 (non-overlapping)	79 (non-overlapping)	13 (partly overlapping)	13 (partly overlapping)
Modulation	MSK	GFSK/DQPSK 8DPSK	DBPSK/DQPSK 16-QAM/64-QAM	BPSK/QPSK/
Miscellaneous	None	Frequency Hopping	Spreading/Code Keying	OFDM

current channel if it is experiencing bad channel conditions. A channel change has to be initiated if the retransmission bandwidth is not able to cope with the packet loss rate over a defined period of time. At the audio sink side (at the wireless headphones), Kleer uses antenna diversity by switching two orthogonal polarized antennas which provides additional interference mitigation in small scale fading for in-vehicular communication. In Table 1 all important details on KLEER are shown and compared to Bluetooth and IEEE 802.11b/g. Figure 2 shows the channel distributions of Bluetooth, IEEE 802.11 and KLEER in the ISM-band.

In the first section, the impact of an IEEE 802.11b/g interferer is discussed and measurement results are presented. The next section analyzes the impact of a Bluetooth interferer in connected and connecting state using an analytical packet error model. Typical Bluetooth states are classified accord-

ing to their impact on a KLEER transmission. It turns out that Bluetooth page state is very critical regarding interference. The following section gives details about the packet timing and the hopping sequence during page state. Furthermore, the characteristic behavior of a page hopping sequence is discussed. In the last section, a statistical analysis is done to carry out a relevance analysis concerning the impact of Bluetooth page state on KLEER.

2   Impact of IEEE 802.11b/g on KLEER

In the network architecture considered in this paper, IEEE 802.11 is using channel 6 with a center frequency of 2.437 GHz. An IEEE 802.11b/g channel occupies a 20 dB-bandwidth of approximately 16/17 MHz. According to the channel map, shown in Fig. 2, four KLEER channels overlap completely and one partly with a single IEEE 802.11 b/g channel. This assumes an IEEE 802.11 bandwidth of 22 MHz according to a more conservative bandwidth definition which is often used in literature. We used KLEER evaluation boards to analyze the impact of an IEEE 802.11 b/g interferer on a KLEER audio streaming. This allows disabling KLEER's dynamic channel switching. By applying an IEEE 802.11 b/g signal on a KLEER transmission we could evaluate the signal to interferer ratio (SIR) at which audio dropouts occurs. For a SIR region of 1 to −20 dB, audio dropouts occurred in up to four KLEER channels (out of 16) in presence of IEEE 802.11b/g traffic. The exact number of disturbed channels (channels where audio dropouts occur) depends on the interferer signal (either IEEE 802.11b or g) and the overlapping spectral power density of the interfering signal in a KLEER channel. Regarding the SIR values, it is important to note that IEEE 802.11b/g is transmitting with a maximum power of 20 dBm compared to KLEER with 1.5 dBm. Therefore a SIR of −20 dB represents almost equal path loss conditions between KLEER victim receiver and KLEER transmitter and IEEE 802b/g transmitter, respectively. Under more unfavorable path loss conditions even more than four KLEER channels can be disturbed. For SIR values between −20 and −30 dB, IEEE 802.11b disturbs five KLEER channels. On the opposite side, KLEER is not affected at all by IEEE 802.11b/g for SIR values higher than 1 dB. Considering the difference in transmitting power between IEEE 802.11 and KLEER, interference is likely in an in-vehicular environment. Therefore KLEER has to avoid IEEE 802.11b/g signals. But if only a limited number of the total 16 channels are affected by IEEE 802.11 signals, KLEER's dynamic channel switching (DSC) is an effective method to avoid the disturbed channels. Our measurements showed that DSC enables KLEER to dynamically change the affected channels during an IEEE 802.11 disturbance with an SIR of −30 dB without audio quality degradation.

3 Impact of Bluetooth on KLEER

KLEER needs a minimum SIR of about 13 dB to prevent audio dropouts in presence of a constantly transmitting Bluetooth GFSK-signal. The SIR difference compared to IEEE 802.11 signals can be explained by the lower spectral bandwidth of Bluetooth. In contrast to IEEE 802.11, Bluetooth communication standard uses Frequency Hopping Spread Spectrum (FHSS) which combines TDMA (Time Division Multiple Access) and FDMA (Frequency Time Division Multiple Access). The TDMA divides the channel in 625 μs slots resulting in 1600 slots s^{-1}. For the connecting state (inquiry or paging) also half slots of 312.5 μs are used. The FDMA is dividing the ISM band in 79 channels of 1 MHz width starting at 2.402 and ending at 2.480 GHz. Each packet is transmitted in a different channel than the previous packet following a pseudo-random hopping sequence. In Bluetooth

specification version 1.2, Bluetooth Special Interest Group (SIG) introduced Adaptive Frequency-hopping (AFH) as additional coexistence mechanism for connected devices (Bluetooth SIG, 2004). When AFH is in use, channels which are classified as "bad" are removed from the hopping sequence. The Bluetooth core specifies a minimum number of 20 RF channels. In presence of an interfering FHSS communication system − like Bluetooth − with uniformly distributed channels over the whole band, a channel switch is unnecessary and even bares risks: a channel change always implies a data overhead and reduces the net data rate. Moreover, the Bluetooth channel classification is influenced negatively by a constantly switching system. In best case, Bluetooth will classify the channels inside the current KLEER channel as "bad" and stop using these channels. Therefore, the interference is only temporary during the Bluetooth AFH-adaptation time. Nevertheless, KLEER's retransmission bandwidth has to be able to compensate the packet loss during an occurring Bluetooth interference to provide overall coexistence. To calculate the resulting packet loss, the packet distribution in time and frequency has to be known. As Bluetooth supports multiple applications in different profiles, the paper focuses on typical in-vehicle scenarios:

- paging/inquiry,

- A2DP audio streaming (ACL connection type),

- hands-free telephony (SCO connection type),

- and connected state without any data transmission.

The measurements were conducted with evaluation boards using an Ellisys Bluetooth sniffer. In Table 2 the average packet distribution in time and frequency is shown.

Comparing the slot rates for both directions as worst-case interference scenario, it shows that paging/inquiry achieves by far the highest slot rate, however, with very short packets of only 68 μs length. In order to make a statement of the interference impact, it is necessary to calculate the resulting packet error rate. According to Golmie (2006), an analytical approach is used to model the occurring interference. Figure 3 shows the timing of the desired packets with respect to the interfering packets seen at the victim's receiver. The model assumes that the desired and interfering packets are sent periodically in a fixed time period of T_{sig} and T_{int}. This simplifies the calculation, but is only valid in case of paging/inquiry and SCO connection. But nevertheless, it also gives a reasonable estimation for the other cases. The packet length is denoted with t_{sig} and t_{int} .

The variable Δt defines the time offset between a desired and an interfering packet. Assuming that Δt is uniformly distributed, the average number that an interfering packet hits a desired packet in time can be calculated, as follows:

$$h_i = \frac{t_{sig} + t_{int,i}}{T_{int,i}}. \tag{1}$$

Table 2. Average packet distribution in time and frequency for different Bluetooth states; data derived from an exemplary measurement using Ellisys Bluetooth sniffer.

State	Packet distribution time			Packet distribution in frequency domain
	Average Slot rate (Master to Slave)	Average Slot rate (Slave to Master)	Average Slot rate (both directions/all packet types)	
Page/Inquiry	1 slot/ID-packet (68 μs)	–	1 slot/ID-packet (68 μs)	Divided in A/B trains with a simultaneous use of 16 channels
A2DP audio streaming	approx. 16 slots/2-DH3-packet (1.4 ms mainly 2-DH3 packets are used)	approx. 8.4 slots/ NULL-packet (126 μs)	approx. 5.5 slots/packet	20–79 channels with AFH
Hands-free	12 slots/2-EV3-packet (392 μs mainly 2-EV3 packets are used) + ca. 40 slots/NULL-packet (NULL 126 μs)	12 slots/packet (mainly 2-EV3 packets [392 μs])	+ ca. 70 slots/packet (NULL packets [126 μs]) approx. 4.8 Slots/packet	20–79 channels with AFH
Connected state without data transmission	40 slots/POLL-packet (126 μs)	40 slots/packet (NULL-packets [126 μs])	20 Slots/packet	20–79 channels with AFH

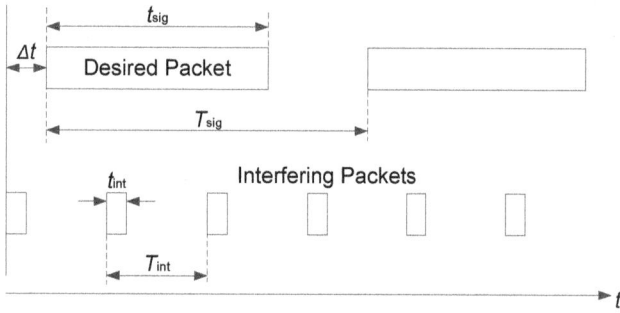

Figure 3. Collisions at the victim's receiver.

Considering different interfering packets $i = 1, 2, 3, .., M$ the total number is a sum of the individual numbers.

$$h = \sum_{i=1}^{M} \frac{t_{\text{sig}} + t_{\text{int},i}}{T_{\text{int},i}} \qquad (2)$$

Assuming that every collision in time and frequency causes a packet error, the packet error rate is:

$$\text{PER} = \begin{cases} h \cdot \dfrac{n_{\text{int}}}{n_{\text{total}}} & \text{if } h \leq \dfrac{n_{\text{total}}}{n_{\text{int}}} \\ 1 & \text{if } h > \dfrac{n_{\text{total}}}{n_{\text{int}}} \end{cases}, \qquad (3)$$

where n_{total} is the total number and n_{int} the number of channels overlapping with the desired signal. The channel distribution is assumed to be uniformly distributed. The above mentioned formula assumes that the collision in time and frequency are independent. Even if not all assumptions of the packet error model are completely valid, it gives a sufficient well estimation of the interference. Using the derived model and assuming a KLEER packet length of 1.3 ms an average

packet error rate can be calculated. For A2DP, Hands-free and no data transmission two different interfering scenarios shall be regarded: in both scenarios, three BT channels are overlapping with a KLEER channel. But one scenario assumes BT to use all 79 channels and the other only 59 channels. The latter represents a situation where AFH avoids 20 channels due to additional IEEE 802.11 interference. In both cases, only co-channel interference with three BT channels inside one KLEER channel is considered. For Bluetooth paging/inquiry 16 channels are used simultaneously in one interval of 10 ms. A characteristic of the paging/inquiry sequence is that the adjacent channels always remain unused. More details about the characteristics of Bluetooth page state are given in the next section. Considering co-channel interference using KLEER's channel bandwidth of 3 MHz three cases have to be distinguished: no interference occurs and one or two channels out 16 are disturbed. The resulting probabilities that a BT packet hits a KLEER packet are shown in Table 3.

The table clearly shows that the critical states are page/inquiry with a possible packet error rate of 13.7 and 27.4 % in average. For Bluetooth connected states, the probabilities lie between 0.2 and 2.8 % which can be compensated by retransmitting the lost packets. The reader could notice that the interference due to paging/inquiry is not relevant as it is a very rare event. This is true for inquiry but not for page state: sometimes it is desirable that a device is able to connect at anytime. For example, in a vehicle the head unit enters periodically the page state every 20–30 s if no device is connected. In such a case, interference from Bluetooth paging has definitely to be excluded. As KLEER's retransmission bandwidth is not able to cope with the packet error rates for noise-equivalent audio signals which cannot be sufficiently compressed. Thus, KLEER's DSC has to be capable to find

Table 3. Average probability that a Bluetooth packet hits a 1.3 ms long KLEER packet in frequency and time for different Bluetooth states.

State	BT channels inside KLEER and total # of channels	Average probability that a BT packet hits a KLEER packet in time and frequency		
		Packet from Master	Packet from Slave	Packet either from Master or Slave
Page	0 of 16	0%	0%	0%
Inquiry	1 of 16	13.7%	0%	13.7%
	2 of 16	27.4%	0%	27.4%
A2DP	3 of 79 (only BT)	1.0%	1.0%	1.0%
audio	3 of 57 (BT & IEEE 802.11)	1.4%	1.4%	2.8%
Hands-free	3 of 79 (only BT)	1.1%	1.0%	2.1%
streaming	3 of 57 (BT & IEEE 802.11)	1.4%	1.3%	2.7%
Connected	3 of 79 (only BT)	0.2%	0.2%	0.4%
state w/o data	3 of 57 (BT & IEEE 802.11)	0.3%	0.3%	0.6%

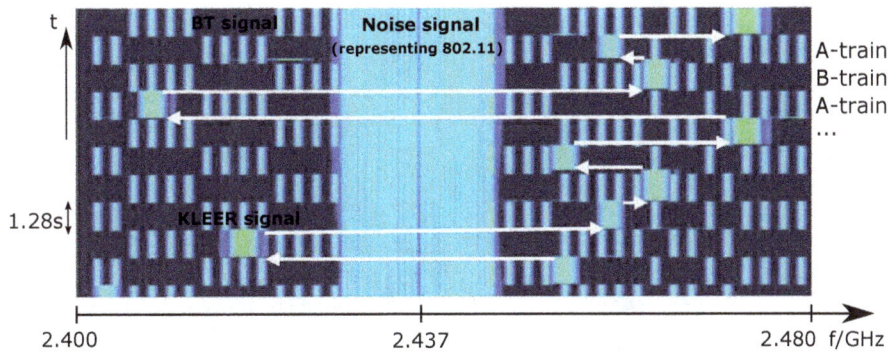

Figure 4. Constant channel switch of a single KLEER transmission in presence of BT paging and IEEE 802.11b/g – measurement done with signal generator and cable setup.

a free channel fast enough without audio degradation. To analyze the coexistence capabilities of KLEER in presence of BT paging/inquiry we generated a BT page state sequence using a vector signal generator. The sequence was generated with a periodic interval of 1.28 s long A and B trains. As BT paging always leaves a region of 16 MHz free, we added a noise signal to fill this gap. This bad case interference scenario was applied to a KLEER transmission. The power ratios are chosen in such a manner that a collision in frequency and time leads very likely to a packet loss. To manage this bad case scenario KLEER has to be able to constantly switch in a free channel. The Fig. 4 shows the measured power density of the described scenario against frequency and time in a waterfall plot.

As shown in Fig. 4, KLEER initiates a channel switch if the Bluetooth page sequence changes from A to B train and vice versa. After every train change, KLEER is able to find an undisturbed channel without any audio degradation. Under these difficult conditions, KLEER's channel switching algorithm is working well.

4 Statistical interference analysis of page (inquiry) sub state

As shown in the last section Bluetooth page (inquiry) sub state is a very serious interferer. For a Bluetooth connection setup the master sends two 68 μs long ID packets each second slot. The ID packets are spaced by 312.5 ms. Each packet is sent on a different frequency according to a short pseudo-random hopping sequence with a total period length of 2^{16} slots ($= 40.96$s). The hopping sequence is determined by the ULAP of the Bluetooth device which is paged. In case of inquiry, the ULAP is usually derived from the general Access Code (GIAC). A total page (inquiry) hopping sequence consists of 32 dedicated wake-up frequencies. These 32 frequencies are divided into two partial sequences of 16 frequencies. The page (inquiry) hopping sequence does not use a hop adaptation. During an interval of 10 ms the master transmits sequentially ID packets at 16 frequencies. The partial sequences – the so called A and B trains – are shifted by half of a page sequence period ($= 40.96$s/2). In modes R1 and R2, the trains are repeated 128 and 256 times (1.28 and 2.56 s) to assure that the slave is able to detect at least one message. After one repetition, it is switched to the other

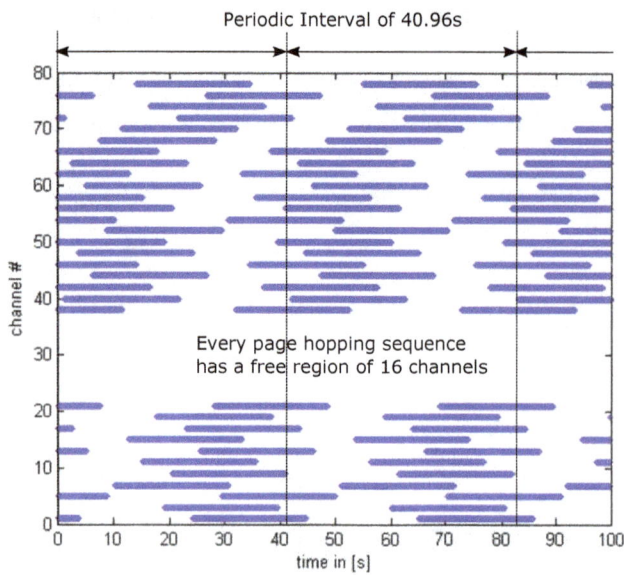

Figure 5. Exemplary Hopping sequence in page state for ULAP = 0x8B6949.

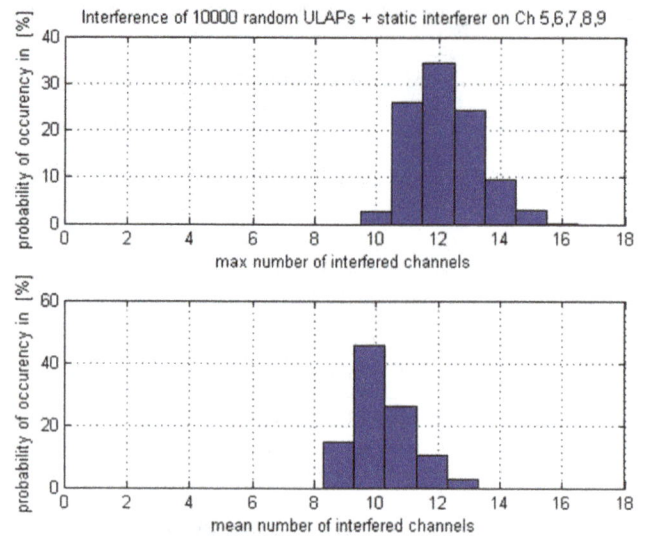

Figure 6. Number of disturbed channel by a Bluetooth page state with probability of concurrency calculated with 10 000 different Bluetooth device addresses.

train. Figure 5 shows the used channels of an exemplary page hopping sequence.

The figure illustrates the characteristics of a page (inquiry) hopping sequence:

- The sequence always leaves a region of 16 channels free.

- There is a gap of at least one frequency between two used channels.

- The page/inquiry hopping sequence repeats every 40.96 s.

- Every 1.28 s one (out of 16) frequencies changes. Between 20.48 s (= 16 × 1.28 s – phase difference between A and B train) two different sets of 16 channels are used.

As described before, every page hopping sequence uses 16 channels per train and has a free region of 16 frequencies. The spectral location of the free region depends on the Bluetooth device address and is uniformly distributed (for random device addresses). Furthermore, we know that an IEEE 802.11b/g signal is able to disturb up to 5 KLEER channels – assuming only co-channel interference. Assuming a bad case scenario, an IEEE 802.11b/g signal lies into the free region of the page hopping sequence. In an absolute worst-case scenario, the 16 frequencies of A or B train are disturbing the remaining 11 KLEER channels outside IEEE 802.11. In this case, all 16 KLEER channels are disturbed either with the IEEE 802.11b/g signal or the paging signal. KLEER would constantly initiate channel switches without the chance to find a free channel – at least during a single page train of 1.28 or 2.56 s. It is important to know the probability of occurrence for this worst case scenario to evaluate

the relevancy. Based on 10 000 calculated page hopping sequences with random ULAP addresses a statistical analysis was performed. For every ULAP address 32 sequences of 16 frequencies were calculated to cover all possibilities over time. In total 320 000 page hopping sequences were calculated and statistically evaluated. Figure 6 shows the probability of occurrence for the number of interfered channels. In the upper figure the probability is given for the maximum number (considering the 32 possible sequences over time for one ULAP address). The lower figure shows average number of interfered channels.

As result it can be stated that the above mentioned worst-case situation is possible: the statistical probability that IEEE 802.11b/g and Bluetooth page state overlap with all 16 KLEER channels is about 0.02 %. In other words, two out of 10 000 randomly generated ULAP addresses are manifesting this behavior. In such a case, the interference on all channels occurs during 1.28 of 40.96 s. It should be emphasized that this worst-case situation only occurs if several factors come together:

- Bluetooth is enabled at the vehicle's head unit (HU) without any connected device.

- IEEE 802.11b/g packet load is high and the SIR is low enough to disturb five KLEER channels.

- The critical sequence is appearing while paging is active.

- A packet collision between a Bluetooth ID packet and a KLEER packet leads very likely to a KLEER packet loss.

Table 4. Possible solutions for paging interference

Solution	Realization	Effect
Increase KLEER's retransmission bandwidth	− Increase net data rate − Decrease amount of transmit data (e.g. using adaptive lossy compression in case of interference)	A channel switch becomes unnecessary with a sufficient retransmission bandwidth
Increase KLEER's number of channels	Decreasing KLEER's channel spacing of 5 MHz	A larger number of channels increases the chance of having non-disturbed channels
Improve SIR at the victim's receiver	− Increase KLEER's Tx power − Decrease Bluetooth Tx power (in page state)	Increasing the power of the desired signal or reducing the power of the interferer signal improves the SIR at the victim's receiver

From a customer view, a page state interference is incomprehensible as Bluetooth is switched off at the mobile device and only IEEE 802.11 as single system is used simultaneously with KLEER. In such a situation the customer cannot understand why he is experiencing audio dropouts. If there is more than one KLEER connection active, of course more than one free channel is needed. At maximum three independent streams (on three channels) are possible. The probability that three or more channels are disturbed during a page state is already 12.4 %. Thus is it certainly necessary to take steps to minimize the interference in such a critical state. Table 4 gives an overview of possible measures for interference reduction: beside of the pure technical aspects, it was important for Daimler to avoid time-consuming costly firmware and hardware changes. Considering the cost and interference aspects, the following two measures were taken: the audio level was lowered digitally on the transmitter's side to improve the audio compression rate. A smaller amount of transmit data increases the retransmission bandwidth. By adjusting the channel switch algorithm to the new retransmission bandwidth, a channel change become more unlikely during page state even if one Bluetooth channel lies inside the KLEER channel. It must be taken into account that an audio level reduction implies an increase of the audio quantization noise. Additionally to the reduction of the audio level, the Bluetooth transmit power was reduced during page/inquiry state to avoid packet loss in case of packet collision. In an in-vehicle situation the distances to the mobile device are very short (typically below 3 m) which usually allows a certain power reduction. First antenna based measurements on random positions inside a vehicle cabin showed that a power reduction of up to 20 dB does not affect the connect ability of Bluetooth devices.

5 Conclusions

In this paper we evaluated the coexistence of KLEER, a proprietary wireless standard in the 2.4 GHz-ISM-band used for audio transmission in full CD quality, in presence of Bluetooth and IEEE 802.11 b/g. KLEER provides a coexistence mechanisms with packet retransmission and dynamic channel switching (DCS). The DCS is able to cope with static (or slowly switching) interferers like IEEE 802.11 b/g without audio degradation. Unavoidable packet loss of a frequency hopping spread spectrum (FHSS) Bluetooth system in connected state can be compensated by retransmitting lost packets. The missing packets are compensated using a buffer management. The investigation showed that a very serious interference scenario consist of a Bluetooth system in page state in combination with an IEEE 802.11 b/g link. This case is particularly critical when the Bluetooth master device enters periodically the page state to allow a connection of slave devices at any time. In an absolute worst-case scenario, all 16 KLEER channels can be disturbed either by Bluetooth ID packets or IEEE 802.11 b/g packets. Two out of 10 000 ULAP addresses are showing this worst-case behavior. Measures are presented to improve the coexistence in this situation. Most of the publications concerning interference are dealing with systems operating in connected state. But this study shows that it is also important to include unconnected states into a full coexistence investigation. Under certain conditions Bluetooth connecting state (paging/inquiry) can be more critical as connected states. The packet rate of Bluetooth paging is approximately five times higher compared to typical connected states as Hands-Free telephony or A2DP streaming. Furthermore, the packets are sent over the whole band (only a region of 16 MHz is free) without channel adaptation (AFH) for interference mitigation. Another important aspect is that a user is not able to understand a Quality-of-service (QoS) degradation during page state interference as it occurs when Bluetooth is switched off at his mobile device. The interference issues of Bluetooth page state should also be viewed from the interferer side. For interference mitigation, methods are needed to avoid interference from other devices sharing the same band – but as second coexistence objective the interference on other devices should be minimized as well. During a single page sequence of 1.28 s (2.56 s) Bluetooth master device sends 2048 (4096) identical 68 bit-packets in a very high rate containing a high amount of redundant information. On the slave's side, this is favorable as the scan time can be minimized and thus energy saved. But for other systems, this could cause serious interference as shown in this paper. An approach to minimize the interference without changing the connecting procedure, is to adaptively reduce the transmit power during the connection state. It has however be assured that the devices are close enough to each other that a reduction in transmit power will not affect the ability to connect devices.

References

Bluetooth SIG: Specification of the Bluetooth System, Core, Version 2.1, available at: http://www.bluetooth.com (last access: 24 March 2015), 2004.

Golmie, N.: Coexistence in Wireless Networks, Cambridge University Press, Cambridge, UK, 2006.

Devries, C. A., Mason, R. D., and Beards, R. D.: RF-to-baseband receiver architecture, US Patent no. 7,539,476, available at: https://www.google.com/patents/US7539476 (last access: 20 March 2015), 2009.

Mason, R. D., Li, R., and DeCruyenaere, J. P. R.: Wireless communications systems and channel-switching method, US Patent App. 12/106,098, available at: https://www.google.com/patents/US20090262709 (last access: 20 March 2015), 2009.

SMSC: ISM Band Coexistence, Document Number KLR0000-WP2-1.2, available at: http://ww1.microchip.com/downloads/en/DeviceDoc/Kleer_ISM_Coexistence.pdf (last access: 24 March 2015), 2007a.

SMSC: Wireless Digital Audio Quality for Portable Audio Application, KLEER KLR0000-WP1-1.4, available at: http://ww1.microchip.com/downloads/en/DeviceDoc/Kleer_AudioQuality.pdf (last access: 24 March 2015), 2007b.

8

A high throughput architecture for a low complexity soft-output demapping algorithm

I. Ali, U. Wasenmüller, and N. Wehn

Microelectronic Systems Design Research Group, University of Kaiserslautern, 67663 Kaiserslautern, Germany

Correspondence to: I. Ali (imran@eit.uni-kl.de)

Abstract. Iterative channel decoders such as Turbo-Code and LDPC decoders show exceptional performance and therefore they are a part of many wireless communication receivers nowadays. These decoders require a soft input, i.e., the logarithmic likelihood ratio (LLR) of the received bits with a typical quantization of 4 to 6 bits. For computing the LLR values from a received complex symbol, a soft demapper is employed in the receiver.

The implementation cost of traditional soft-output demapping methods is relatively large in high order modulation systems, and therefore low complexity demapping algorithms are indispensable in low power receivers. In the presence of multiple wireless communication standards where each standard defines multiple modulation schemes, there is a need to have an efficient demapper architecture covering all the flexibility requirements of these standards. Another challenge associated with hardware implementation of the demapper is to achieve a very high throughput in double iterative systems, for instance, MIMO and Code-Aided Synchronization.

In this paper, we present a comprehensive communication and hardware performance evaluation of low complexity soft-output demapping algorithms to select the best algorithm for implementation. The main goal of this work is to design a high throughput, flexible, and area efficient architecture. We describe architectures to execute the investigated algorithms. We implement these architectures on a FPGA device to evaluate their hardware performance. The work has resulted in a hardware architecture based on the figured out best low complexity algorithm delivering a high throughput of 166 Msymbols/second for Gray mapped 16-QAM modulation on Virtex-5. This efficient architecture occupies only 127 slice registers, 248 slice LUTs and 2 DSP48Es.

1 Introduction

In transmitter, a constellation mapper takes groups of bits and maps them to particular constellation points. A specific magnitude and phase represents a certain combination of bits in the transmitted symbol. Due to distortion by the wireless channel, an error occurs in the position of each transmitted constellation point. In the receiver, the phase and magnitude of each received symbol is extracted, and a decision is made about what combination of bits the transmitter sent.

In the receiver, bit level demapping can be performed such that the output of demapper is "hard", i.e., either a logical value 1 or 0. Alternatively, the demapper output can be "soft"; a soft-value indicating the probability, that the modulated bit associated with a given demapper output is to be of logical value 1 or 0. The soft-output (LLR) of the kth bit c_k in noisy received symbol sequence r is

$$\mathrm{LLR}(c_k|r) = \ln \frac{p(c_k = 1|r)}{p(c_k = 0|r)} \qquad (1)$$

where p denotes probability.

If modulating bits are uncoded or algebraic coded such as RS-codes or BCH codes, the demapper output is typically hard. If modulating bits are coded with a convolutional, LDPC, or Turbo-Code encoder the demapper output must be soft in order to yield superior performance. Consequently, soft-output demappers are an integral part of many modern communication receivers.

Optimal soft-output demapping algorithms involve computationally complex functions such as logarithmic and exponential functions, and thus are not well suited for hardware implementation. On the other hand, suboptimal methods significantly reduce the computational complexity by adopting simplified functions. However, they still require to calculate distances between the received symbol and all constellation points Li et al. (2009); Su et al. (2011); Lin et al. (2010); Ryoo et al. (2003); Li and Shi (2014) and Lee et al. (2011). The computational complexity of suboptimal methods is further reduced in so-called low complexity soft-output demapping algorithms. A large number of works in this domain have focused on theoretical/simulation aspects of the algorithms aiming to attain superior frame error rate (FER) performance. Little attention has been paid to the actual implementation of such algorithms that look to deliver more than 100 Msymbols/second.

Usually the demapping function is executed only once on each burst in the receiver. However, the double iterative systems such as MIMO and Code-Aided Synchronization engage the demapper in their outer iteration. Consequently, the demapping function is executed multiple times on each burst in the receiver. Accordingly, the double iterative systems require very high throughput demapper. In this regard, consider the needs of gateway of Second Generation Digital Video Broadcasting Interactive Satellite System which is typically 20 Msymbols/second DVB (2014). In case of utilizing code-aided synchronization, typically 8 outer iterations are performed (see Fig. 1) to achieve the desired communications performance as reported in Ali et al. (2014). In such system, the demapper must deliver a throughput of 8-times 20 Msymbols/second. Therefore, we decided to set a minimum 160 Msymbols/second throughput specifications for this work.

In the presence of multiple wireless communication standards where each standard defines multiple modulation schemes, there is a need to have single demapper architecture covering all the flexibility requirements of these standards. We focus on popular Gray mapping M-PSK and M-QAM modulation schemes in this work which are specified in many wireless communication standards. The architectures reported in Altera (2007); Park et al. (2008); and Jafri et al. (2010) support multiple modulation schemes. However, they do not satisfy the throughput requirement outlined above.

Based on the results of a thorough literature search and deep analysis, we find the following algorithms having remarkable reduced complexity without compromising on quality of communications performance: (1) the algorithm reported in Lin et al. (2010) identifies the two required constellation points to compute one LLR in a very simple way and (2) The algorithms reported in Tosato and Bisaglia (2002); Ryoo et al. (2003); Kim et al. (2006); Arar et al. (2007); and Sun and Zeng (2011) are quite similar to each other and provide a very simple approach to compute LLR. We call this approach as decision threshold algorithm in this sequel.

The computational complexity of the aforementioned algorithms have been examined in the literature by counting number of operations required which is not a sufficient measure to derive realistic complexity for hardware implementation. Instead, hardware complexity metrics are: throughput,

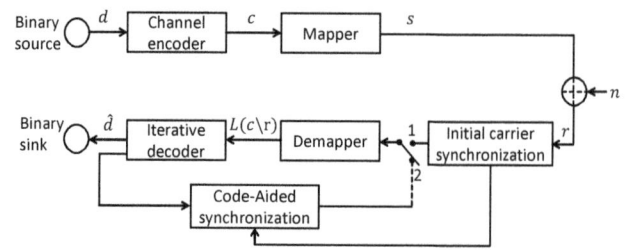

Figure 1. Baseband model of iterative channel decoding based system.

latency, resource utilization, and power. We investigate the hardware performance of the considered algorithms by realizing FPGA/ASIC implementations under the constraint of above specified throughput. At the conclusion of our work, we identify a demapping algorithm having the lowest implementation complexity.

This paper is structured as follows. In Sect. 2, we describe the system model as well as optimal and traditional suboptimal algorithms while Sect. 3 explains low complexity suboptimal algorithms. The communications performance of the algorithms is shown in Sect. 4. The hardware architectures and their implementation complexity are compared in Sect. 5, and Sect. 6 concludes this work.

2   System model

The system model comprising channel encoder, iterative channel decoder, mapper, demapper, initial carrier synchronization, and Code-Aided synchronization is shown in Fig. 1. The "Channel encoder" processes binary signal d and produces the encoded signal c. Then, the "Mapper" block maps M coded bits $c_0, c_1, \ldots, c_{M-1} \in \{0, 1\}$ to a complex symbol s using the mapping function $s = \text{map}(c_0, c_1, \ldots, c_{M-1})$. The Additive White Gaussian Noise (AWGN) channel adds noise n in the signal. Discrete-time baseband signal at the receiver can be represented as

$$r(k) = s(k) \cdot e^{j(2\pi k f_0 T + \Phi)} + n(k) \qquad k = 0, 1, \ldots, K-1 \quad (2)$$

where $s(k)$ is the transmitted signal, K is the length of the received signal, T is the symbol duration, f_0 is frequency offset, Φ is phase offset, and $n(k)$ is a sequence of complex white Gaussian noise samples with variance σ^2.

In the receiver after performing automatic gain control, frame detection, timing synchronization, and initial carrier synchronization (phase/frequency) the resulting data sequence is transferred to "Demapper". The demapping module demodulates the complex channel symbols and extracts M soft-outputs for a received symbol using a log likelihood ratio calculation. The "Iterative decoder" estimates the transmitted bits using soft-input from the "Demapper". The soft-output of "Iterative decoder" is used by "Code-Aided synchronization" to further compensate the frequency and phase offset from the received burst. Afterwards, the newly corrected burst is passed to "Demapper" and subsequently the next iteration of "Iterative decoder" is performed. Hence, in this double iterative system demapping function is performed after each iteration of the decoder. After presenting the system model, we discuss about the demapping algorithms. The soft-output demapping methods are classified into two major categories optimal and suboptimal which are explained in the subsequent sections.

2.1 Optimal soft demapping

For M-ary modulation scheme, the demapper needs to calculate log-likelihood ratios on the coded bits $c_0, c_1, \ldots, c_{M-1}$ for each incoming received symbol. The channel information of the coded bit c_k conditioned on the received symbol r can be calculated as follows.

$$LLR(c_k|r) = \ln \frac{\sum_{i=0}^{2^{M-1}-1} \exp\left[-\frac{1}{2\sigma^2}\left(r - s_{k(1,i)}\right)^2\right]}{\sum_{i=0}^{2^{M-1}-1} \exp\left[-\frac{1}{2\sigma^2}\left(r - s_{k(0,i)}\right)^2\right]} \quad (3)$$

where σ^2 is variance of AWGN channel, $s_{k(1,i)}$ and $s_{k(0,i)}$ represent the constellation points whose kth bits are one and zero respectively and M represents the number of bits in one modulated symbol. In 16-QAM modulation, four bits constitute a symbol so in this case $M = 4$.

It can be clearly seen in Eq. (3) that the optimal demapping method involves logarithmic and exponential functions to compute LLR. Because of these computationally complex mathematical operations, the optimal demapping method is not suitable for hardware implementation. This computational complexity is reduced in suboptimal demapping methods which is described in the following section.

2.2 Suboptimal soft demapping

In order to eliminate the logarithmic and exponential functions in Eq. (3), the suboptimal algorithms adopt an approximation. Since the sum term in Eq. (3) is dominated by the largest term, it can be simplified as reported in Robertson et al. (1995). The simplification can be formally expressed as

$$\ln \sum_j \exp(-a_j) \approx \max(-a_j) = \min(a_j) \quad (4)$$

where $a_j >= 0$. With this approximation, LLR can be computed as follows.

$$LLR(c_k|r) = \frac{1}{2\sigma^2} \cdot \left[\min_i(r - s_{k(0,i)})^2 - \min_i(r - s_{k(1,i)})^2\right] \quad (5)$$

where $i = 0, 1, \ldots, 2^{M-1} - 1$.

It is evident from Eq. (5) that the suboptimal demapping algorithm significantly reduces the computational complexity by avoiding logarithmic and exponential functions as opposed to the optimal algorithm. Despite this simplification,

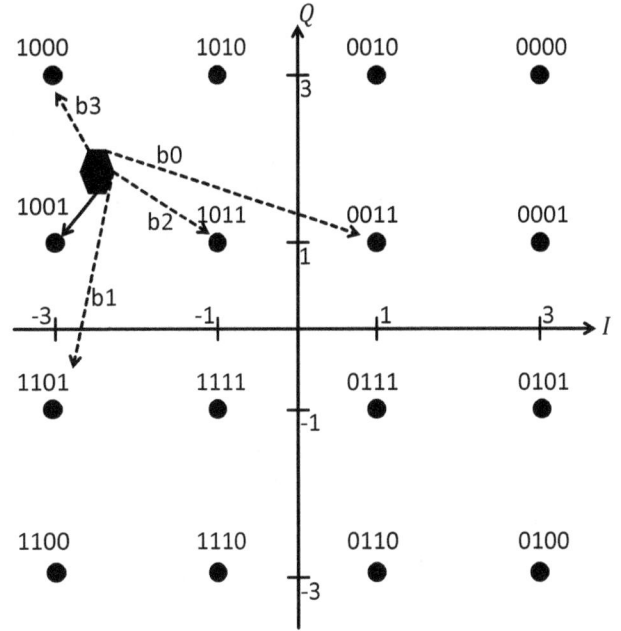

Figure 2. Lin algorithm illustration for Gray mapped 16-QAM modulation. The black color circles denote constellation points whereas the black color hexagon represents the received symbol

the suboptimal demapping algorithm involves computation of all possible Euclidean distances and then an exhaustive search to determine the two nearest constellation points. This complexity is considerably prominent in high order modulation schemes. This computational complexity can be further reduced by adopting some simple techniques as explained in the subsequent section.

3 Low complexity suboptimal demapping algorithms

This section explains two low complexity suboptimal demapping algorithms which are applicable to popular Gray mapped modulation schemes: (1) Lin algorithm and (2) decision threshold algorithm.

3.1 Lin algorithm

The algorithm described in Lin et al. (2010) does not compute all possible Euclidean distances as opposed to the traditional sub-optimal demapper. Instead, it identifies two constellation points $s_{k(0,i)}$ and $s_{k(1,i)}$ of Eq. (5) that are at minimum distance from the received symbol followed by computation of only two squared distances. This identification is carried out using very simple mathematical operations. This technique is explained with an example in Fig. 2.

The magnitude of the received symbol in the complex plane is $(-2.5, 1.5)$ in the considered example. At first, the Cartesian coordinates of the received symbol are rounded to the coordinates of its nearest constellation point (NCP). The magnitude of real part of the received signal is -2.5 which is closer to -3 as compared to -1. Similarly, the imaginary part of the received symbol is 1.5 which is closer to 1 instead of 3. Therefore, the NCP of the received symbol is $(-3,1)$ in the constellation diagram. The corresponding bit mapping of the NCP is "1001". This NCP is used to compute the first squared distance from the received symbol. Afterwards, four nearest constellation points are identified with respect to this computed CP where kth bit of the formers is flipped corresponding to the kth bit of the latter. We call the former constellation points as Flipped Constellation Points (FCPs).

For the considered case, the first bit b_0 (MSB) in the NCP (1001) is "1". Now we want to compute the FCP whose first bit is "0", i.e., flipped with respect to the first bit of NCP. This can be accomplished by using a simple transformation $x'x1x$. In this transformation term x' means flip the corresponding bit, x represent no change in the corresponding bit, and 1 means replace the corresponding bit by 1. Using this transformation, the computation of FCP for MSB of NCP (1001) results in "0011" which is highlighted in Fig. 2 for bit b_0. The second squared distance is computed between the received symbol and this FCP. Similarly, the remaining three FCPs are computed corresponding to the second $b1$, third $b2$ and fourth $b3$ bits using the transformations $xx'x1$, $xxx'x$ and $xxxx'$ respectively.

Finally, LLR of one bit is computed by using the two calculated squared distances, i.e., squared distance between the received symbol and the NCP, and squared distance between the received symbol and its corresponding FCP. Remark, the first computed squared distance can be utilized for LLR calculation of the remaining three bits of the received symbol. In short, for a 16-QAM modulated symbol five squared distances are computed to calculate LLR of four bits, whereas for the same case the traditional sub-optimal demapper needs to compute 16 squared distances.

It is very important to mention that the abovementioned mathematical transformations to compute FCP are specific to the mapping scheme shown in Fig. 2. If the mapping scheme is changed, these mathematical transformations to compute FCP also need to be modified accordingly. Furthermore, this technique is applicable to only Gray mapping modulation schemes, including PAM, PSK, and square QAM. Under these constraints, the algorithm computes the distances which are exactly needed in Eq. (5) as claimed in Lin et al. (2010).

3.2 Decision Threshold Algorithm

In order to reduce the computational complexity, this algorithm adopts a simple decision threshold comparison mech-

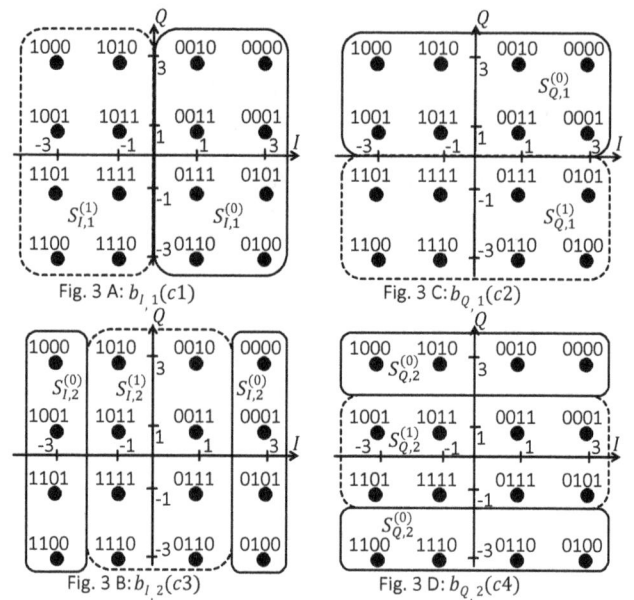

Figure 3. Partitions of the 16-QAM constellation used in decision threshold algorithm

anism, and thus a single distance computation accomplishes the soft demapping.

In the case of 16-QAM constellation, the partitions $(S_{I,k}^0, S_{I,k}^1)$ are shown for the generic in-phase components of the complex signal $b_{I,k}$ (c1, c2) in Fig. 3a and b. The partitions $(S_{Q,k}^0, S_{Q,k}^1)$ for the quadrature component of the complex signal $b_{Q,k}$ (c3, c4) are shown in Fig. 3c and d. The MSB c1 is always 1 in the left half section and 0 in the right half section (see Fig. 3a). The second bit c2 is always 1 in the lower half section and 0 in the upper half section (see Fig. 3c). The third bit c3 is always 1 in the middle section and 0 in the outer section (see Fig. 3b). The LSB bit c4 is always 1 in the middle section and 0 in the outer section (see Fig. 3d). The decision threshold algorithm exploits this property of Gray mapping and provides very simple expression to calculate the LLR.

As discussed that components of the complex signal are delimited by either horizontal or vertical boundaries. Therefore, the two symbols within the two subsets, nearest to the received signal, always lie in the same row if the partition boundaries are vertical (bits $b_{I,1}$ and $b_{I,2}$ in Fig. 3a and b) or in the same column if the boundaries are horizontal (bits $b_{Q,1}$ and $b_{Q,2}$ in Fig. 3c and d). The same observation holds true for 8-PSK and 64-QAM constellations. As a consequence, the LLR of the constituting bits in 16-QAM modulation can be derived as follows.

$$LLR(c_1|r) = \frac{2}{\sigma^2} \cdot \left[x \right] \qquad (6)$$

$$LLR(c_2|r) = \frac{2}{\sigma^2} \cdot \left[y \right] \qquad (7)$$

$$LLR(c_3|r) = \frac{2}{\sigma^2} \cdot \left[|x| - \frac{C+D}{2} \right] \qquad (8)$$

$$LLR(c_4|r) = \frac{2}{\sigma^2} \cdot \left[|y| - \frac{C+D}{2} \right] \qquad (9)$$

where two positive constants C and D represent the magnitudes of I and Q components of 16-QAM symbol which are 1 and 3 in this example. The terms x and y represent the distances of real and imaginary parts of the received symbol from the origin respectively in the complex plane. Regarding term $\frac{2}{\sigma^2}$, the detailed derivation can be found in Ryoo et al. (2003). The above described equations show that computing $LLR(c_1|r)$ and $LLR(c_2|r)$ require no distance calculation, whereas the computation of $LLR(c_3|r)$ and $LLR(c_4|r)$ require calculating two absolute values and two simple subtractions. It is worth mentioning that the expressions Eqs. (13) to () are specific to the mapping scheme shown in Fig. 3. If the mapping scheme is changed, these mathematical expressions also need to be modified accordingly. Furthermore, this technique is applicable to only Gray mapping modulation schemes.

For a given constellation diagram of Gray coded 8-PSK modulation, LLR of the constituting bits can be computed as follows.

$$LLR(c_1|r) = \frac{2}{\sigma^2} \cdot \left[x \right] \qquad (10)$$

$$LLR(c_2|r) = \frac{2}{\sigma^2} \cdot \left[y \right] \qquad (11)$$

$$LLR(c_3|r) = \frac{2}{\sigma^2} \cdot \left[|x| - |y| \right] \qquad (12)$$

For a given constellation diagram of Gray coded 64-QAM modulation, LLR of the constituting bits can be computed as follows.

$$LLR(c_1|r) = \frac{2}{\sigma^2} \cdot \left[x \right] \qquad (13)$$

$$LLR(c_2|r) = \frac{2}{\sigma^2} \cdot \left[-|x| + 4 \right] \qquad (14)$$

$$LLR(c_3|r) = \frac{2}{\sigma^2} \cdot \left[-\Big||x| - 4\Big| + 2 \right] \qquad (15)$$

$$LLR(c_4|r) = \frac{2}{\sigma^2} \cdot \left[y \right] \qquad (16)$$

$$LLR(c_5|r) = \frac{2}{\sigma^2} \cdot \left[-|y| + 4 \right] \qquad (17)$$

$$LLR(c_6|r) = \frac{2}{\sigma^2} \cdot \left[-\Big||y| - 4\Big| + 2 \right] \qquad (18)$$

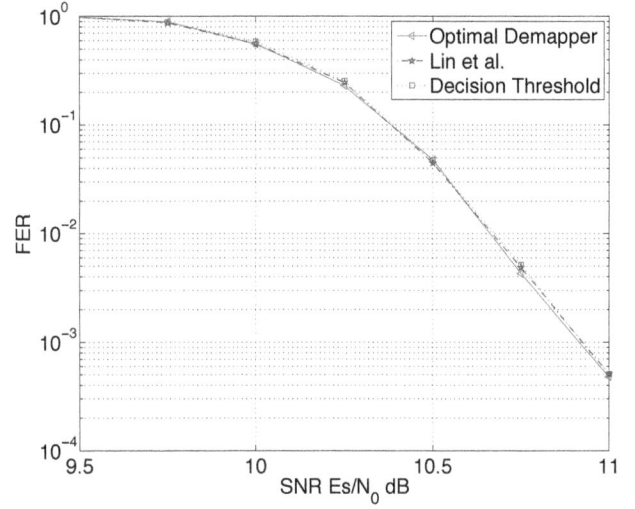

Figure 4. FER performance of Turbo-Code decoder after 8 iterations. The length of 16-QAM modulated burst is 536 symbols and code rate is 3/4. Three different demapping algorithms are applied: 1- optimal, 2- Lin algorithm, and 3- decision threshold algorithm

4 Communications performance

We compare the communications performance of optimal algorithm, Lin algorithm and decision threshold algorithm in Fig. 4. The simulations were carried out with bit true models of the hardware units to take into account quantization losses. We used 9 bit quantization each for input real and imaginary component, 6 bit for $\frac{2}{\sigma^2}$, and 6 bit for output LLR. We used a 16-state duo-binary Turbo-Code decoder in our simulations having Max-Log-Map with 0.75 extrinsic scaling factor, 8 iterations, and 7 bit for the extrinsic LLR. Both initial carrier synchronization and Code-Aided synchronization are performed to compensate the phase and frequency offsets. The FER graph clearly shows that the performance of all investigated algorithms is nearly identical. In short, by setting appropriate value of $\frac{2}{\sigma^2}$ the investigated suboptimal algorithms show similar communications performance to that of optimal algorithm. Remark that the simplified mathematical expressions Eqs. (6) to (9) adopted in decision threshold algorithm are equivalent to Eq. (5) adopted in Lin algorithm.

5 Hardware performance

In this section, we describe the architectures for abovementioned low complexity suboptimal demapping algorithms and compare their implementation performance. We used synthesizable VHDL to model the architectures.

5.1 Architecture for Lin algorithm

We present the architecture for Lin algorithm in Fig. 5. This architecture is described to support only 16-QAM modula-

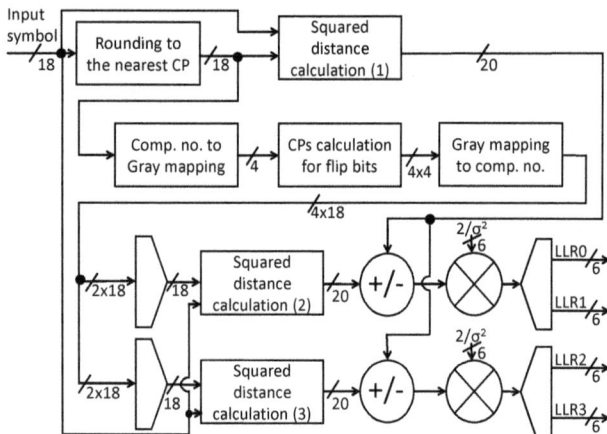

Figure 5. Proposed architecture for Lin algorithm.

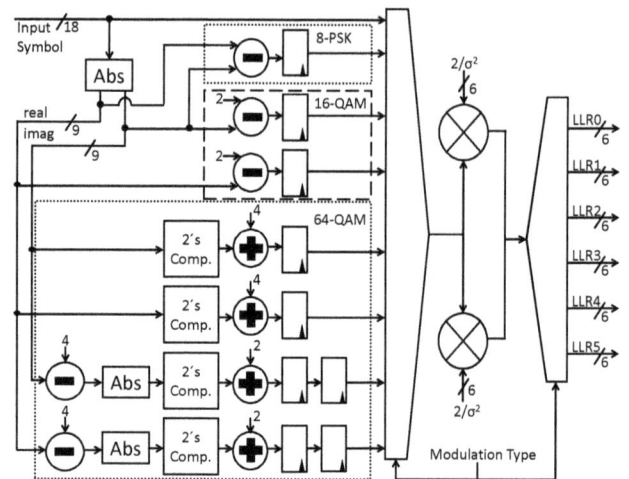

Figure 6. Proposed architecture for decision threshold algorithm

tion scheme for evaluation purpose but in reality the Lin algorithm can be applied to all Gray mapped M-PSK and M-QAM modulation schemes.

In the architecture, the rounding of the input received symbol towards the NCP is carried out in "Rounding to the nearest CP" block according to the procedure explained in Sect. 3.1. Then first squared distance is computed between the NCP and the received input symbols in "Squared distance calculation (1)" block to implement one term of Eq. (5). This result is used either first or second term of this equation depending upon the value of the corresponding bit. The NCP which is a complex number, is mapped to a predefined Gray code in "Comp. no. to Gray mapping" block. The resulting Gray code of the NCP is used to compute the nearest FCP. With respect to 16-QAM modulation scheme, four FCPs are computed. This operation is performed in "CPs calculation for flip bits". The resulting Gray codes of four FCPs are converted into complex numbers in "Gray mapping to comp. no.". The squared distances are computed between the resulting complex numbers of four FCPs and the received symbol in "Squared distance calculation (2)" and "Squared distance calculation (3)". To save and reutilize the hardware units (multiplier and adder), we compute only two squared distances at a time and therefore we use a multiplexer and a demultiplexer at the input and output of the multipliers.

The results of "Squared distance calculation (1)", and either "Squared distance calculation (2)" or "Squared distance calculation (3)" are used to compute LLR of each bit. All in all, only two squared distances are used to compute LLR of each bit. Finally, $\frac{2}{\sigma^2}$ is multiplied to compute LLR. The quantization bitwidths adopted in this work are mentioned in the figure. We compute two LLRs per clock cycle to achieve the aforementioned throughput.

5.2 Architecture for decision threshold algorithm

We present the architecture for decision threshold algorithm in Fig. 6. The proposed architecture provides flexibility to

Table 1. FPGA (Virtex-5) Place-and-Route results comparison.

Parameter	Jafri et al. (2009)	Proposed 1	Proposed 2
Algorithm	conventional suboptimal	Lin algorithm	decision threshold algorithm
Flexibility Modulation Support	BPSK to 256-QAM	16-QAM	BPSK, QPSK, 8-PSK, 16-QAM, 64-QAM
Mapping	Gray, non-Gray	Gray	Gray
Frequency (MHz)	156	312	333
Throughput (Msymbols/sec) 16-QAM	26	156	166
Resources			
Slice Regs	1596	288	127
Slice LUTs	2627	279	248
DSP48Es	6	8	2
BRAM	8	0	0

support following modulation schemes: BPSK, QPSK, 8-PSK, 16-QAM, and 64-QAM. In the case of BPSK and QPSK the input received symbol is directly used for LLR calculation. For the other mentioned modulation schemes, LLRs of first two bits are computed directly from the input received symbol because they do not involve any arithmetic operations. The computation of LLRs of the other bits involve simple arithmetic operations like absolute value, addition, subtraction and 2's complement. We calculate two LLRs per clock cycle to achieve the desired throughput.

5.3 FPGA implementation results

We compare the implementation results of our proposed architectures with a state-of-the-art demapper Jafri et al. (2010) in Table 1. As the latter design is implemented on a Xilinx Virtex-5 FPGA, so we used the same FPGA device (xc5vlx330-2ff1760) for implementation of our architectures to make a fair comparison.

The implementation results show that our proposed architectures achieve a much higher clock frequency and consequently deliver almost 6 times higher throughput besides occupying almost 10 times less resources as compared with state-of-the-art implementation. These results also show that the architecture based on decision threshold algorithm has less implementation complexity than that of Lin algorithm. The former saves 56 % slice registers, 12 % slice LUTs and 75 % DSP48Es than the latter. In summary, decision threshold algorithm has the lowest implementation complexity among the investigated architectures.

5.4 ASIC implementation results

Because the architecture described for decision threshold algorithm shows the lowest implementation complexity on FPGA, we selected this architecture for ASIC implementation. We implemented it on a 65 nm low power CMOS library. We used Synopses tools to perform Synthesis and, P&R. This efficient design occupies only 0.006 mm^2 area (2886 gates) after P&R and with worst case process parameters (1.1V, 125 °C). The design achieves a high clock frequency of 645 MHz, and therefore it delivers a very high throughput of 322 Msymbols/second with 16-QAM modulation. The design consumes only 3.85 mW power at nominal case.

6 Conclusions

Our investigation reveals that the decision threshold algorithm is a clear winner among the investigated demapping algorithms from the point of view of communications and implementation performance. The communications performance achieved by this algorithm costs only a tiny fraction of the computational effort required to achieve the same communications performance using the optimal and traditional sub-optimal algorithms. We have presented a very high throughput, area efficient, low power, and flexible architecture based on this algorithm. Our proposed architecture delivers almost 6 times higher throughput and requires about 10 times less resources on a FPGA as compared with state-of-the-art implementation.

References

Digital Video Broadcasting (DVB): Secocond Generation DVB, Interactive Satellite System (DVB-RCS2), Part 2: Lower Layers for Satellite standard, ETSI EN 301 545-2 V1.2.1, 2014.

Ali, I., Wasenmüller, U., and Wehn, N.: Hardware Implementation Issues of Carrier Synchronization for Pilot-Symbol Assisted Bursts: A Case Study for DVB-RCS2, in: 8th IEEE International Conference on Signal Processing and Communication Systems, Gold Coast, Australia, doi:10.1109/SPC.2013.6735092, 2014.

Altera: Constellation Mapper and Demapper for WiMAX, Tech. rep., http://www.altera.com/literature/an/an439.pdf, May 2007.

Arar, M., Amours, C. D., and Yongacoglu, A.: Simplified LLRs for the Decoding of Single Parity Check Turbo Product Codes Transmitted Using 16QAM, Research Letters in Communications, p. 4, doi:10.1155/2007/53517, 2007.

Jafri, A. R., Baghdadi, A., and Jezequel, M.: Rapid design and prototyping of universal soft demapper, in: IEEE International Symposium on Circuits and Systems (ISCAS), 3769–3772, 2010.

Kim, K. S., Hyun, K., Yu, C. W., Park, Y. O., Yoon, D., and Park, S. K.: General Log-Likelihood Ratio Expression and its Implementation Algorithm for Gray-Coded QAM Signals, ETRI Journal, 28, 291–300, 2006.

Lee, J. H., Sunwoo, M. H., Kim, P. S., and Chang, D.-I.: Low complexity soft-decision demapper for DVB-S2 using phase selection method, in: Proceedings of the 5th International Conference on Ubiquitous Information Management and Communication, p. 45, ACM, 2011.

Li, J. and Shi, Y.: Simplified Soft-output Demapper Based on a Linear Transformation Technique for M-ary PSK., Sensors & Transducers, 181, 1726–5479, 2014.

Li, M., Nour, C., Jego, C., and Douillard, C.: Design of rotated QAM mapper/demapper for the DVB-T2 standard, in: IEEE Workshop on Signal Processing Systems, SiPS 2009, 018–023, doi:10.1109/SIPS.2009.5336265, 2009.

Lin, D., Xiao, Y., and Li, S.: Low Complexity Soft Decision Technique for Gray Mapping Modulation, Wireless Personal Communications, 52, 383–392, 2010.

Park, J. W., Sunwoo, M. H., Kim, P. S., and Chang, D.-I.: Low complexity soft-decision demapper for high order modulation of DVB-S2 system, in: International SoC Design Conference (ISOCC '08), vol. 02, II–37–II–40, 2008.

Robertson, P., Villebrun, E., and Hoeher, P.: A comparison of optimal and sub-optimal MAP decoding algorithms operating in the log domain, in: Communications, 1995. ICC '95 Seattle, IEEE International Conference on "Gateway to Globalization", vol. 2, 1009–1013, doi:10.1109/ICC.1995.524253, 1995.

Ryoo, S., Kim, S., and Lee, S. P.: Efficient soft demapping method for high order modulation schemes, in: CDMA International Conference (CIC), Seoul, Korea, 2003.

Su, J., Lu, Z., Yu, X., and Hu, C.: A novel low complexity soft-decision demapper for QPSK 8PSK demodulation of DVB-S2 systems, in: International Conference of Electron Devices and Solid-State Circuits (EDSSC), 1–2, doi:10.1109/EDSSC.2011.6117701, 2011.

Sun, X. and Zeng, Z.: A Novel Simplified Log-Likelihood Ratio for Soft-Output Demapping of CMMB, in: Advances in Computer Science and Education Applications, vol. 202, 350–356, Springer, Berlin, Heidelberg, 2011.

Tosato, F. and Bisaglia, P.: Simplified soft-output demapper for binary interleaved COFDM with application to HIPERLAN/2, in: IEEE International Conference on Communications (ICC), vol. 2, 664–668, 2002.

9

Networks of high frequency inhomogeneous transmission lines

F. Ossevorth[1], **H. G. Krauthäuser**[1], **S. Tkachenko**[2], **J. Nitsch**[2], and **R. Rambousky**[3]

[1]TU Dresden, Elektrotechnisches Institut, Helmholtzstraße 9, 01069 Dresden, Germany
[2]Otto-von-Guericke-Universität Magdeburg, Lehrstuhl für Elektromagnetische Verträglichkeit, 39016 Magdeburg, Germany
[3]Bundeswehr Research Institute for Protective Technologies and NBC Protection, Humboldtstraße 100, 29633 Munster, Germany

Correspondence to: F. Ossevorth (fabian.ossevorth@tu-dresden.de)

Abstract. It is well known from classical transmission line theory, that transmission lines can be folded into impedances and thereby used in an electrical network setting. But it is also possible to create large networks of transmission lines consisting of tubes and junctions. The tubes contain the transmission lines and the junctions consider the mutual influences of the adjacent tubes or the terminals. The calculation of the currents and voltages at the junctions can be performed with the help the BLT-equation. So far this method is not applicable for nonuniform transmission lines described in a full wave method, because the lack of a distinct voltage gives no possibility for junctions. Junctions only make sense, when the considered network offers the possibility to propagate a TEM-Mode. If this requirement is fullfilled, nonuniform transmission lines could be included in an electrical network. This approach is validated in this paper in form of numerical simulations as well as measurements.

1 Introduction

In classical transmission line theory the input impedance of any line can be calculated at any position. Through this method the line itself can be used in an electrical network to calculate the necessary power at the load. By using the input impedance transmission lines can also be branched. Since voltage and current are distinct at every position along the line branch points can be included to combine transmission lines and to build complex networks. These branch points are called junctions and consider the mutual influences of the branched transmission lines. Such a network can be computed with the aid of the BLT-equation, as shown in Tesche et al. (1997, Sect. 6).

When the transmission line is not homogeneous the classical theory is no longer suitable. With a full wave approach all mutual effects of the line are considered, a reduction of the line through the input impedance is not accurate. For the impedance a distinct voltage is necessary, but in a full wave method only potentials exist. But under certain circumstances a nonuniform line can also be reduced, similar to a classical transmission line. It is also possible to include nonuniform lines with branches. This will be shown in the following sections.

The first section gives a brief overview of the Transmision Line Super Theory, a full wave method. In the following section parts of a nonuniform line are combined into an impedance similar to the classical theory. The results will show, that such an approach is valid. Following the collapsing of the line, a simple network with one branch will be investigated and treated numerically.

2 Transmission Line Super Theory

Transmission Line Super Theory (TLST) is a concise analytic method to obtain the potential and current along nonuniform wires, as described in Nitsch and Tkachenko (2010). These wires are approximated as thin wires with infinite conductivity.

Given are n wires above an infinitely conducting ground plane. The scalar potential ϕ has the form

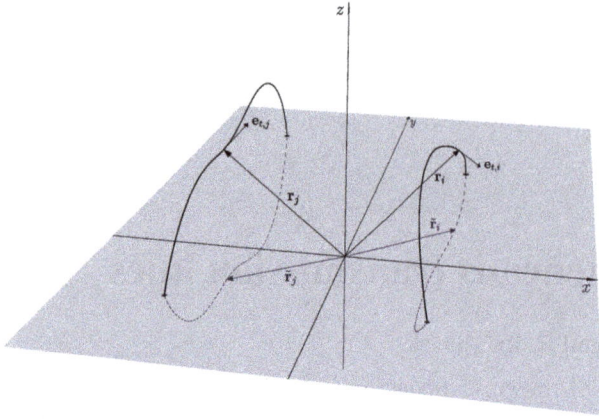

Figure 1. Arbitrary wires above groundplane.

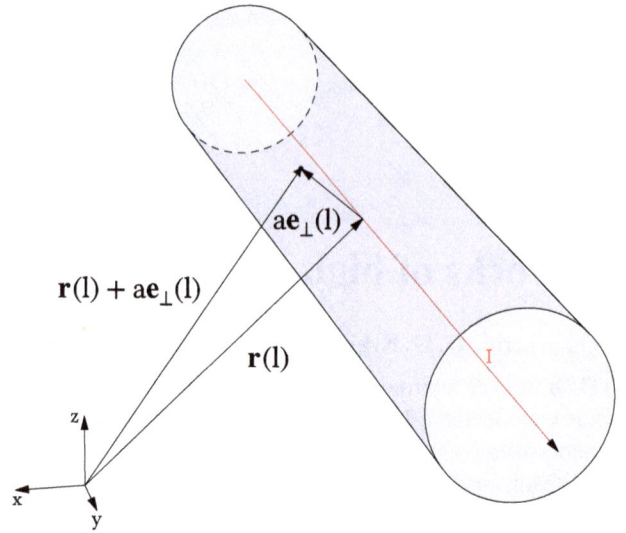

Figure 2. Point of evaluation.

with radius a of the wire, Fig. 2.

If the radius a is very small, then the tangential vector on the surface of the wire is very close to the tangential vector of the center of the wire. Thus, the boundary condition can be simplified to

$$\boldsymbol{e}_{t,i}(l_i) \cdot \boldsymbol{E}_i(\boldsymbol{r}_i(l_i)) = 0 \qquad (8)$$

and it follows an integral equation for ϕ and I

$$\frac{d\phi_i(l_i)}{dl_i} + j\omega\frac{\mu_0}{4\pi}\sum_{j=1}^{n}\int_0^{L_j} I_j(l'_j)\Big(\boldsymbol{e}_{t,i}(l_i)\cdot\boldsymbol{e}_{t,j}(l'_j)G_{ij}(l_i,l'_j,k)$$

$$-\boldsymbol{e}_{t,i}(l_i)\cdot\widetilde{\boldsymbol{e}}_{t,j}(l'_j)\widetilde{G}_{ij}(l_i,l'_j,k)\Big)\,dl'_j = \boldsymbol{E}^{\mathrm{exc}}\cdot\boldsymbol{e}_{t,i}(l_i)$$

$$\frac{d\phi_i(l_i)}{dl_i} + j\omega\frac{\mu_0}{4\pi}\sum_{j=1}^{n}\int_0^{L_j} I_j(l'_j)\boldsymbol{G}_L(l_i,l'_i,k)dl'_j$$

$$= \boldsymbol{E}^{\mathrm{exc}}\cdot\boldsymbol{e}_{t,i}(l_i). \qquad (9)$$

With the help of this equation and Eq. (1) one obtains a coupled differential equation with varying coefficients $\boldsymbol{P}$ similar to the classical transmission line equations

$$\frac{d}{dl}\begin{pmatrix}\boldsymbol{\phi}(l)\\\boldsymbol{I}(l)\end{pmatrix} + j\omega\begin{pmatrix}\boldsymbol{P}_{11} & \boldsymbol{P}_{12}\\\boldsymbol{P}_{21} & \boldsymbol{P}_{22}\end{pmatrix}\begin{pmatrix}\boldsymbol{\phi}(l)\\\boldsymbol{I}(l)\end{pmatrix} = \begin{pmatrix}\boldsymbol{v}^{\mathrm{exc}}(l)\\\boldsymbol{0}\end{pmatrix}, \qquad (10)$$

as shown in Nitsch and Tkachenko (2010). Solving this equation gives the potential and current for every conductor with respect to the matricant, Gantmacher (1960), which depends on the common parameter l

$$l_i = \frac{L_i}{L}l \quad 0 \le l \le L, \qquad (11)$$

$$\phi = -\frac{1}{j\omega 4\pi\varepsilon_0}\sum_{i=1}^{n}\int_0^{L_i}\frac{\partial I_i(l'_i)}{\partial l'_i}\Big(G(l_i,l'_i,k) - \widetilde{G}(l_i,l'_i,k)\Big)\,dl'_i$$

$$= -\frac{1}{j\omega 4\pi\varepsilon_0}\sum_{i=1}^{n}\int_0^{L_i}\frac{\partial I_i(l'_i)}{\partial l'_i}G_C(l_i,l'_i,k)dl'_i, \qquad (1)$$

where $G(l_i,l'_i,k)$ is Green's function in free space

$$G(l_i,l'_i,k) = \frac{e^{-jk|\boldsymbol{r}_i(l_i)-\boldsymbol{r}'_i(l'_i)|}}{|\boldsymbol{r}_i(l_i) - \boldsymbol{r}'_i(l'_i)|} \qquad (2)$$

and $\widetilde{G}(l_i,l'_i,k)$ the mirrored Green's function

$$\widetilde{G}(l_i,l'_i,k) = \frac{e^{-jk|\boldsymbol{r}_i(l_i)-\widetilde{\boldsymbol{r}}'_i(l'_i)|}}{|\boldsymbol{r}_i(l_i) - \widetilde{\boldsymbol{r}}'_i(l'_i)|}, \qquad (3)$$

Figure 1. These functions depend on the natural arc length l_i, the integration variable l'_i and on $k = \frac{2\pi f}{c}$, with the speed of light c and frequency f.

The vector potential $\boldsymbol{A}$ is also needed

$$\boldsymbol{A} = \frac{\mu_0}{4\pi}\sum_{i=1}^{n}\int_0^{L_i} I(l'_i)\big(\boldsymbol{e}_{t,i}(l'_i)G(l_i,l'_i,k) - \widetilde{\boldsymbol{e}}_{t,i}(l'_i)\widetilde{G}(l_i,l'_i,k)\big)\,dl'_i, \qquad (4)$$

with $\boldsymbol{e}_{t,i}$ the tangential vector of wire i

$$\boldsymbol{e}_{t,i} = \frac{\partial\boldsymbol{r}_i(l_i)}{\partial l_i} \qquad (5)$$

evaluated on the surface of the wire. Inserting the equation for $\boldsymbol{A}$ (Eq. 4) into the equation for the electrical field

$$\boldsymbol{E} = \boldsymbol{E}^{\mathrm{sc}} + \boldsymbol{E}^{\mathrm{exc}} = -\nabla\phi - j\omega\boldsymbol{A} + \boldsymbol{E}^{\mathrm{exc}} \qquad (6)$$

gives a general equation for the E-field. On the surface of the wire, the boundary condition is

$$\boldsymbol{n}_i \times \boldsymbol{E}_i(\boldsymbol{r}_i(l_i) + \boldsymbol{e}_{\perp,i}(l_i)a) = \boldsymbol{0} \qquad (7)$$

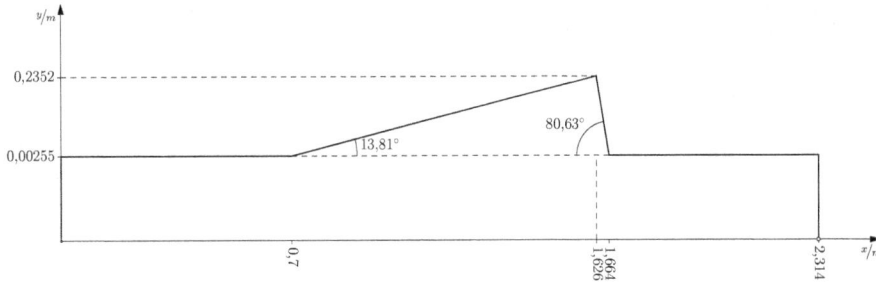

Figure 3. Geometry of a nonuniform transmission line.

giving

$$\begin{pmatrix} \boldsymbol{\phi}(l) \\ \boldsymbol{I}(l) \end{pmatrix} = \boldsymbol{M}_{l_0}^l \begin{pmatrix} \boldsymbol{\phi}(l_0) \\ \boldsymbol{I}(l_0) \end{pmatrix} + \int\limits_{l_0}^{l} \boldsymbol{M}_{l_0}^{l'} \begin{pmatrix} \boldsymbol{v}^{\mathrm{exc}}(l') \\ \boldsymbol{0} \end{pmatrix} dl'. \qquad (12)$$

The matricant is a square matrix of dimension $2n \times 2n$

$$\boldsymbol{M}_{l_0}^l = \begin{pmatrix} \boldsymbol{\phi}_1(l) & \boldsymbol{\phi}_2(l) \\ \boldsymbol{I}_1(l) & \boldsymbol{I}_2(l) \end{pmatrix} \qquad (13)$$

containing two linear independent solution vectors for $\boldsymbol{\phi}$ and $\boldsymbol{I}$.

So far, the parameters are unknown. They can be determined through a perturbation approach, also shown in Nitsch and Tkachenko (2010), based on an already known solution for the matricant from classcial transmission line theory

$$\boldsymbol{P} = -\frac{1}{j\omega} \frac{d}{dl} \boldsymbol{M}_{l_0}^l \left(\boldsymbol{M}_{l_0}^l \right)^{-1} \qquad (14)$$

giving an analytic solution for the parameters

$$\boldsymbol{P}(l) = \frac{\mu_0}{4\pi} \begin{pmatrix} \int_{l_0}^{L} \boldsymbol{G}_L(l,l',k) \cdot \mathrm{diag}(e^{-jkl_i(l')})dl' & \int_{l_0}^{L} \boldsymbol{G}_L(l,l',k) \cdot \mathrm{diag}(e^{jkl_i(l')})dl' \\ \frac{4\pi k}{\omega\mu_0} \mathrm{diag}(e^{-jkl_i(l')}) & -\frac{4\pi k}{\omega\mu_0} \mathrm{diag}(e^{jkl_i(l')}) \end{pmatrix}$$
$$\begin{pmatrix} k\int_{l_0}^{L} \frac{\boldsymbol{G}_C(l,l',k)\cdot\mathrm{diag}(e^{-jkl_i(l')})}{\omega 4\pi\varepsilon_0}dl' & -k\int_{l_0}^{L} \frac{\boldsymbol{G}_C(l,l',k)\cdot\mathrm{diag}(e^{jkl_i(l')})}{\omega 4\pi\varepsilon_0}dl' \\ \mathrm{diag}(e^{-jkl_i(l')}) & \mathrm{diag}(e^{jkl_i(l')}) \end{pmatrix}^{-1}. $$
$$(15)$$

After evaluating the parameters the differential Eq. (10) can be solved with the help of the product integral, Gantmacher (1960)

$$\boldsymbol{M}_{l_0}^l = \lim_{\Delta l_k \to 0} \prod_k e^{\boldsymbol{P}\Delta l_k}. \qquad (16)$$

Various numerical methods for the solution of this integral have been investigated in Steinmetz (2006), following these investigations a Runge–Kutta method of 4th order gives very accurate results both in computation time and accuracy of the solution.

3   Network description of a nonuniform transmission line

One essential part of a network description of transmission lines is the concept of the input impedance. When it is possible to collapse parts of a wire into an input impedance and use this impedance as termination for the rest of the wire, then the prerequisites for more complex networks are met.

An experiment was conducted in order to proof simulation data with measurements. A thin wire with radius 0.35 mm was fixed as in Fig. 3 above a conducting ground plane.

The nonuniform part in the center of the wire is flanked by uniform parts to ensure the propagation of a TEM-mode along these parts of the line. The actual structure of the wire is shown in Fig. 4.

For the measurements a network analyzer was used, connected to the line from below the ground plane. As can be seen from Fig. 5 the results of the measurements for the input impedance and the calculation match up to 200 MHz. The following deviation in the measurements is due to sideeffects of the experiment, e.g. the coupling to surrounding apertures, which is not considered in the calculations. But especially the dielectric material seems to have a considerable effect on the final results. Nevertheless, the results of a method of moments simulation and a TLST method show very good agreement. With these correct results for the complete line, new calculations for the input impedance can be compared to now proofed results.

At first the line was cut into two parts in the last half of the line and thus partitioned into a front and back part, which were calculated separately, Fig. 6. This means, that the mutual influences between the two parts were neglected, which is an essential property of a network characteristic.

Thereafter the input impedance is calculated from Eq. (17), taking into account the terminal impedance Z_L and the matricant of the last part of the line $\boldsymbol{M}_{3,L_2}^L$

$$i(L) \begin{pmatrix} Z_L \\ 1 \end{pmatrix} = \boldsymbol{M}_{3,L_2}^L \begin{pmatrix} \phi(L_2) \\ i(L_2) \end{pmatrix}. \qquad (17)$$

The separation of the line took place in the uniform part of the line, where a TEM-mode propagates, giving an explicit

(a) Structure

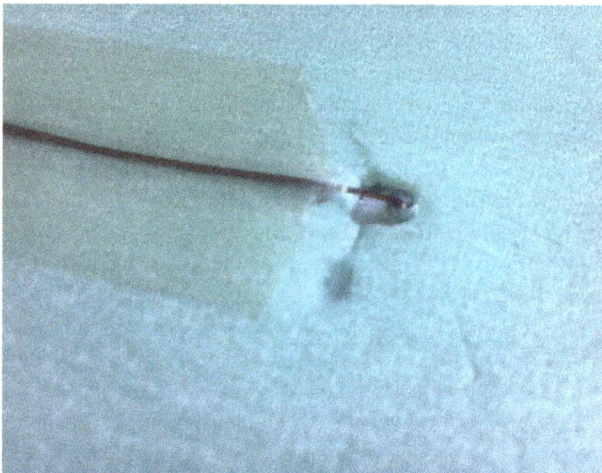

(b) Connection

Figure 4. Photographs of the real wire.

voltage and therefore the ratio between $\frac{\phi(L_2)}{i(L_2)}$ is valid giving

$$Z_{i,2} = \frac{\phi(L_2)}{i(L_2)} = \frac{M_{3,12} - Z_L M_{3,22}}{Z_L M_{3,21} - M_{3,11}}. \tag{18}$$

When cutting the line into two parts and calculating both lines separately one makes an error, when using a full wave method to compute potential and current. Since a TEM-mode is dominant on these parts of the line the parameters of the separated lines should reflect this effect. But when using the full wave method the lines end at the cutted parts and therefore the parameters also change at the end of the line. This can be avoided in the calculation of Eq. (15) by taking the lines longer than they actually are. Now, the lines seem to overlap the junction and the parameters become constant at the position where the real line ends. In this way the parameters are matched to the TEM-mode. The results show an excellent agreement which can be seen from Fig. 7.

Table 1. Layout of the network structure.

Part	Description	
1	front part	0–0.318 m
2	junction	classical transmission lines of length 0.1 m
3	nonuniform part	0.518–2.314 m
4	branch	0.324 m

4 Branched network

In the previous section it was shown that a nonuniform line can be calculated piecewise, when the separation of the wire took place in the uniform parts. Through this method it is shown that a nonuniform line with uniform parts can be treated as a simple network. A more complex network would include junctions too. In extending the line from the previous section about an additional wire rectangular to the line, Fig. 8, a branch point is inserted and a more complex network created.

The whole structure is now divided into four parts, Table 1. At first, each structure except the junction itself is calculated with the TLST. After that the resulting matricants are prepared for a classical electrical network solver. For that the matricants are transformed into admittance form. It is necessary to pay attention to the specific definition for the admittance transformation for each network solver. In the present case the following definition is used, Fig. 9,

$$i(L) = -i_2, \; i(0) = i_1 \text{ und } v(L) = v_2, \; v(0) = v_1, \tag{19}$$

and the matricants are transformed with

$$\overline{Y} = \begin{pmatrix} -M_{12}^{-1} M_{11} & M_{12}^{-1} \\ M_{22} M_{12}^{-1} M_{11} - M_{21} & -M_{22} M_{12}^{-1} \end{pmatrix}. \tag{20}$$

Now each part, except for the junction, is in admittance form, which gives a reduced scheme as shown in Fig. 10. This scheme is then used for the network solver in form of Listing 1. A voltage source of 1 V with a resistance of 50 Ω is connected through the knots 0, 1 and 2. The current in this branch is measured with a voltage source of 0 V. The junction is modeled with the nodes 4, 5, 6, 8 using three classical transmission lines with the network solver's own functions. Every line is terminated with 50 Ω which reflects the connection of the networkanalyzer used for measurements. The calculated, simulated and measured results are shown in Fig. 11. The calculation and the simulation with a method of moments show good agreement. Again the measurements agree up to 400 MHz and diverge for higher frequencies. The effects of the insulation are not considered in both the calculation and the simulation. Nonetheless, the plots show, that networks with branches can be calculated with the TLST. It is required that the branches occur in the uniform parts of a line, where a TEM-Mode can dominate and classical transmission line theory is applicable.

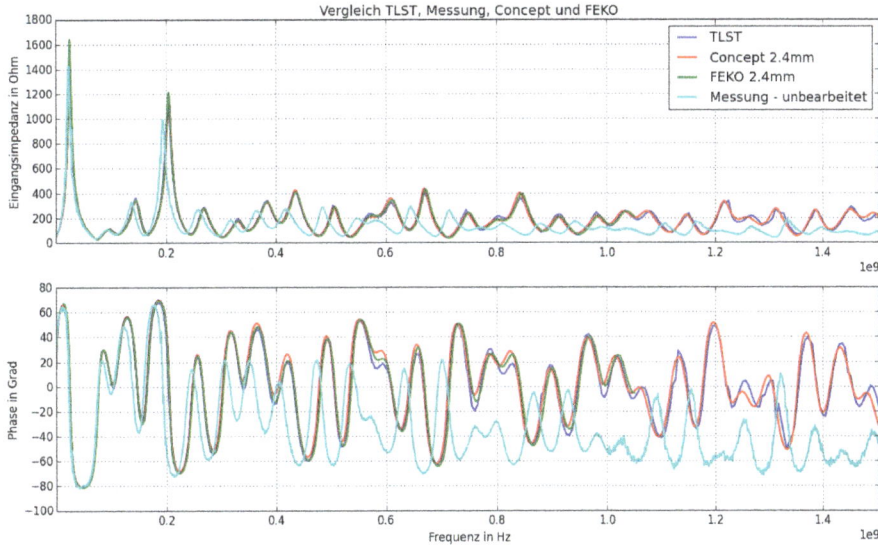

Figure 5. Magnitude and phase of the input impedance of the whole line.

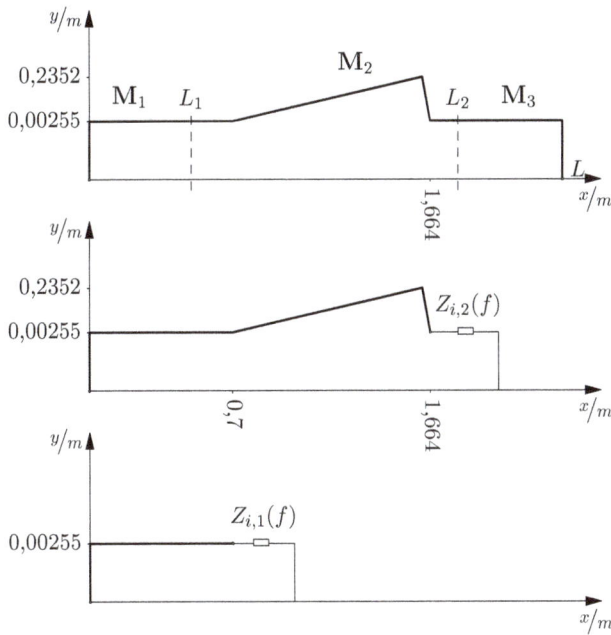

Figure 6. Calculation of input impedances

```
*Branched Network
v1  1  0  1
R1  1  2  50
v2  2  3  0
x1  pexl_ymat 3 4 Matricants_1st_seg.y
Tl1  4  5  wire_over_ground 0.1 0.25e-3 2.4e-3
Tl2  5  6  wire_over_ground 0.1 0.25e-3 2.4e-3
x2  pexl_ymat 6 7 Matricants_2nd_seg.y
Tl3  5  8  wire_over_ground 0.1 0.25e-3 2.4e-3
x3  pexl_ymat 8 9 Matricants_3rd_seg.y
R2  7  0  50
R3  9  0  50
.ac lin 280 1e6 1.5e9
.print wis_network_junction.res 2 @v2
```

Listing 1. Script for network solver.

5 Conclusions

It could be shown in this article that nonuniform transmission lines can be divided into separate parts and calculated independently from each other. This gives the possibility for a network abstraction of nonuniform lines which can be extended to complex networks. The requirement for this approach of separating the transmission line is, that the line has uniform parts on which a TEM-mode can propagate. It is along these parts of the line that the wire can be separated only. Because of the TEM-mode distinct voltages exist which allow the application of classical network theory on these parts. Apart from separating one transmission line only, there was also a branched transmission line investigated. The computed results for the single parts have been used in a network solver as admittance matrices. The overall result of the input impedance has been compared to results of a simulation with a method of moments as well as with measurements and has shown a good agreement. These results confirm the usability of the Transmission Line Super Theory for networks. Further studies are necessary to apply the theory to more complex networks with multiple conductors and to reduce the computational effort.

(a) ABCD-parameters of the matricant

(b) Y-parameters

Figure 9. Current and voltages on twoports.

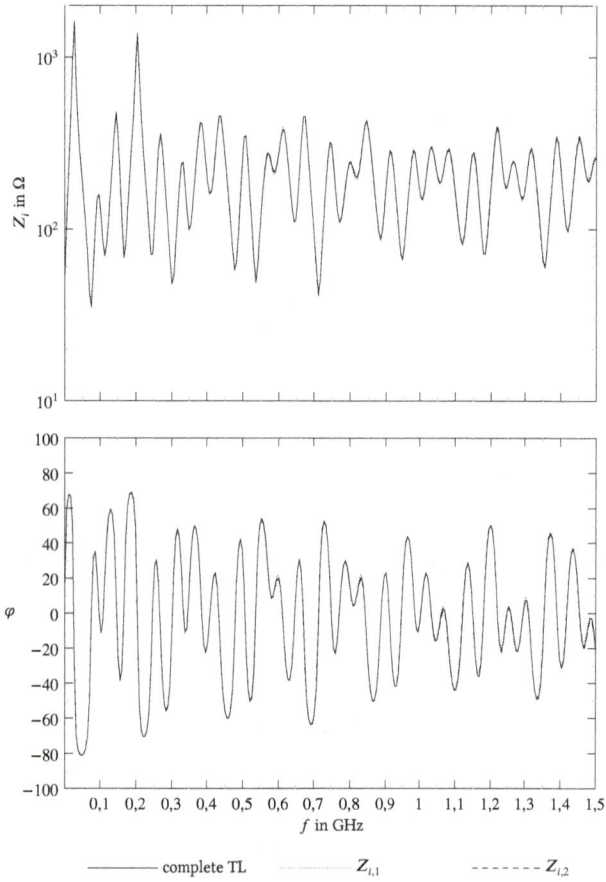

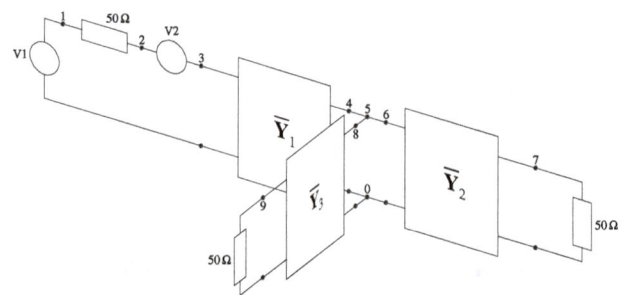

Figure 7. Results of the input impedance for the separated line with an asymptotic approach at the cutted ends.

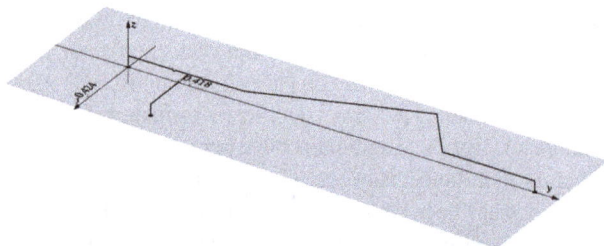

Figure 8. Network with junction

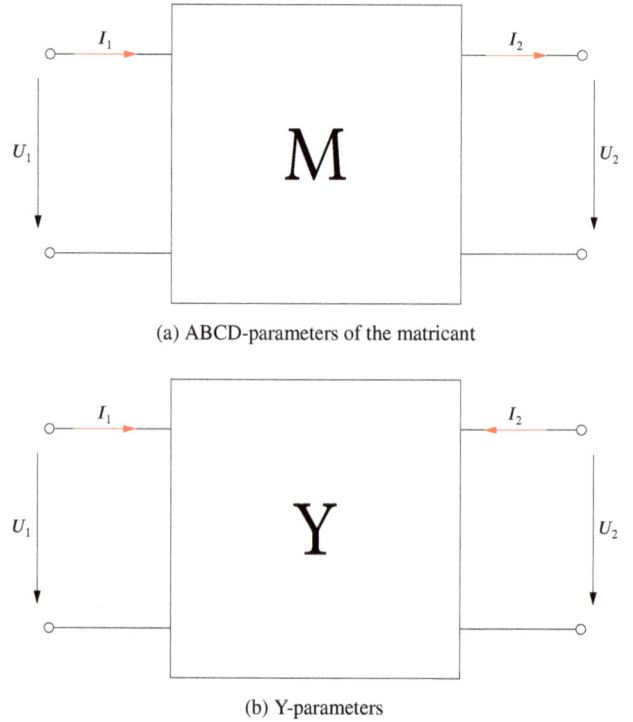

Figure 10. Layout of the network with admittance matrices.

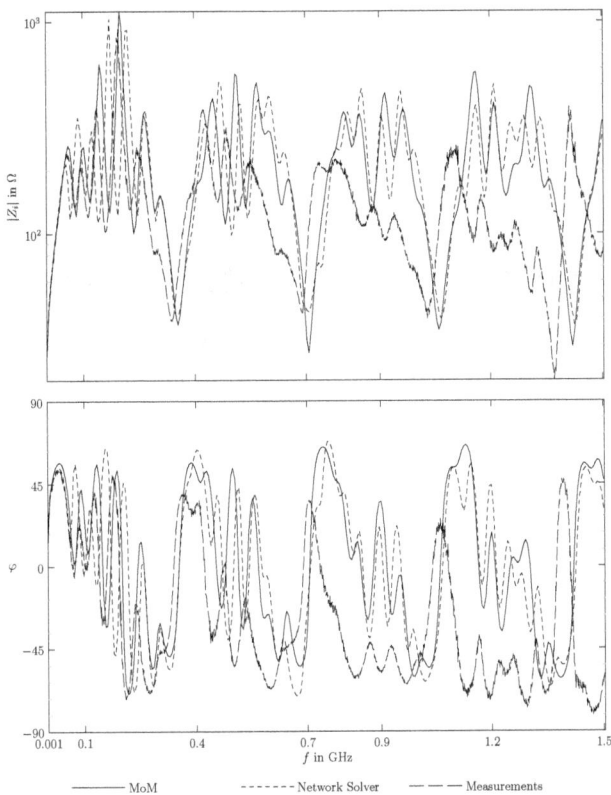

Figure 11. Input impedance on knot 3.

Acknowledgements. The research was funded by the Wehrwissenschaftliches Institut für Schutztechnologien, Munster, Germany, under the contract number E/E590/DZ005/AF119.

References

Gantmacher, F.: The Theory of Matrices, no. Bd. 2 in Chelsea Publishing Series, American Mathematical Society, 1960.

Nitsch, J. and Tkachenko, S.: High-Frequency Multiconductor Transmission-Line Theory, Found. Phys., 40, 1231–1252, doi:10.1007/s10701-010-9443-1, 2010.

Steinmetz, T.: Ungleichförmige und zufällig geführte Mehrfachleitungen in komplexen technischen Systemen, PhD thesis, Otto-von-Guericke-Universität Magdeburg, 2006.

Tesche, F. M., Ianoz, M. V., and Karlsson, T.: EMC Analysis Methods and Computational Models, John Wiley & Sons, 1997.

10

Impedance spectra classification for determining the state of charge on a lithium iron phosphate cell using a support vector machine

P. Jansen[1], D. Vergossen[1], D. Renner[1], W. John[2], and J. Götze[3]

[1]Audi Electronics Venture GmbH, Gaimersheim, Germany
[2]SiL GmbH – Paderborn/Leibniz Universität Hannover, Hanover, Germany
[3]Technische Universität Dortmund (AG DAT), Dortmund, Germany

Correspondence to: P. Jansen (patrick.jansen@audi.de)

Abstract. An alternative method for determining the state of charge (SOC) on lithium iron phosphate cells by impedance spectra classification is given. Methods based on the electric equivalent circuit diagram (ECD), such as the Kalman Filter, the extended Kalman Filter and the state space observer, for instance, have reached their limits for this cell chemistry. The new method resigns on the open circuit voltage curve and the parameters for the electric ECD. Impedance spectra classification is implemented by a Support Vector Machine (SVM). The classes for the SVM-algorithm are represented by all the impedance spectra that correspond to the SOC (the SOC classes) for defined temperature and aging states. A divide and conquer based search algorithm on a binary search tree makes it possible to grade measured impedances using the SVM method. Statistical analysis is used to verify the concept by grading every single impedance from each impedance spectrum corresponding to the SOC by class with different magnitudes of charged error.

1 Introduction

The exact determination of the state of charge (SOC) of a cell, especially of a lithium iron phosphate cell, is a challenging task in signal processing. The requirements as regards the accuracy of the determined SOC are very significant in the automotive industry to ensure that the electrochemical storage device operates in a reliably and efficiently mode.

The exact SOC in automotive applications, for hybrid as well as conventional electrical supply systems on board a vehicle, is a very important information for the energy management system (EMS). The EMS needs to know how much energy is still left in the battery, and how much energy can be charged back. This essential knowledge of the energy level in the storage device defines the whole operational strategy of the EMS. The SOC indicates critical states such as deep discharge or overcharge. These levels of extremely high or low SOC can lead to irreversible damage in the battery (Piller et al., 2001). The task of the EMS is to avoid these critical states in any case to enable a high endurance.

In this context a specific definition of the SOC is needed. The most common definition for the SOC is the ratio between the difference of the rated capacity C_n and the charge balance Q_b to the rated capacity C_n. The SOC is 1 when state of charge FULL is reached and 0 after a net discharge of the rated capacity (Sauer et al., 1999).

$$\text{SOC} = \frac{C_n - Q_b}{C_n} \quad (1)$$

This definition ignores the problem of the battery aging, as the capacity that can be delivered by a battery may change in the course of its life due to problems such as the loss of charge acceptance of the active materials on either of the electrodes, changes in the physical properties of the electrolyte or corrosion on the current conductors (Piller et al., 2001). This aging behavior is called state of health (SOH). Together the temperature behavior and the SOH have the biggest influence on the SOC.

The aim of this work is to propose a method to determine the SOC of a lithium iron phosphate cell, in an automotive application under load conditions, with a specific SOH and a defined temperature from the frequency domain data. The reference cell used for the impedance spectroscopy was in mint condition, so the SOH was approximately 100 % when the impedance spectroscopy was carried out.

2   Methods for determining the state of charge

The following are existing methods for determining the SOC independent of the battery type:

- discharge test,

- Ah balance,

- open circuit voltage,

- Kalman filter,

- state space observer,

- artificial neuronal network,

- machine learning.

Some of the methods are limited in their range of functionality. The discharge test and Ah balancing for instance are not suitable solutions for an automotive on-board application. The most common time domain based methods for on-board SOC determination are equivalent circuit diagram (ECD) (see Fig. 1) based methods such as the Kalman Filter, the extended Kalman Filter and the state space observer. The mathematical basis of those methods is a state space model from the ECD (Lee et al., 2007; Codeca et al., 2008; Plett, 2004).

$$\begin{bmatrix} SOC_{k+1} \\ U_{RC_{1k+1}} \\ \vdots \\ U_{RC_{nk+1}} \end{bmatrix} = \begin{bmatrix} 1 & 0 & \cdots & 0 \\ 0 & 1 - \frac{\Delta t}{R_{RC_1} \cdot C_{RC_1}} & 0 & \vdots \\ \vdots & 0 & \ddots & \vdots \\ 0 & \cdots & \cdots & 1 - \frac{\Delta t}{R_{RC_n} \cdot C_{RC_n}} \end{bmatrix}$$

(2)

$$\cdot \begin{bmatrix} SOC_k \\ U_{RC_{1k}} \\ \vdots \\ U_{RC_{nk}} \end{bmatrix} + \begin{bmatrix} \frac{\Delta t}{C_n} \\ \frac{\Delta t}{C_{RC_1}} \\ \vdots \\ \frac{\Delta t}{C_{RC_n}} \end{bmatrix} \cdot I_{Batt_k}$$

$$\hat{U}_{Batt_k} = \begin{bmatrix} \frac{U_{OCV_k}}{SOC_k} & U_{RC_{1k+1}} \dots U_{RC_{nk+1}} \end{bmatrix}$$

$$\cdot \begin{bmatrix} SOC_k \\ U_{RC_{1k}} \\ \vdots \\ U_{RC_{nk}} \end{bmatrix} + R_i \cdot I_{Batt_k}$$

(3)

In this model based on the ECD, the open circuit voltage (OCV) can be calculated to correct the SOC of the state space Ampere hour integrator to compensate for untraceable capacity losses. The quality of the model's dynamic behavior is important when it comes to the accuracy of the calculated OCV. That dynamic behavior depends on the number of RC-circuits of the ECD (see Fig. 1). The RC-circuits represent

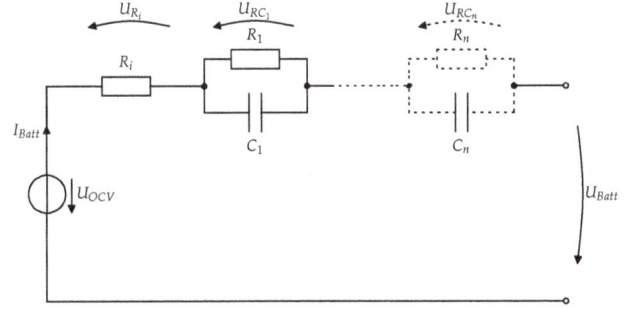

Figure 1. Equivalent circuit diagram.

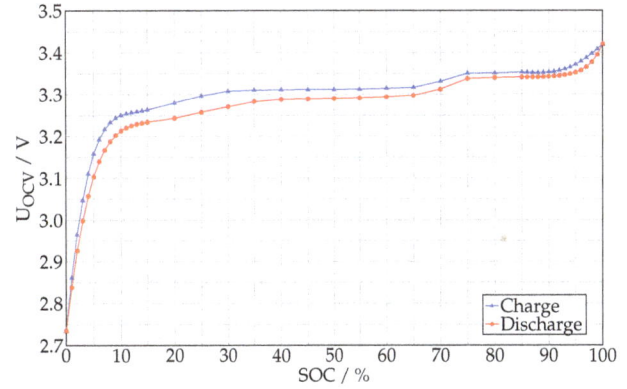

Figure 2. Open circuit voltage curve.

several different time based chemical reactions such as diffusion or the charge carrier movement.

The ECD based methods are suitable for many different types of chemistries but they reach their limits when it comes to lithium iron phosphate chemistry. There are two main reasons for this, in particular for lithium iron phosphate cells: the extremely smooth plateau of the OCV curve in the middle SOC range (see Fig. 2) and the accuracy of the ECD. The OCV calculated from the terminal voltage highly depends on the accuracy of the ECD parameters. For that reason OCV based methods are only capable of estimating the SOC in the middle range of the OCV curve with a monadic percentage precision with very substantial effort in terms of measurement.

$$1\,mV \cong 1\,\%_{SOC}$$

(4)

Recent theoretical work on methods in the field of machine learning, use support vector Regression or support vector machine (SVM) on time domain based data such as voltage, current and temperature to determine the SOC (Weng et al., 2013; Álvarez Antón et al., 2013; Hu et al., 2014). Another approach in this case is to use the frequency domain data to determine the SOC. The basis of this method will be provided in this paper.

3 Principle and methodology

The SVM is a binary classifier from the field of machine learning theory. The SVM is a support vector learning algorithm for pattern recognition, with the aim of classifying quantities with certain attributes and grade unknown samples to one of two classes. There are other methods for classification problems such as the nearest neighbor decision (NND) and its derivatives such as the k_n-nearest neighbor decision and the distance-weighted k_n-nearest-neighbor decision. NND-methods use euclidean metrics to evaluate the distance between the sample and its classified single-nearest neighbor or k_n-nearest neighbors. An *a priori* assumption of the underlying statistics of the training data as for a Bayes classifier is not necessary. Therefore the big advantages of this method are the simplicity and performance. A disadvantage in this regard is the fact that, as sets of training data increase, the classification probability decreases (Cover and Hart, 1967; Dudani , 1976). Compared to the SVM method, NND-methods have higher costs in terms of memory space to store the entire volume of training data and in terms of runtime because all the training data has to be evaluated to grade a single sample. SVMs on the other hand resign on the storage of the training data. Their aim is to detect a pattern within the training data via its class-specific attributes so that the data can be intersected by a hyperplane based on support vectors and the training data can be separated without errors (Cortes and Vapnik, 1995).

3.1 Support vector machine

A binary support vector classifier such as the SVM is based on a class of linear hyperplanes,

$$(\underline{w} \cdot \underline{x}) + b = 0, \quad \underline{w} \in \Re^N, b \in \Re \tag{5}$$

to separate a number of elements into two specific classes, based on class specifying attributes – for instance color, shape or other metadata – using a hyperplane. The hyperplane is the shortest orthogonal line connecting the convex hulls of this two classes, and intersects them half way. The optimum hyperplane has a symmetrical maximum margin to both convex hulls. The normal vector $\underline{w}$ with a threshold b defines the linear hyperplane and its margin of the two classes, so that a grading of a new unknown element $\underline{x}$ is possible. The so-called support vectors $\underline{x}_k$ are specific objects of the training data. That are the elements of the convex hulls closest to the margin (see Fig. 3). The SVM is also applicable to non linear separable data, by using the so-called "Kernel Trick" to transform the data into a high-dimensional feature space where the data is linear separable.The kernel depends on several usable funktions, for instance a polynomial or a radial basis function, to evaluate the hyperplane that separates the data in the feature space. A suitable kernel function has to be chosen specifically for the training data.

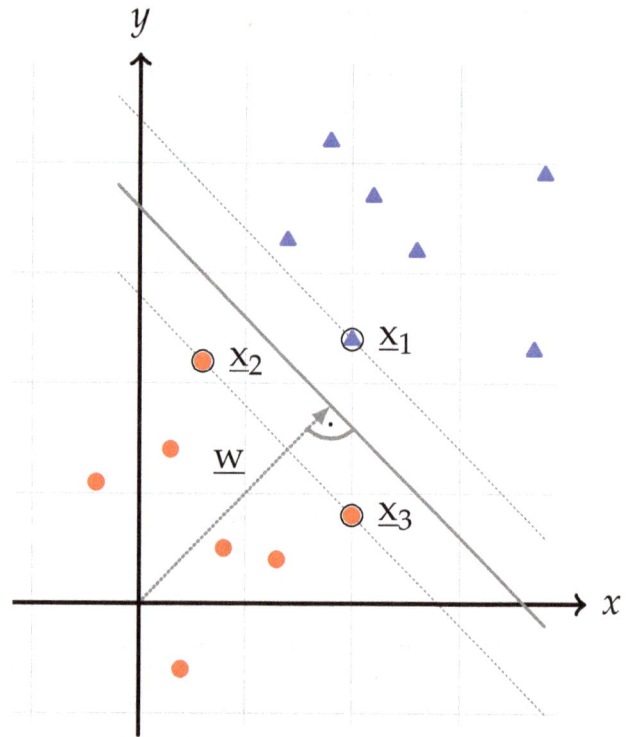

Figure 3. Support vector machine in 2-D space (Hearst et al., 1998).

The training data is required as the basis for the classification, where every element $\underline{x}_i$ of a quantity is affiliated to one class by its class label y_i.

$$(\underline{x}_1, y_1), \ldots, (\underline{x}_l, y_l) \in \Re^N \times \{\pm 1\} \tag{6}$$

The class label is defined by ± 1, so that all elements of one class are labeled by $+1$, and all elements of the other class are labeled by -1. Based on this information in a first step the SVM classification function, based on a linear kernel, is capable of evaluating the optimal hyperplane to separate the two classes.

The corresponding SVM decision function for linear separable data

$$f_{(x)} = \text{sgn}((\underline{w} \cdot \underline{x}) + b) \tag{7}$$

leads in a second step to a grading of an new unknown element $\underline{x}$ to one of the classes with the return of its affiliation $\{\pm 1\}$.

In case of non linear separable data the normal vector $\underline{w}$ is a representation of a linear combination of support vectors $\underline{x}_k$ from the training data, the corresponding class labels y_k and the lagrange multipliers $\underline{v}_k$ (Hearst et al., 1998).

$$\underline{w} = \sum_{k}^{l} \underline{v}_k y_k \underline{x}_k \tag{8}$$

As this efficient learning algorithm has simple correspondence to a linear method in a high-dimensional feature space

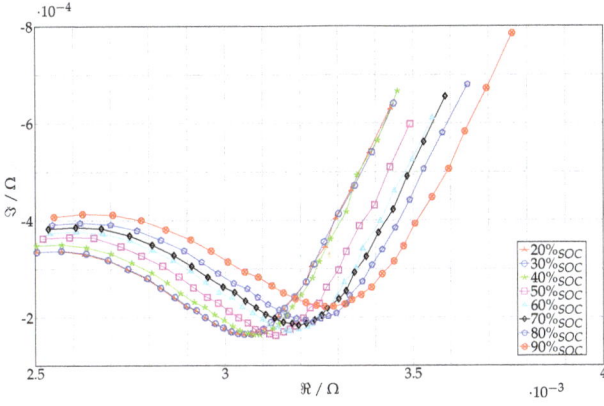

Figure 4. Impedance spectra frequency domain cutout 80–0.1 Hz.

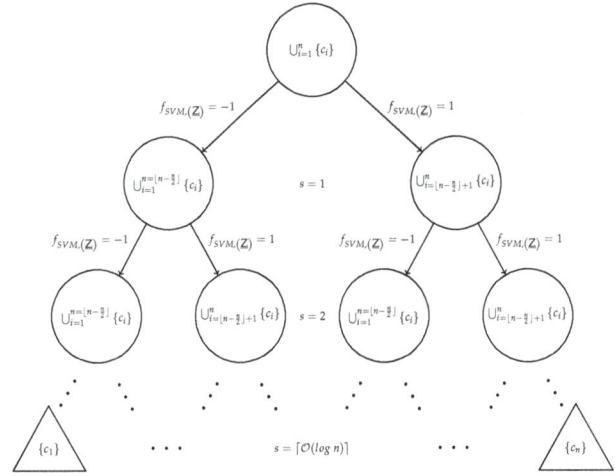

Figure 5. Binary search tree.

that is non-linearly related to its input space, it is straightforward to analyze it mathematically. We are dealing here with a classification algorithm, where a superset of elements is separated into two power sets or classes where single new elements are graded to one of those classes. A SVM is only capable of a single binary decision regarding whether the applicable element belongs to one power set or the other.

3.2 Impedance spectra classification

Determining the SOC via impedance spectra classification using SVM is an alternative method to achieve this aim. This method resigns an electric ECD with components such as the OCV curve and the element parameters of the electric network.

The task of determining the SOC of a lithium iron phosphate cell can be achieved with an optimal classifier such as the SVM by grading measured impedances to a certain class. The classes for the SVM are represented by all the impedance spectra for different SOC levels, corresponding to defined temperature and aging states, generated by an impedance spectroskopy, are the foundation of this classification method (see Fig. 4). The data of those spectra represents the training data of the SVM classification function – based on a polynomial kernel function – that is used to calculate the hyperplanes that separate all impedance spectra to their nearest neighbors. As noted above, the SVM is only capable of a binary decision, therefore with more than two classes a separation of every impedance spectra to its nearest neighbor has to be realized by a hyperplane via SVM. So n classes will yield to $n-1$ hyperplanes to be evaluated by the SVM to separate all spectra from each other.

The SVM decision function can only make binary decisions so that all the SVM decisions have to be rated and contextualized. The most efficient way to do so is to create a graph to arrange the hyperplanes. The whole quantity of the impedance spectra elements or the superset, the root of the graph, is separated by the median of the hyperplanes repre-

sented by their specific linear combinations,

$$\underline{\widetilde{w}}_{\text{med}} = \sum_{k}^{l} \underline{\widetilde{v}}_{k,\text{med}} y_k \underline{\widetilde{x}}_{k,\text{med}} \tag{9}$$

of lagrange multipliers $\underline{\widetilde{v}}_{k,\text{med}}$, corresponding class labels y_k and the support vectors $\underline{\widetilde{x}}_{k,\text{med}}$ of the represented SOC class $\{c_i\}$.

$$\underline{\widetilde{x}}_{k,\text{med}} \in \bigcup_{i}^{n} \{c_i\} \tag{10}$$

The two new nodes of the graph represent the two roughly equal power sets of the above superset. The separation of the generated power sets can be repeated recursively down to the power set elements representing a single impedance spectrum or SOC class $\{c_i\}$. The resulting graph corresponds to a binary search tree, where the root is the whole superset of all elements from the impedance spectra with the nodes as a power set of its parent superset and the leaves representing the single SOC classes (see Fig. 5).

This binary search tree can easily be parsed by a binary tree search algorithm where the edges of the graph represent the binary decisions of the SVM decision function. So a binary tree search algorithm such as the divide and conquer search algorithm, applied to the afore mentioned binary search tree, makes it possible to grade measured impedances $\underline{Z}_i$ using the SVM decision function,

$$f_{\text{SVM},(\underline{Z}_i)} = \text{sgn}\left(\sum_{k=1}^{l} \underline{\widetilde{v}}_{k,\text{med}} \cdot y_k \langle \underline{Z}_i, \underline{\widetilde{x}}_{k,\text{med}} \rangle^d + b \right) \tag{11}$$

where d indecates the degree of the used polynomial kernel function.

$$k\left(\underline{Z}_i, \underline{\widetilde{x}}_{k,\text{med}}\right) = \langle \underline{Z}_i, \underline{\widetilde{x}}_{k,\text{med}} \rangle^d \tag{12}$$

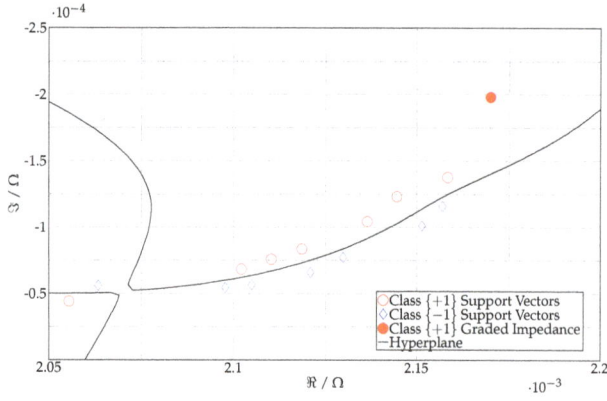

Figure 6. Impedance grading using SVM.

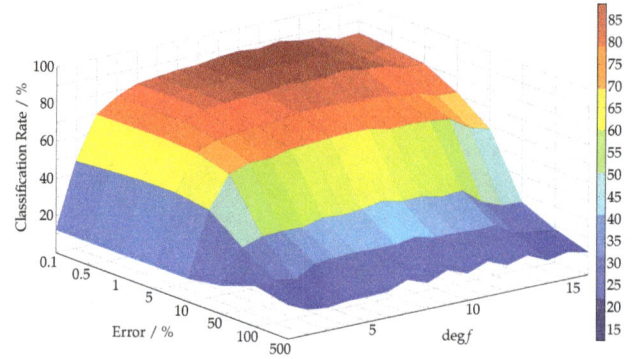

Figure 7. Statistic analysis of positive classification rates for single impedances depending on polynomial degree and impedance error.

3.3 Impedance grading

The SOC of the cell can now be determined by grading at least one on-vehicle measured impedance $\underline{Z}$ from the cell under load conditions. The impedance spectra of the relevant defined SOC levels therefore have to be classified, as described above. The measured impedance now has to be graded to a single SOC class – SOC specific impedance spectrum – to determine the SOC of the cell. The binary decision of the SVM decision function can only prove for two classes to which class, separated by the hyperplane, the measured impedance belongs (see Fig. 6). By using a divide and conquer search algorithm on the binary search tree with the SVM decision function as a key criterion, the SOC can be determined by multiple binary decisions along the search tree.

This divide and conquer algorithm starts at the median hyperplane of all separated impedance curves, that separates this superset into two roughly equal power sets. The binary decision, of the SVM decision function, whether the measured impedance belongs to the power set on one side of the hyperplane or the other decreases the quantity of relevant SVM decisions by half. The remaining power set after the decision containing the measured impedance will therefore also be divided by its median hyperplane into two subsidiary power sets. By continuing this recursive structure the SVM decisions on the binary search tree ultimately grade the measured impedance to a single dedicated SOC class. This class represents the SOC of the cell for the measured impedance. This implementation of this binary SVM based tree search algorithm makes an optimal execution time of $\mathcal{O}(log n)$ for the grading of one measured impedance possible.

3.4 Statistical Verification

Trials with measured impedance spectra have demonstrated that the new concept for grading impedances using SVM is effective for determining the SOC. After classifying the impedance spectra with the SVM classification function, an error ϵ, in both directions ($\Im$ and $\Re$), consisting of normally distributed random noise $X_{(\omega)} \sim \mathcal{N}\left(\mu, \sigma^2\right)$ charged with different magnitudes m,

$$X : \Omega \to \Re | X_{(\omega)} = \frac{1}{\sigma\sqrt{(2\pi)}} e^{-\frac{1}{2}\left(\frac{\omega-\mu}{\sigma}\right)^2},\qquad(13)$$

$$\epsilon = X_{(\omega)} \cdot m, \left\{m \in \Re | 10^{-7} \leq m \leq 10^{-4}\right\},\qquad(14)$$

where ω is a random value, the mean value of $\mu = 0$ and the variance of $\sigma^2 = 1$, where added to each impedance of a impedance spectrum. These error charged impedance objects are the specification for the SOC determination algorithm, to clarify that they would be graded correctly to their origin impedance spectrum.

The classification rate of this trial is highly dependent on two different factors. The first important factor is the polynomial degree of the kernel, which defines the separation accuracy of the hyperplanes between the impedance spectra. Another important aspect is the magnitude of the charged error of the original impedances of the spectra.

A linear representation of the hyperplanes is highly inefficient, because of the very low classification rates (see Fig. 7). The optimal polynomial degree would be around 5–8 as a trade-off between accuracy and execution time. Another feature is the high tolerance to the variance of error. The binary SVM based tree search algorithm is capable of grading impedances charged with a variance of error up to 10 %, with an accuracy of 60 %, for a single impedance. The statistical accuracy with 30 graded impedance objects of one SOC class rises up to 80 %. A classification rate of 90 % can then be achieved by decreasing the variance of error below 1 % (see Fig. 8).

Statistical evaluation demonstrates that the concept of a binary SVM based tree search is capable of determining the SOC of a lithium iron phosphate cell in the middle SOC range.

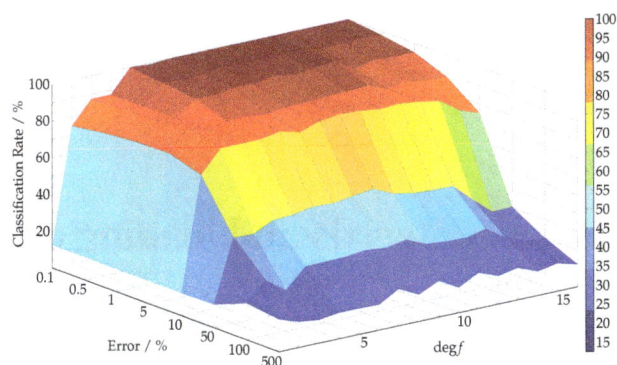

Figure 8. Statistic analysis of positive classification rates for the SOC determination.

4 Conclusions and future work

The binary SVM based tree search approach is a new method for determining the SOC of a lithium iron phosphate cell in the middle SOC range. The trial data demonstrated that, that the SOC can be determined with a certainty of 60 % for one measured impedance, and a maximum error variance of 10 %. The accuracy of the classification rate can increase to 90 % depending on two factors, the variance of error of the measured impedance and the polynomial degree of the hyperplane function. Therefore the requirements for a on-board impedance spectrum measurement can be circumvented. For an in-vehicle application, it is important to be able to identify different impedances at certain frequencies. To calculate impedances for several frequencies out of the time domain by a Fast Fourier Transformation, would be one method for this application (Klotz et al., 2011). Taking the results of the impedance grading method into account, it is possible to identify the requirements for the impedance calculation.

Future topics for research in this regard are the analysis of impedances from time domain data for classification purposes to determine the on-board SOC and the comparison with other methods such as the ECD based Kalman Filter or the state space observer, NND and its derivatives and artificial neuronal networks for SOC determination to identify the advantages and disadvantages of the different methods to combine them into a hybrid SOC determination algorithm. It will also be important to incorporate aging detection to update the hyperplanes in an enhanced machine learning method based on the above binary SVM search tree algorithm. Updating the hyperplanes will ensure that the SOC is correctly determined as the cell ages over its lifetime.

Acknowledgements. This contribution was developed within the scope of the project Drive Battery 2015 (Intelligente Steuerungs- und Verschaltungskonzepte für modulare Elektrofahrzeug-Batteriesysteme zur Steigerung der Effizienz und Sicherheit sowie zur Senkung der Systemkosten – AEV-Subproject: Optimierung des Energiemanagements von Fahrzeugen mit Lithium-Ionen Starter- und Bordnetzbatterien) which is funded by the BMWi (Bundesministerium für Wirtschaft und Energie) under the grant number 03 ET6003 I. The responsibility for this publication is held by the authors only.

References

Ávarez Antón, J. C., García Nieto, P. J., Viejo, C. B., and Vilán Vilán, J. A.: Support Vector Machines Used to Estimate the Battery State of Charge, IEEE Trans. Power Electro., 28, 5919–5926, 2013.

Codeca, F., Savaresi, S., and Rizzoni, G.: On battery State of Charge estimation: A new mixed algorithm, IEEE Intl. Conf Contr., 102–107, 2008.

Cortes, C. and Vapnik, V.: Support-vector networks, Machine Learning, Kluwer Academic Publishers, Dordrecht, the Netherlands, 20, 273–297, 1995.

Cover T. M. and Hart P. E.: Nearest Neighbor Pattern Classification, IEEE T. Inform. Theory, IT-13, 1, 21–27, 1967.

Dudani S. A.: The Distance-Weighted k-Nearest-Neighbor Rule, IEEE T Syst. Man. Cyb., SMC-13, 325–327, 1976.

Hearst, M., Dumais, S., Osuna, E., Platt, J., and Schölkopf, B.: Support vector machines, IEEE Intell. Syst. App., 13, 18–28, 1998.

Hu, J., Hu, J., Lin, H., Li, X., Jiang, C., Qiu, X., and Li, W.: State-of-Charge Estimation for Battery Management System Using Optimized Support Vector Machine for Regression, J. Power Sour., 269, 682–693, doi:10.1016/j.jpowsour.2014.07.016, 2014.

Klotz, D., Schönleber, M., Schmidt, J., and Ivers-Tiffée, E.: New approach for the calculation of impedance spectra out of time domain data, Electrochim. Acta, 56, 8763–8769, 2011.

Lee, J., Nam, O., and Cho, B.: Li-ion battery SOC estimation method based on the reduced order extended Kalman filtering, J. Power Sour., 174, 9–15, 2007.

Piller, S., Perrin, M., and Jossen, A.: Methods for state-of-charge determination and their applications, J. Power Sour., 96, 113–120, 2001.

Plett, G. L.: Extended Kalman filtering for battery management systems of LiPB-based HEV battery packs – Part 1: Background, J. Power Sour., 134, 252–261, 2004.

Sauer, D. U., Bopp, G., Jossen, A., Garche, J., Rothert, M., and Wollny, M.: State of Charge – What do we really speak about?, International Telecommunications Energy Conference (INTELEC), Copenhagen, Denmark, 6–8 June, 1999.

Weng, C., Cui, Y., Sun, J., and Peng, H.: On-board state of health monitoring of lithium-ion batteries using incremental capacity analysis with support vector regression, J. Power Sour., 235, 36–44, 2013.

11

Analysis of the effect of different absorber materials and loading on the shielding effectiveness of a metallic enclosure

S. Parr[1], **H. Karcoon**[1], **S. Dickmann**[1], and **R. Rambousky**[2]

[1]Faculty of Electrical Engineering, Helmut-Schmidt-University/University of the Federal Armed Forces Hamburg, Germany
[2]Bundeswehr Research Institute for Protective Technologies and NBC Protection (WIS) Munster, Germany

Correspondence to: S. Parr (stefan.parr@hsu-hh.de)

Abstract. Metallic rooms as part of a complex system, like a ship, are necessarily connected electromagnetically via apertures and cables to the outside. Therefore, their electromagnetic shielding effectiveness (SE) is limited by ventilation openings, cable feed-throughs and door gaps. Thus, electronic equipment inside these rooms is susceptible to outer electromagnetic threats like IEMI[1]. Dielectric or magnetic absorber inside such a screened room can be used in order to prevent the SE from collapsing at the resonant frequencies. In this contribution, the effect of different available absorber materials is compared, as well as other properties like weight and workability. Furthermore, parameter variations of the absorber as well as the effect of loading in form of metallic and dielectric structures on the SE are analyzed.

1   Introduction

Inside metallic, empty, perfectly conducting and cuboid rooms or enclosures, resonances are excited by external fields at the frequencies

$$F_{(m,n,p)} = \frac{c}{2}\sqrt{\left(\frac{m}{a}\right)^2 + \left(\frac{n}{b}\right)^2 + \left(\frac{p}{d}\right)^2} \quad (1)$$

with a, b and d: dimensions, c: speed of light in free space, m, n and p: positive integers, one of which may be zero (Dawson et al., 2001). This results in large spectral and spatial variations in field levels up to 45 dB (Izzat et al., 1998). Electronic equipment inside the room can thus be impaired due to the high field strengths at the resonant frequencies. The SE at the resonances can be improved by lining the inner walls with absorbing material (Olyslager et al., 1999).

---

[1]Intentional Electromagnetic Interference

In this contribution, two studies are carried out. First, different absorber materials are compared with respect to their damping property, flammability, workability, weight and price. Therefore, a cuboid screened enclosure with dimensions of 40 cm is analyzed in the frequency range of 400 to 1000 MHz via simulation and measurement. It is illuminated by a TEM wave which couples in through an aperture at its front side. The inner back side is tiled with a layer of different absorber materials: polyurethane-carbon foam, ferrite tiles and two different types of ferrite composite absorber. The SE is determined via simulation and measurement. For the simulation, knowledge about the dielectric and magnetic properties of the absorber materials is necessary. This is provided by a reflection and transmission measurement inside a coaxial transmission line.

In the second study, the SE of a full size room with a door gap as aperture is investigated numerically. As the field strength inside the resonator is spatially varying at the resonant frequencies, the SE is calculated via the mean energy density. Different parameter variations of the absorber like thickness and number are carried out in the frequency range from 40 to 100 MHz. Furthermore, the impact of metallic and dielectric loading inside the enclosure on SE is determined.

2   Comparison of different absorbing materials

2.1   Permittivity and permeability of the absorber

First, the influence of absorber inside a screened enclosure on the SE is analyzed for different materials. In order to carry out numeric calculations, the complex permittivity and per-

Table 1. Minimum electric shielding effectiveness of the enclosure with absorber.

Absorber material (thickness)	ferrite (5.5 mm)	polyurethane-carbon (20 mm)	silicon-ferrite (2 mm)	polyethylene-ferrite (4 mm)	polyethylen-ferrite (2 mm)
Measurement	30 dB	24 dB	17 dB	30 dB	28 dB
Simulation	21 dB	20 dB	14 dB	34 dB	32 dB

Table 2. Comparison of different properties of the absorber materials.

	ferrite 5.5 mm	polyurethane-carbon 20 mm	silicon-ferrite 2 mm	polyethylene-ferrite 4 mm	polyethylene-ferrite 2 mm
Price	$400\,€\,m^{-2}$	$120\,€\,m^{-2}$	$1139\,€\,m^{-2}$	$2831\,€\,m^{-2}$	$1910\,€\,m^{-2}$
Damping quality for 400 to 1000 MHz	good	fair	poor	good	good
Flammability	not inflammable	inflammable (flame-resistant version available)	not inflammable	not inflammable	not inflammable
Workability	poor	good	good	good	good
Weight per area	$27.8\,kg\,m^{-2}$	$0.96\,kg\,m^{-2}$	$6.3\,kg\,m^{-2}$	$14.8\,kg\,m^{-2}$	$7.4\,kg\,m^{-2}$

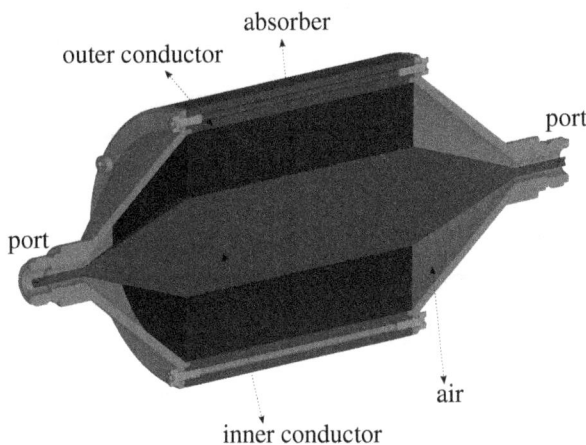

Figure 1. Coaxial line experiment setup for determination of the complex permittivity.

Figure 2. Absorber samples for the coaxial line experiment. From left to right: polyurethane-carbon, ferrite, silicon-ferrite, polyethylene-ferrite.

meability

$$\varepsilon_r = \varepsilon_r' - j\varepsilon_r'' \quad (2)$$
$$\mu_r = \mu_r' - j\mu_r'' \quad (3)$$

of the absorber have to be known. They are determined via a coaxial line experiment measuring the reflection and transmission coefficients with a network analyzer. The setup of the coaxial line with the absorber inserted is shown in Fig. 1. The analyzed absorber materials include polyurethane-carbon foam as an dielectric absorber, ferrite tiles, silicon-ferrite and polyethylene-ferrite as magnetic ab-

sorber. The pre-cut absorber samples are shown in Fig. 2. The permittivity and permeability are calculated using the Nicolson-Ross-Weir (NRW) algorithm (Nicolson and Ross, 1970) from the S parameters. A measurement of the empty line yields reasonable results for permittivity and permeability of air up to 1 GHz. Above, higher order modes spoil the measurement (Ihsan et al., 2011). The extracted values for ferrite and polyurethane-carbon are shown in Fig. 3.

2.2 Shielding Effectiveness of the resonator with and without absorber

A screened enclosure in form of a cube with dimensions of approximately 40 cm and a circular aperture with a radius r_0 of 15 mm at its front side is analyzed. The thickness of the

Ferrite

Ferrite

Polyurethane foam + Carbon

Polyurethane foam + Carbon

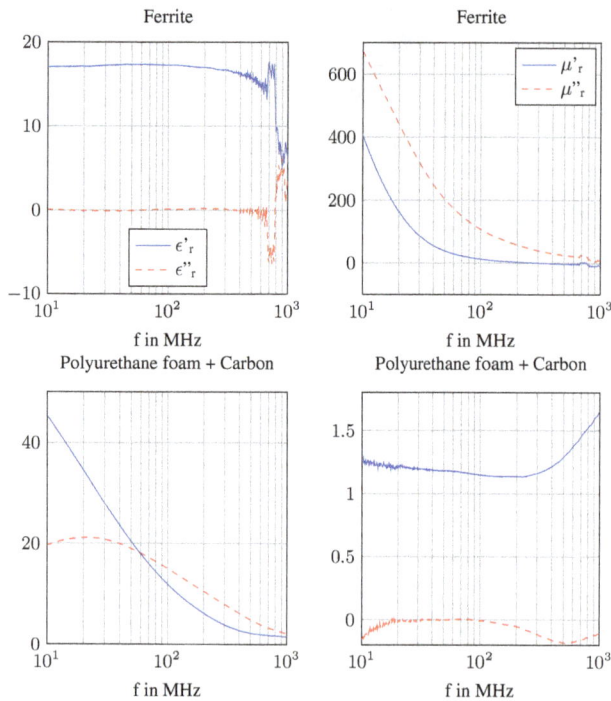

Figure 3. Permittivity and Permeability of ferrite and polyurethane-carbon absorber.

Table 3. Minimum wavelength inside the absorber in the frequency range 40 to 100 MHz.

Material	λ_{min} in m
vacuum	3
polyurethane-carbon	0.63
ferrite	0.069
silicon-ferrite	0.70
polyethylen-ferrite	0.38

absorber covering the inner back wall depends on the used material. The box is illuminated with a TEM wave and its electric shielding effectiveness SE_{el} is defined as

$$SE_{el} = 20 \times \log_{10} \frac{E_0}{E_1} \text{ in dB.} \tag{4}$$

with E_0: electric field in absence of the shield, E_1: electric field inside the shield. SE_{el} is determined via measurement and simulation in the frequency range of 400 MHz to 1 GHz. The lower limit is chosen in such a way that the first resonance at 529 MHz is covered. A GTEM cell is used as a source for a TEM electromagnetic wave for the measurement (Parr et al., 2012). The simulation is done using the Finite-Element-Method within the software FEKO. The minimum of SE_{el} of the empty resonator at the analyzed frequencies is at 10 dB. It is improved by the different absorbers reaching values shown in Table 1.

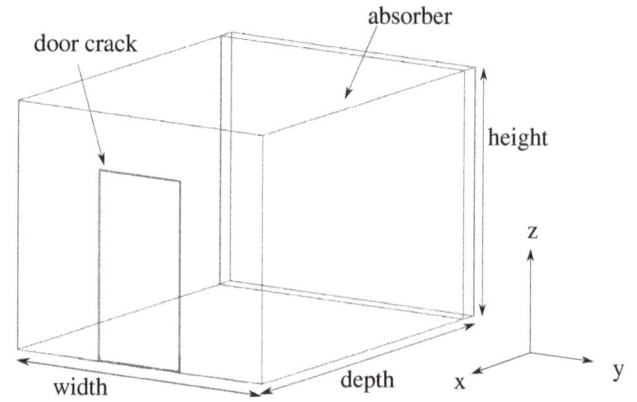

Figure 4. Screened room with door gap.

Table 4. Minimum value for SE_{em} with different absorber thicknesses e.

e	0 cm	5 cm	10 cm	20 cm
SE_{min}	−26 dB	−21 dB	−12 dB	−2 dB

2.3 Consideration of other absorber properties

Not only the ability to improve the SE, but also other absorber properties like flammability, workability, weight and price have to be considered for practical purposes. They are given for the different materials in Table 2. Ferrite has the best damping properties, but lacks workability and is heavy, whereas polyurethane-carbon absorber shows an overall decent performance.

3 Parameter variations of the absorber and loading

3.1 Model for the numerical calculations

In this section, different parameter variations of the absorber geometry inside a screened room are carried out, and the effect of dielectric and metallic structures inside is considered. Therefore, the SE of a screened room with dimensions 4 m (depth) × 3 m (width) × 2.6 m (height) as shown in Fig. 4 is analyzed via FEM simulation. As aperture a round about door gap is assumed, representing a shielded door, that is not closed properly. The incoming TEM wave travels in x direction and its electric field is polarized 45° to the z axis, in order to excite all modes. As the field levels inside the room vary significantly in space at the resonant frequencies, the shielding effectiveness SE_{em} is calculated via the mean energy density w of the electromagnetic field:

$$SE_{em} = 10\log_{10} \frac{\overline{w_0}}{\overline{w_1}} \text{ dB} \tag{5}$$

with $\overline{w_0}$: mean electromagnetic energy density in absence of the shield and $\overline{w_1}$: mean electromagnetic energy density

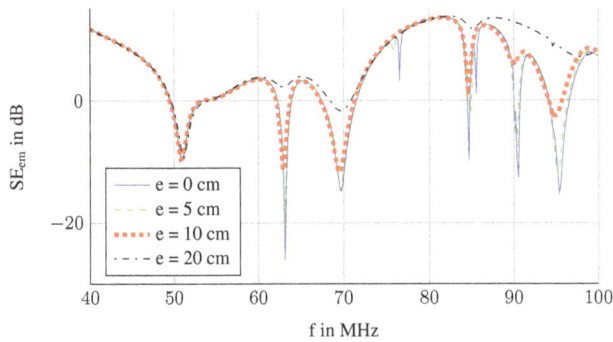

Figure 5. CEM results for SE$_{em}$ for different polyurethane-carbon absorber thicknesses e.

Table 5. Minimum value for SE$_{em}$ with different number of walls lined with 10 cm absorber.

walls lined with absorber	without absorber	rear wall	rear wall and ceiling	ceiling, rear and side walls
SE$_{min}$	-26 dB	-12 dB	-6 dB	-3 dB

Figure 6. Screened room with metallic structures (terminals).

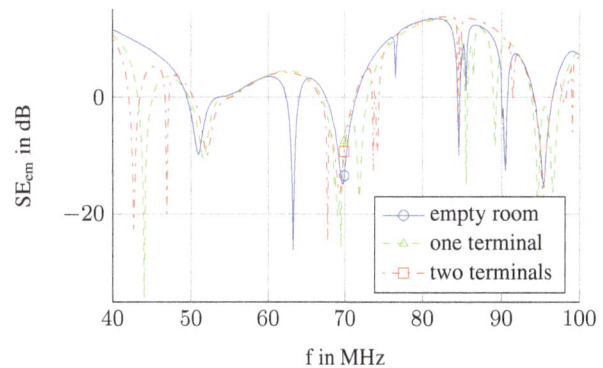

Figure 7. CEM results for SE with and without terminals inside the room.

inside the shield. It is calculated as the average value over 240 points with a spacing of 0.5 m. An adaptive frequency sampling is chosen with a minimum frequency increment of 150 kHz in the range of 40 to 100 MHz, covering the first resonance of the room at 62.5 MHz.

Due to its high permittivity and/or permeability, the absorber region is meshed more densely. The minimum wavelength inside the absorber in the analyzed frequency interval for the different materials is shown in Table 3. Not only the absorber region is meshed densely but also the adjacent metallic surfaces of the enclosure. These are numerically solved with the Methods of Moments, which leads to high time and memory consumption in the case of ferrite.

3.2 Parameter variations of the absorber

As absorber material polyurethane-carbon is chosen. At first, the thickness of the absorber layer e at the rear of the room is varied. The CEM[2] results for SE$_{em}$ for values for e of 5, 10 and 20 cm are shown in Fig. 5. The minimum SE$_{em}$ without absorber is -26 dB at 62.5 MHz, which corresponds to the first resonance of the room (110). The resonance at 51 MHz is caused by the door gap and is therefore not affected by the absorber. SE$_{min}$, the minimum value for SE$_{em}$ in the analyzed frequency region for different absorber thicknesses is shown in Table 4.

Furthermore, the number of walls that are lined with absorber is varied. As a result, the resonant frequencies of the room shift slightly, as the resonator gets electrically larger. The minimum value for SE$_{em}$ is shown in Table 5.

[2]Computational Electromagnetics

3.3 Effect of dielectric and metallic structures inside the room

In order to distinguish between the effects of dielectric and metallic structures inside the room on SE, both cases are analyzed separately. At first, metallic structures in the form of two cuboids, named here terminals, are inserted in the room model as depicted in Fig. 6. The numeric results for SE$_{em}$ with none, one and both terminals are shown in Fig. 7. As a result, the metallic structures shift the resonant frequencies and cause additional resonances in the low frequency region.

Next, the effect of dielectric structures in form of persons is considered, that are modeled as columns with electromagnetic properties of human muscle. As a reasonable assumption, two persons are modeled inside the room (Fig. 8). The numeric results in Fig. 9 show, that the (110) and (210) resonances are completely damped, because their electric field is polarized in z direction, parallel to the columns in the centre of the room. The value for SE$_{em}$ at other resonances however is lower with the dielectrics inside the room.

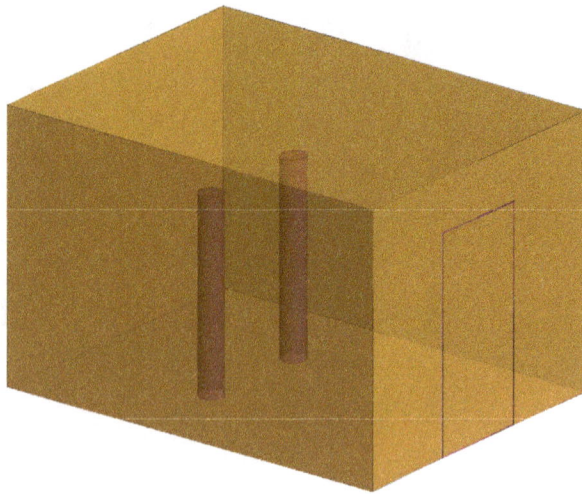

Figure 8. Screened room with dielectric structures (persons).

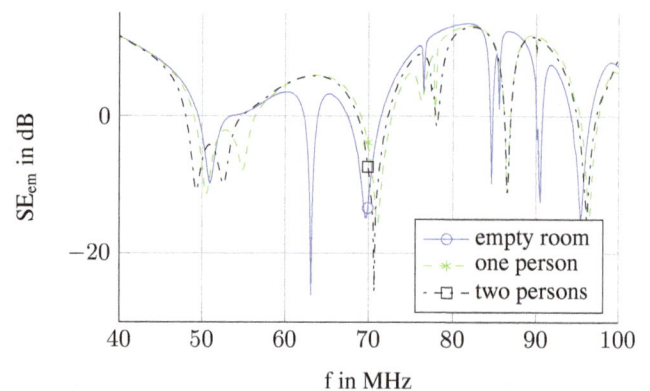

Figure 9. CEM results for SE with and without persons inside the room.

4 Conclusions

In the framework of an electromagnetic analysis of a complex system with consideration of resonant room and enclosure structures, two studies have been caried out. First, various absorber materials have been compared with respect to their effect on the SE in the frequency range of 400 to 1000 MHz. Ferrite and a composite ferrite absorber have the best damping properties, improving the minimum SE from 10 to 30 dB, while polyurethane-carbon foam has significant advantages in price, weight and workability. Then, the effect of absorber on the resonances of a metallic room has been analyzed, and the improvement in SE quantified with different parameter variations. It shows, that a 10 cm polyurethane-carbon layer at the rearside of the room improves the minimum value for SE from -26 to -12 dB. Finally, the effect of loading on the resonance behavior has been considered. Metallic structures inside the room cause additional resonances below the first room resonance, and therefore reduce the SE. The effect of dielectric structures on SE depends on the electric field distribution of the resonant modes. Both, metallic and dielectric loading, result in a slight shift of the resonant frequencies. The results show, that the susceptibility of a complex system to an outer electromagnetic threat in the form of IEMI can be reduced by using absorber inside the resonant structures.

Edited by: F. Gronwald

References

Dawson, L., Dawson, J. F., Marvin, A. C., and Welsh, D.: Damping resonances within a screened enclosure, IEEE Trans. Electromagn. Compat., 43, 45–55, 2001.

Ihsan, Z., Lubkowski, G., Adami, C., and Suhrke, M.: Characterization of the absorbing material used in EMC experiments, in: Proc. EMC Europe 2011 York, 774–777, 2011.

Izzat, N., Craddock, I. J., Hilton, G. S., and Railton, C. J.: Analysis and realisation of low-cost damped screened rooms, IEE Proceedings-Science, Measurem. Technol., 145, 1–7, 1998.

Nicolson, A. M. and Ross, G. F.: Measurement of the Intrinsic Properties of Materials by Time-Domain Techniques, IEEE Trans. Instrum. Meas., 19, 377–382, 1970.

Olyslager, F., Laermans, E., De Zutter, D., Criel, S., De Smedt, R., Lietaert, N., and De Clercq, A.: Numerical and experimental study of the shielding effectiveness of a metallic enclosure, Electromagnetic Compatibility, IEEE Transact., 41, 202–213, 1999.

Parr, S., Dickmann, S., and Rambousky, R.: Damping resonances of a screened enclosure using absorbing material, in: Electromagnetic Compatibility (EMC EUROPE), 2012 International Symposium on, 1–5, doi:10.1109/EMCEurope.2012.6396853, 2012.

12

Statistical sensor fusion of ECG data using automotive-grade sensors

A. Koenig, T. Rehg, and R. Rasshofer

BMW Group Research and Technology, Munich, Germany

Correspondence to: A. Koenig (alexander.ak.koenig@bmw.de)

Abstract. Driver states such as fatigue, stress, aggression, distraction or even medical emergencies continue to be yield to severe mistakes in driving and promote accidents. A pathway towards improving driver state assessment can be found in psycho-physiological measures to directly quantify the driver's state from physiological recordings. Although heart rate is a well-established physiological variable that reflects cognitive stress, obtaining heart rate contactless and reliably is a challenging task in an automotive environment. Our aim was to investigate, how sensory fusion of two automotive grade sensors would influence the accuracy of automatic classification of cognitive stress levels. We induced cognitive stress in subjects and estimated levels from their heart rate signals, acquired from automotive ready ECG sensors. Using signal quality indices and Kalman filters, we were able to decrease Root Mean Squared Error (RMSE) of heart rate recordings by 10 beats per minute. We then trained a neural network to classify the cognitive workload state of subjects from heart rate and compared classification performance for ground truth, the individual sensors and the fused heart rate signal. We obtained an increase of 5 % higher correct classification by fusing signals as compared to individual sensors, staying only 4 % below the maximally possible classification accuracy from ground truth. These results are a first step towards real world applications of psycho-physiological measurements in vehicle settings. Future implementations of driver state modeling will be able to draw from a larger pool of data sources, such as additional physiological values or vehicle related data, which can be expected to drive classification to significantly higher values.

1 Introduction

Over the last years, the automotive industry has seen a surge in driver assistance systems (DAS) to increase security and comfort of the driver. Empowered by an increasing number of sensors such as radar, lidar, ultrasound or video based systems, the vehicle can estimate a model of the current state of the environment, its objects and properties, such as speed or orientation. Meanwhile, the DAS know very little about the state of the driver. Information on the current cognitive, emotional or medical state of the driver is, however, of high importance: according to a study by the German Automobile Club ADAC, tired drivers are responsible for one in four traffic fatalities in Germany (ADAC, 2014). In addition to fatigue, driver states such as stress, aggression, distraction or even medical emergencies continue to be yield to severe mistakes in driving and promote accidents. The vehicle should therefore know whether or not the driver is tired, aggressive, stressed or distracted to be able to adapt its DAS accordingly. Commercially available systems in vehicles currently focus on fatigue estimations and rely on measures of steering wheel interaction and break/acceleration commands of the driver to detect this single driver state. While fatigue estimation is already an important step towards increased safety, these systems must be improved in two ways: faster detection of driver states and a larger number of reliably detected driver states.

A pathway towards improving driver state assessment can be found in psycho-physiological measures to directly quantify the driver's state from physiological recordings, thereby removing the proxy of deducting the driver's state from steering wheel dynamics and pedal interactions (Koenig et al., 2014). One key measure of the psycho-physiological state of a driver is the Electrocardiogram (ECG) and variables derived from it, such as heart rate or heart rate variability. Heart rate is a particularly well established physiological variable that reflects cognitive stress of subjects (Mehler et al., 2012; Mulder et al., 2000). One problem of previous work on automatically classifying the cognitive stress level of a driver from heart rate information was that recordings were typically done using medical grade, adhesive wet electrodes to record ECG.

For security and comfort reasons, any estimation of a driver's state can only happen through contactless sensors that seamlessly integrate into a vehicle. Obtaining ECG data contactless and reliably to deduct heart rate is a challenging task in an automotive environment. Galvanic and capacitive sensors in the steering wheel, seat and seatbelt can record ECG. Galvanic sensors, built into a steering wheel for example, have the limitation that the driver has to have both hands on the steering wheel to obtain a differential recording between left and right hand from which the ECG can be extracted. Capacitive sensors, located in the seat or the seatbelt, have limited signal availability when electric currents generated through friction motion pollute the signal. These currents induce local electric fields stronger than the ECG signal and therefore render a useful signal recording impossible whenever the subject moves. In addition, thick clothing can prevent sufficient capacitive coupling between human and sensor. So, while galvanic and capacitive sensors are available automotive-ready, they have limited reliability due to movement artifacts.

Sensor fusion of different sources of ECG derived signals promises higher signal availability and therefore improved driver state assessment. Our aim was to investigate, how the use of automotive grade ECG sensors would influence the accuracy of automatic classification of cognitive stress levels in drivers, and how sensory fusion algorithms could be employed to improve this accuracy. Previous research, mostly performed in the medical field, has focused on fusing existing signals of different kinds, such as blood pressure, ECG and oxygen saturation (SpO_2) to obtain more robust estimates of heart rate (Feldman et al., 1997). Only Li and colleagues looked at fusing recordings of two medical-grade ECG signal sources to obtain higher reliability of heart rate estimates (Li et al., 2008). However, so far no one has investigated how the fusion could improve heart rate recordings in an automotive setting and how classification of driver states might benefit from such sensor fusion.

In this publication, we report our experiments on inducing cognitive stress in subjects and estimating stress levels from their heart rate signals, acquired from automotive ready ECG sensors. We recorded ECG data of five subjects using automotive-grade sensors in the seat and steering wheel during a cognitively challenging task and employed statistical sensor fusion algorithms, namely Kalman filters, to increase reliability of ECG estimates. We trained a neural network to classify the cognitive workload state of subjects from heart rate and compared classification performance for ground truth, the individual sensors and the fused heart rate signal.

Table 1. The cognitive task used to induce cognitive stress levels in subjects. The stimulus (instruction by the experimenter) and the required answer of the subject are shown in the first and second line, respectively.

Stimulus by experimenter	4	8	2	6	0	8	5	2	5
2-back response	–	–	4	8	2	6	0	8	5

2 Methods

2.1 Experimental setup

We aimed at quantifying how well currently available automotive sensors are suited to automatically classify cognitive stress from physiological data. We chose to focus on heart rate as a physiological recording, as heart rate was shown to very well reflect the cognitive stress level of drivers (Mehler et al., 2012). Classifying cognitive stress levels from medical-grade heart rate recordings is well established in literature (Mehler et al., 2012; Koenig et al., 2011). Using only automotive-grade sensors, we wanted to quantify how strongly sensor fusion algorithms could influence the classification accuracy. We hypothesized that we should be able to obtain higher availability of heart rate readings by fusing several automotive-grade sensors, thereby improving classification performance of cognitive stress.

We induced cognitive stress through a well-established memory task, the two-back task (Table 1), which requires subjects to verbally repeat a sequence of numbers between 0 and 9 at a delay of two numbers (see). The experiment consisted of only two conditions of each two minutes length: a baseline condition without cognitive stress and one with cognitive stress.

Heart rate data was computed from ECG data, recorded from five healthy subjects (2f and 3m, 23 years ±2 years) at BMW Group Research and Technology, Munich, Germany, using standard beat detection algorithms. Automotive grade sensor systems were embedded into a driver seat and a steering wheel. To create realistic conditions, subjects were instructed to move in the seat and take their hands on and off the steering wheel as they felt comfortable. Ground truth ECG data was obtained from a medical grade three point leads attached through wet electrodes directly to the subjects' chest using a Gtec g.USBAmp system (Gtec, Graz, Austria) in combination with recording software written in Matlab Simulink 2010 (The Mathworks, Natick, MA, USA) at 512 samples per second. The seat sensor consists of capacitive electrodes in the sitting area and backrest that measures the? ECG as a potential difference between electrodes. The steering wheel sensor consists of a set of Plessey sensors (Plessey Semiconductors, Plymoth, UK) that galvanically record the potential difference between left and right hand. Both auto-

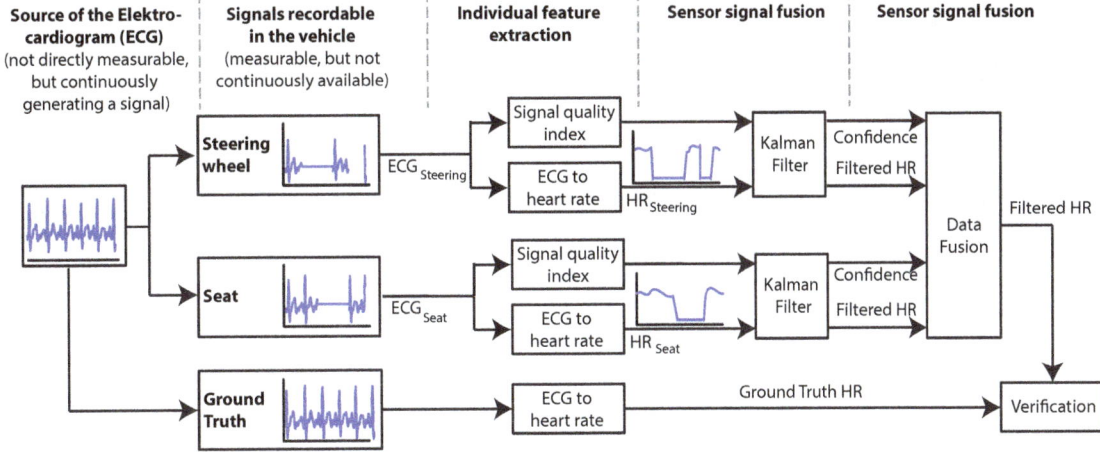

Figure 1. Smoothing and filtering system for the ECG data obtained from seat and steering wheel sensors.

motive sensors sampled data at 1 kHz. While the ground truth data is very robust towards noise induced through movement or friction currents, the seat sensor loses signal when it loses direct contact with the driver, for example when the driver moves to adjust his/her sitting position or looks at the side mirrors. The steering wheel can only sense ECG when the driver has both hands on the wheel. Heart rate was extracted from individual ECG signals as $HR = 60\,s\,/\,RR$ in beats per minute using a standard peak detection algorithm, where RR was the time in between two consecutive R waves, the most prominent peak in the ECG signal.

2.2 Data filtering and fusion

Steering wheel and seat sensors experienced data loss due to loss of contact to the human. To filter the individual HR signals, we followed the approach of Li2008 and used Kalman filters in combination with a Signal Quality Index (SQI), which adapted the Kalman Gain at runtime depending on the current signal quality. We then used the Kalman filtered signals of both, steering wheel and seat, and selected the signal that had the higher Kalman Filter confidence (see Fig. 1). The Kalman Filter used $A = 1$ as its state space matrix with no input B. C was set to 1, D to 0. As heart rate fluctuates by several beats per minute over time (referred to as Heart Rate Variability), the only driver to our system was the process noise, which we estimated beforehand to be 8.6 from the variance of a standard heart rate recording.

The SQI was computed as a combination of three different measures: kurtosis of ECG, running variance of heart rate and absolute values of heart rate. Kurtosis (kSQI) of the ECG signal was calculated using a moving window with a window size of 0.5 s which was found to be optimal for both seat and wheel signals. kSQI was computed as

$$kSQI = \frac{1}{n}\sum_{i=1}^{n}\left(\frac{x_i - \bar{x}}{s}\right)^4 \tag{1}$$

where x is the heart rate data, $\bar{x}$ is the mean of x and s is the standard deviation of x. If kurtosis was found to be above a threshold of 80, the signal was quantified as too noisy. Variance of heart rate was used as a second indicator of signal quality (varSQI) and computed from a 4 s window, symmetrically around the current data point. As the window was placed symmetrically around the current data point, we introduced a delay of 2 s during computation of SQI.

Usually heart rate variance, the fluctuation of heart rate at rest, lies around ± 2.5 bpm (Koenig et al., 2010). A variance of more than 40 was found to indicate that the original ECG signal had been too noisy. The third indicator for signal quality was the absolute value of heart rate (aSQI): heart rates above 200 bpm or below 30 bpm are physiologically not possible. Only if both, the absolute value and either variance or kurtosis SQI indicate good signal quality the measurement could be trusted.

$$kSQI = \begin{cases} 0 \text{ if } K > 80 \\ 1 \text{ if } K \leq 80 \end{cases} \tag{2}$$

$$varSQI = \begin{cases} 0 \text{ if VAR} > 40 \\ 1 \text{ if VAR} \leq 40 \end{cases} \tag{3}$$

$$aSQI = \begin{cases} 0 \text{ if } 30 < A < 200 \\ -5 \text{ otherwise} \end{cases} \tag{4}$$

the SQI was then computed as

$$SQI = kSQI + varSQI + aSQI \tag{5}$$

According to the SQI, the Kalman filters R matrix, the covariance matrix of the measurements, was adjusted, which directly influenced the Kalman gain. If the SQI > 0 the value of R was set to 0.5. If it was below zero, R was set to e^5. The filter always trusts the signal with higher SQI and uses its inner state transition matrix for the estimation if both measurements are noisy. The computations of SQI delayed the signal by 2 s, which was found to be tolerable.

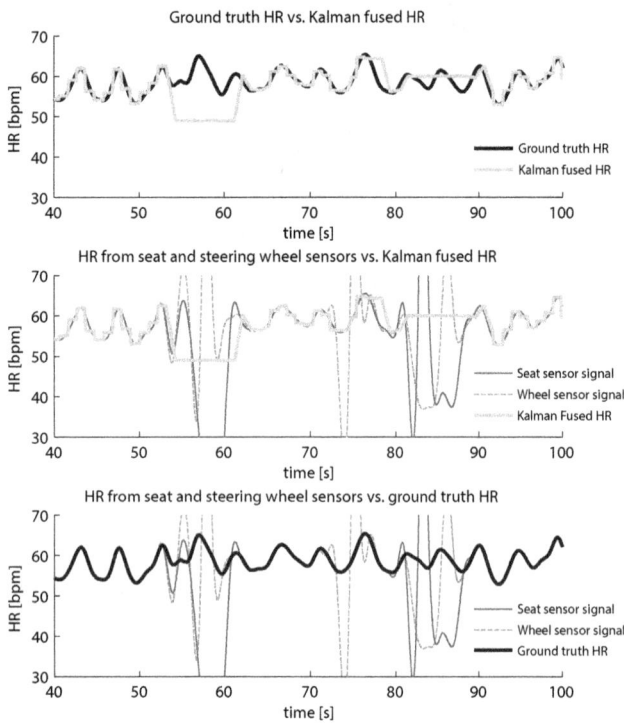

Figure 2. Exemplary subject, 60 s of data. The top panel shows the comparison between the ground truth heart rate data and the Kalman filter fused data, based upon the individual sensory data from seat and steering wheel. When both sensors fail to record reliable data, the Kalman filter switches to its internal model. The middle panel shows the comparison between the Kalman fused data and the individual sensory data. The bottom panel shows the comparison between ground truth and the individual sensory data.

2.3 Quantitative data analysis

To quantify improvement of our approach in terms of signal availability, we compared ground truth heart rate to the single sensor heart rate readings of steering wheel and seat and to the fused signal via computation of root mean square error (RMSE). In addition to quantifying the exact average improvement through RMSE, we wanted to assess if our approach could improve future efforts in driver state modeling. Our previous results indicated that fluctuations of ±2.5 beats per minute are within the normal boundaries of heart rate variability (Koenig et al., 2011a) during constant conditions of cognitive stress. We therefore computed the percent availability as the time during which the sensory signal deviated by less than 2.5 beats per minutes from the ground truth, assuming that any deviation smaller than 2.5 beats per minute from ground truth would not result in miss-classification of cognitive stress levels derived from changes in heart rate.

Table 2. Average performance of individual sensors and improvements of heart rate estimations through data fusion as measured by percent deviation from ground truth and Root Mean Square Difference (RMSD) between sensory signal and ground truth.

	Steering wheel alone	Seat alone	Fused data
Percent deviation less than ±2.5 beats per minute from ground truth	37 ± 18	51 ± 4	61 ± 9
RMSE (bpm)	20 ± 6	14 ± 1	5 ± 1

Table 3. Classification results of cognitive stress from heart rate.

	Ground truth	Steering wheel alone	Seat alone	Fused data
Percent correctly classified	77 ± 12	68 ± 4	61 ± 10	73 ± 11

2.4 Classifying cognitive stress

Heart rate was used as a measure for cognitive stress. We trained a neural network (Neural Network Toolbox, Matlab, The Mathworks) with heart rate estimates as input and performed leave one out classification for all five subjects, i.e. we trained the classifier on all but the ith subject data and classified the data of subject i with it. The neural network used a hidden layer with 20 neurons and a 70, 15 and 15 % split of the training data for training, validation and test respectively. These parameters were identified empirically. Four different sets of input data were provided to train the net:

- ground truth to evaluate the maximally possible correct classification with perfect, unpolluted data

- only heart rate data extracted from the seat sensor

- only heart rate data extracted from the steering wheel sensor

- fused heart rate data obtained from both, seat and steering wheel sensor, as described above

Leave one out classification then allowed computing the percent correctly classified.

3 Results

The fusion algorithm allowed dynamic switching between sensor sources. As seen in Fig. 2 the Kalman Filter used its internal model of $A = 1$ when both sensory signals were unavailable. As soon as one of the sensors was available again, the algorithms switched immediately back to trust its sensor readings.

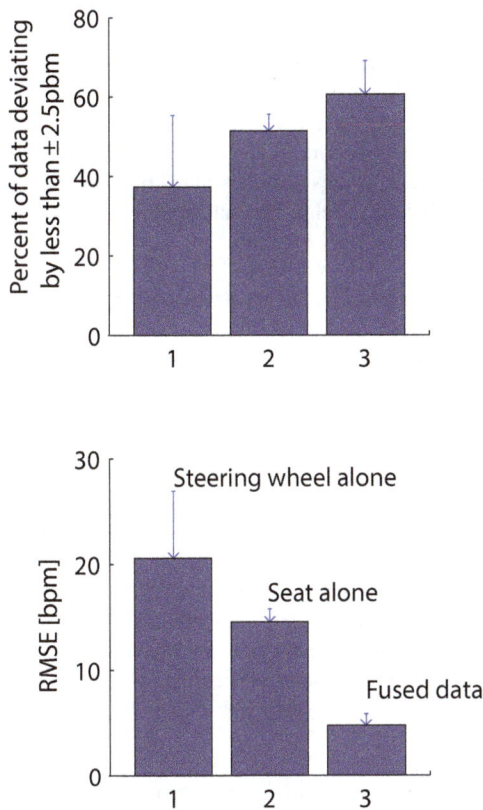

Figure 3. Aggregate data of all five subjects. The top panel shows mean and standard error of percent of all data that was found to deviate by less than ±2.5 beats per minute from ground truth for the steering wheel sensor alone, the seat sensor alone and the fused data. The bottom panel shows mean and standard error of the root mean square error (RMSE) between the ground truth and the respective sensors.

Through our sensor fusion concept, we were able to improve availability of heart rate information during our tests from 37 % for the steering wheel sensor and 51 % for the seat sensor by 10 to 61 % total signal availability (as compared to 100 % availability of ground truth data). The availability of the steering wheel sensor between subjects varied greatly by ±18 %, depending on whether or not the subjects were keeping their hands on the wheel (Table 2). Signal fusion decreased RMSD by 15 and 10 bpm respectively from steering wheel and seat alone.

The classifier used heart rate data as input to classify whether or not the subject was experiencing cognitive stress or not. It has to be noted that 50 % correct classification would denote chance, as we only differentiated between two classes. The best possible classification, given ground truth data, was found to be 77 %. Using only heart rate from seat data, classification dropped to 68 %, a decrease by 9 % (Table 3). Steering wheel data alone only provided correct classification in 60 % of cases. By fusing the signals, we could reach a total of 73 %, only 4 % lower than with ground truth.

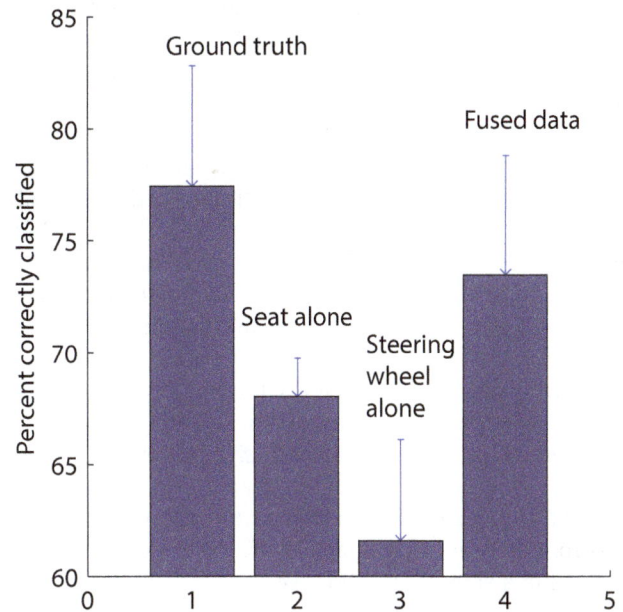

Figure 4. Results of neural network based classification of cognitive stress levels of subjects. Ground truth allowed 77 % correct classification, which was almost reached by fused data. ECG steering wheel and ECG seat sensor proved to be worse an average of 7 and 10 %, respectively.

4 Discussion

We investigated the effects of heart rate fusion algorithms on classification of driver states using automotive-grade sensors. The results presented in this paper are a first step towards real world applications of psycho-physiological measurements in vehicle settings. An ever increasing number of driver assistance systems need to understand the current cognitive, emotional or vigilance state of the driver to adapt their support functionality, for example, by adapting warning thresholds.

Previous experiments on automatic classification of subjects cognitive stress levels featured a larger array of sensory input, such as additional physiological signals and task performance data (Koenig et al., 2011a; Mehler et al., 2012; Mulder et al., 2000). Additional physiological values often recorded include Galvanic Skin Response (GSR), a measure for stress and arousal (Boucsein, 2005; Dawson et al., 2007), a measure for cognitive workload or breathing frequency, a measure for stress and mental effort (Carrol et al., 1986; Suess, 1980). In previous experiments, when using only physiological recordings of health subjects, we were only able to classify cognitive workload of subjects at 38 % correct classification for a 4 class problem, where chance level was 25 % (Koenig et al., 2011b). With the additional input of task performance indicators, the correct classification of this 4-class problem rose to 84 %. In this light, our classification results of 73 % correct classification of cognitive stress levels looks very promising, given that ground truth only provided 77 % correct classification. In comparison, the best single sensor was at 68 %, which corresponds to a 5 % increase in correct classification through the use of filtering software.

Future implementations of driver state modeling will be able to draw from a larger pool of data sources, such as additional physiological values or vehicle related data. Already now, fatigue is quantified from steering wheel and pedal interactions alone. While vehicle based driver state classification has the disadvantage of requiring long calibration times during which the vehicle learns its internal model, this data is constantly available during driving. Combining such available data with data from future sensory systems such ECG sensors, eye trackers and driver cameras can be expected to drive ground truth classification to significantly higher values.

It would not have to be necessary to employ Kalman Filters for this problem, as simple if-then rules could have performed in a similar way. However, one major advantage of our approach is the possibility to extend the system by additional sources of ECG or heart rate recordings. Possible are, for example, heart rate estimates extracted from camera images (Wu et al., 2012), which could be recorded from a driver camera system or additional capacitive sensors in the seat belt. These signals could then be fed as additional input sensory data into the Kalman filter.

Future fusion algorithms will need to not only fuse heart rate signals computed from noisy ECG sources, but perform fusion on the ECG level directly. Probability based approaches, such as particle filters, will be used to estimate RR intervals, which would result in a binary representation of the CG signal. From an estimated RR signal, additional parameters such as Heart Rate Variability which was shown to correlate with cognitive load and emotional stress levels (Malik and Camm, 1990).

Apart from only quantifying drivers' stress levels or tiredness and reacting to it, future developments call for manipulation of the driver state. Active modification, as for example through light, sound, scent or climate control, will allow the vehicle to de-stress the driver or wake him or her up. The basis for these applications will be a reliable driver state monitoring system.

Acknowledgements. The authors thank their colleague Justus Jordan for his support.

References

ADAC e.V.: Müdigkeit im Straßenverkehr – unterschätzt, verkannt, tödlich, Artikelnummer 2831141, available at: www.adac.de/infotestrat/ratgeber-verkehr, 2014.

Boucsein, W.: Electrodermal measurement, in: Handbook of human factors and ergonomics methods, edited by: Stanton, N., Hedge, A., Brookhuis, K., Salas, E., and Hendrick, H., London, CRC Press, 2005, 18-11–18-18, 2005.

Carrol, D., Turner, J. R., and Prasad, R.: The effects of level of difficulty of mental arithmetic challenge on heart rate and oxygen-consumption, Internat. J. Psychophysiol., 4, 167–173, 1986.

Dawson, M. E., Schell, A. M., and Filion, D. L.: Handbook of Psychophysiology, 3rd Edn., 2007.

Feldman, J. M., Ebrahim, M. H., and Bar-Kana, I.: Robust Sensor Fusion Improves Heart Rate Estimation: Clinical Evaluation, J. Clin. Monit., 13, 379–384, 1997.

Koenig, A., Omlin, X., Zimmerli, L., Sapa, M., Krewer, C., Bolliger, M., Mueller, F., and Riener, R.: Psychological state estimation from physiological recordings during robot-assisted gait rehabilitation, J. Rehabil. Res. Dev., 48, 367–385, 2011a.

Koenig, A., Novak, D., Pulfer, M., Omlin, X., Perreault, E., Zimmerli, L., Mihelij, M., and Riener, R.: Real-time control of cognitive load in neurological patients during robot-assisted gait training, Trans. Neural. System. Rehabilitat. Engin., 19, 453–464, 2011b.

Koenig, A., Decke, R., and Rasshofer, R. H.: Fahrerzustandsmodellierung, Elektronik automotive, 33–37, 2014.

Li, Q., Mark, R. G., and Clifford, G. D.: Robust heart rate estimation from multiple asynchronous noisy sources using signal quality indices and a Kalman filter, Physiol. Meas., 29, 15–32, 2008.

Malik, M. and Camm, A. J.: Heart rate variability, Clin. Cardiol., 13, 570–576, 1990.

Mehler, B., Reimer, B., and Coughlin, J.: Demand From a Working Memory Task, An On-Road Study Across Three Age Groups Sensitivity of Physiological Measures for Detecting Systematic Variations in Cognitive Demand from a Working Memory Task, An On-road Study Across Three Age Groups, Human Fact., 54, 396–412, 2012.

Mulder, G., Mulder, L., Veldman, J., and van Roo, A.: A psychophysiological approach to working conditions, in: Engineering psychophysiology: Issues and applications, edited by: Backs, R. W. and Boucsein, W. M., Lawrence Erlbaum Associates, 139–159, 2000.

Suess, W. M., Alexander, A. B., Smith, D. D., Sweeney, H. W., and Marion, R. J.: The effects of psychological stress on respiration – a preliminary-study of anxiety and hyperventilation Psychophysiology, 17, 535–540, 1980.

Wu, H., Rubinstein, M., Shih, E., Guttag, J., Durand, F., and Freeman, W. T.: Eulerian Video Magnification for Revealing Subtle Changes in the World, ACM Transact. Graph., 31, 2012.

13

Charge pump design in 130 nm SiGe BiCMOS technology for low-noise fractional-N PLLs

M. Kucharski and F. Herzel

IHP, Im Technologiepark 25, 15236 Frankfurt (Oder), Germany

Correspondence to: F. Herzel (herzel@ihp-microelectronics.com)

Abstract. This paper presents a numerical comparison of charge pumps (CP) designed for a high linearity and a low noise to be used in a fractional-N phase-locked loop (PLL). We consider a PLL architecture, where two parallel CPs with DC offset are used. The CP for VCO fine tuning is biased at the output to keep the VCO gain constant. For this specific architecture, only one transistor per CP is relevant for phase detector linearity. This can be an nMOSFET, a pMOSFET or a SiGe HBT, depending on the design. The HBT-based CP shows the highest linearity, whereas all charge pumps show similar device noise. An internal supply regulator with low intrinsic device noise is included in the design optimization.

1 Introduction

Fractional-N phase-locked loops (PLL) are widely used in radio frequency (RF) and high-speed digital applications. As opposed to integer-N architecture they avoid the trade-off between low phase noise and fine frequency step, which makes them especially attractive for radar systems discussed by Pohl et al. (2012), wireless sensor nodes as outlined by Ussmuller et al. (2009) or wireless base stations, see Osmany et al. (2013). In a PLL, a phase detector (PD) compares the reference phase with the divided output of a voltage-controlled oscillator (VCO). In integrated PLLs, the PD is usually composed of a phase-frequency detector (PFD) and a charge pump (CP). The CP circuit is used to inject into the low-pass filter (LPF) a current that is proportional to the phase difference at the PFD input. The PLL CP is essentially a switchable current source, in contrast to charge pumps known from power electronics where they serve as DC/DC converters. In a fractional-N PLL the divider ratio is often modulated by means of a sigma-delta modulator (SDM). Unlike the topolo-

gies employing accumulators, the SDM shifts the quantization noise to frequency offsets above the loop bandwidth. Unfortunately, this noise is folded down to low frequencies if the PD is nonlinear. This causes in-band phase noise and fractional spurs as discussed by De Muer and Steyaert (2003), Riley et al. (2003), Pamarti et al. (2004), Chien et al. (2004), Arora et al. (2005), and Hedayati and Bakkaloglu (2009). Besides this, the thermal device noise in the CP must be minimized for a low in-band phase noise as outlined by Levantino et al. (2013).

This paper compares different types of CPs with respect to linearity and device noise, where either an nMOSFET, a pMOSFET or a SiGe HBT are the crucial devices in the steady state. Since the investigated CP topologies suffer from a low output resistance, a low-noise voltage regulator is suggested in order to minimize the effect of supply noise.

2 PLL architecture for a low phase noise

In order to reduce CP noise, the PFD may drive two parallel CPs (Fig. 1). In this architecture, a fine-tuning loop comprising a LPF is combined with a coarse-tuning loop in parallel. The latter includes only a capacitor to ground between the CP and the VCO. A conventional LPF can be extended by adding a resistive voltage divider, so the fine tuning voltage is kept roughly at a constant level. This results in a stable VCO gain over the whole tuning range, which is preferable in order to keep the loop bandwidth at the same level. The fine-tuning varactor diode in the VCO can be made small, which improves the phase noise due to lower sensitivity. Also, the fine-tuning loop provides stability of the PLL. To preserve the wide tuning range a coarse-tuning loop must be applied. The coarse control voltage sweeps the entire tuning range and determines the VCO frequency. As shown by

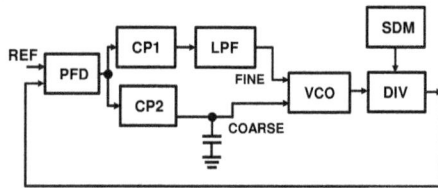

Figure 1. PLL architecture with two parallel tuning loops.

Herzel et al. (2003), if the dual-loop is correctly designed, a relatively large coarse loop capacitor does not deteriorate the settling speed of the PLL significantly.

The PD composed of a PFD and the two CPs converts the PLL phase error to currents. The PD nonlinearity has a strong effect in the presence of high-frequency quantization noise generated by SDM. It appears that this noise is folded down from large frequency offsets to the in-band region of the spectrum due to self-mixing in the nonlinear PD. Thus, mixed-down quantization noise cannot be filtered and deteriorates the close-in phase noise performance.

In modern fractional-N PLLs the main in-band noise sources include the reference buffer, PFD and the CP among which the latter is often the main contributor. It is due to CP device noise and already mentioned quantization noise folding. Since a fractional-N PLL often has wide loop bandwidth, the thermal noise of the CP becomes an issue.

As outlined by Herzel et al. (2010) large gate-source overdrive voltages $V_{ov} = V_{GS} - V_{th}$ of the CP transistors may improve PD linearity and reduce device noise. Figure 2 shows a CP architecture where V_{ov} is maximized. The most critical transistor in Fig. 2 is M1, since it switches in the steady state. The pMOSFET M2 is active only during the settling of the PLL. In the steady state, a constant current I_{OS1} flowing onto the filter capacitance is delivered by the pMOSFET M4. Its value equals the average current flowing through M1. Therefore, the offset current I_{OS1} defines the ON time of the CP and the duty cycle in the steady state. The output of CP1 for VCO fine tuning is DC biased by a resistive voltage divider to stabilize the VCO gain and the phase noise spectrum. Let us assume that due to PVT variations the VCO frequency is shifted. In a standard PLL topology this would change the DC value of the VCO control voltage. Since in a typical integrated VCO the gain varies by a factor of three or more over the tuning range, the loop bandwidth would change drastically. To prevent this, the DC value of the fine tuning voltage can be fixed using our PLL architecture with two parallel CPs and a voltage divider for the fine-tuning loop. This approach has been discussed and verified experimentally by Osmany et al. (2013). If the DC value $VDD \times RB1/(RB1 + RB2)$ is close to the position of the VCO gain maximum, the PLL is robust with respect to variations of the device parameters with process, supply voltage and temperature (PVT). Because RB1 and RB2 are part of the loop filter they change the PLL transfer function. This makes the system more suscepti-

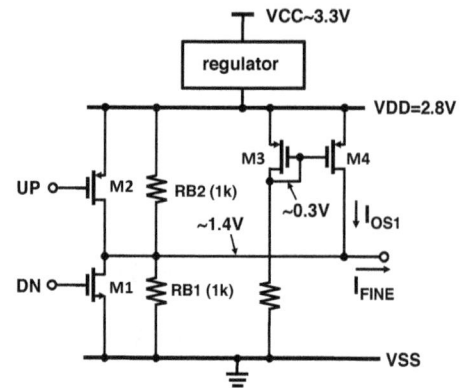

Figure 2. Charge pump CP1 for VCO fine tuning with DC output biasing and DC offset current.

ble to low frequency noise of the VCO. If the values are too small the VCO flicker noise will be significant. As shown by Herzel et al. (2010), the values should be around 1 kΩ in case of HBT-based VCO and around 10 kΩ for MOS-based VCO. Since we will use SiGe-HBTs for the VCO in the future, we have used resistor values as small as RB1 = RB2 = 1 kΩ in this design. The typical PD characteristic shows strong nonlinearity around zero transition due to so-called *dead zone* region. It is caused by the finite speed of logic gates in the PFD. It also suffers from gain mismatch between the UP and DN current pulses. Both problems can be solved by utilizing a DC offset current source added at the outputs of both CPs. This shifts the steady state operating point of the PD to a region, where either M1 or M2 transistor responds to the phase changes at the PD input. This idea was originally suggested by Chien et al. (2004), where a DC current to ground was introduced. Osmany et al. (2013) have employed this architecture in a fractional PLL, where a robust low-noise performance was achieved over a wide tuning range. Depending on the sign of the offset current, either nMOSFETs or pMOSFETs are changing their output current pulses with time in the steady state. In a SiGe BiCMOS process, HBTs can be employed for CP design. They are promising owing to their fast switching speed. Unfortunately, they introduce shot noise which may easily exceed the thermal device noise of MOSFET-based CPs.

The CP in Fig. 2 is relatively susceptible to supply noise. Therefore, it should be stabilized by an internal voltage regulator. Figure 3 shows a regulator which transforms a supply voltage VCC of about 3.3 V into a stable voltage of VDD = 2.8 V. High-voltage MOSFETs with moderate gate lengths were employed for reliability. A bandgap reference (BGR) according to Brokaw (1974) was used to generate a temperature stable voltage of 1.1 V. The high-frequency supply noise is reduced by the low-pass filter between BGR and CMOS amplifier. This may be important if CP and VCO are operated from the same supply, since the PLL loop filter is bypassed then. Relatively large gate widths were used to re-

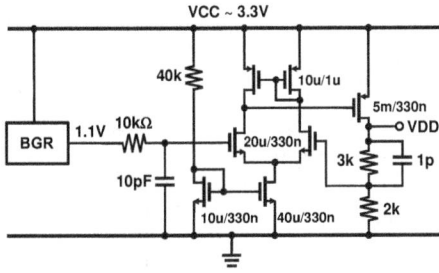

Figure 3. Low-noise voltage regulator for generation of a stable internal supply voltage of VDD = 2.8 V.

duce the device noise and to make the output voltage VDD independent of the load.

3 Charge pump figures of merit

3.1 Linearity

For an ideal phase detector the average CP output current I is a linear function of the phase deviation ϕ from the steady-state average phase error ϕ_0 at the PFD input. The linear PD gain is then given by

$$K_{PD} = I_{CP}/(2\pi), \tag{1}$$

where I_{CP} is the CP current in the ON state. For our PLL architecture with DC phase offset, the strong nonlinearities at low phase errors are avoided. Far from the origin, the PD input–output characteristic is smooth and can be approximated by a low-order polynomial. A parabolic fit may be enough here as shown by Herzel et al. (2010). Then we obtain

$$I = K_{PD}\phi + \frac{\beta}{2} K_{PD} \phi^2, \tag{2}$$

where β represents the PD nonlinearity. The quantity β is the normalized curvature of the PD input–output characteristic at the PD operating point and is given by

$$\beta = \frac{d^2 I/d\phi^2}{dI/d\phi}. \tag{3}$$

For a binary weighted CP, the quantity β does not depend on the CP current, but only on the CP architecture. As shown by Herzel et al. (2010) the in-band phase noise due to SDM noise folding in the nonlinear PD is proportional to β^2 provided that a DC offset current is employed at the CP output for PD linearization. In other words, a reduction of β by a factor of two reduces this phase noise contribution by 6 dB.

For the calculation of β from the PD characteristic the phase error range should be centered at the desired static phase ϕ_0 error which is typically between 36 and 72°. The ϕ range for curve fitting should be relatively narrow for a high accuracy but has to cover the phase excursions in the steady state.

3.2 Device noise

Commercial phase noise meters usually display the two-sided power spectral density (PSD) of the PLL output phase which is numerically close to the single-sideband phase noise. Therefore, we use the *two-sided* output current PSD of the CP current S_{CP} ($A^2 Hz^{-1}$) for characterizing the CP device noise. The in-band phase noise PSD due to CP current noise is then given by

$$S_\phi = S_{CP} \frac{N^2}{|K_{PD}|^2}, \tag{4}$$

where N is the feedback divider ratio. Note that circuit simulation tools usually output the one-sided current PSD ($2 \times S_{CP}$ in our notation) which must be divided by two for our purpose.

Usually, S_{CP} is proportional to the CP current I_{CP} in the ON state. In fractional-N PLLs the noise-optimum loop bandwidth is typically between 300 kHz and 1 MHz. The largest contributions to the PLL jitter stem from the phase noise spectrum around the loop bandwidth. Since the noise corner frequency scales with the charge pump duty cycle as follows from Eq. (15) of Herzel et al. (2010), the CP noise corner is typically below the loop bandwidth. Therefore, CP flicker noise is usually less important than white noise (thermal, shot) in the CP.

Due to the long correlation time of $1/f$ noise compared to the sampling period at the PD input, the flicker noise PSD scales with α_{CP}^2 where α_{CP} is the CP duty cycle. This was also discussed by Lacaita et al. (2007). Since white CP noise PSD scales with α_{CP} only, the CP noise corner frequency is proportional to α_{CP}. For these noise sources, S_{CP} is proportional to the CP duty cycle $\alpha_{CP} = T_{ON}/T_s$, where T_{ON} is the CP activation time and T_s is the sampling period at the PFD input. The static phase error at the PD input is related to the CP duty cycle by $\phi_0 = 360° \times \alpha_{CP}$. In order to obtain a device noise figure of merit (FOM) which characterizes a CP *architecture* and is independent of I_{CP} and α_{CP}, we define a FOM by

$$FOM = \frac{S_{CP}}{I_{CP}\alpha_{CP}}, \tag{5}$$

which is given in units of $A Hz^{-1}$.

3.3 Power supply rejection ratio

In addition to device noise, a CP may be affected by noise generated in other circuit blocks on the same die. In order to minimize the effect of this noise, the internal supply voltage VDD should be stabilized with respect to variations of the unregulated supply voltage VCC. Let us assume that a sinusoidal modulation of amplitude V_m and frequency f_m is added to the DC value of VCC. In this case, a signal of the same frequency but lower amplitude will appear at the internal supply voltage VDD. At the CP output the amplitude will be further attenuated to a value V_{out}. The power supply rejection ratio of the regulated CP is defined as

$$PSRR(f_m)(dB) = 20\log(V_m/V_{out}). \qquad (6)$$

4 Charge pump design

The CP design was performed in a 130 nm SiGe BiCMOS technology described by Rücker et al. (2010). This technology features high-performance HBTs with peak transit frequencies f_T of 240 GHz, maximum oscillation frequencies f_{max} up to 330 GHz, and breakdown voltages BV_{CEO} of 1.7 V along with high-voltage HBTs ($f_T = 50$ GHz, $f_{max} = 130$ GHz, $BV_{CEO} = 3.7$ V). Short-channel MOSFETs with 130 nm gate length are available for digital processing. Moreover, high-voltage MOSFETs with a minimum gate length of 330 nm are available. These MOSFETs are better suited for CP design since a larger voltage range can be used for VCO tuning, compared with the short-channel MOSFETs. For the used technology the threshold voltage V_{th} is about 0.6 V for the high-voltage MOSFETs, which results in overdrive voltages as large as 2.7 V for a supply voltage of 3.3 V.

In the following, we consider three CPs. The first one is the nMOS-based CP shown in Fig. 2. Here, only the nMOS-FET M1 changes its output current pulses in the steady state, while a constant UP current is delivered by the current mirror. The regulator derives a stable internal supply voltage of VDD = 2.8 V from a global supply voltage of VCC ≈ 3.3 V. The gate potential of the pMOSFETs in the current mirror is as low as 0.3 V to minimize their noise contribution *for a given current*. Note that both the CP and the current mirror transistors are in triode region. The current mirror with its output transistor in triode region provides a stabilized current with respect to variations of the threshold voltage V_{th}. In triode region the drain current depends linearly on the overdrive voltage $V_{GS} - V_{th}$, whereas in saturation it is a square-law characteristic. However, the drain current in this configuration depends strongly on the drain-source voltage, but this is not critical here due to the resistive voltage divider at the CP output. Eventually, the value of the DC offset current (and the resulting PD offset) can always be adjusted to compensate for PVT variations by using binary weighted current sources. In order to stabilize the static phase error, a classical cascode CP and an offset current source with a high output resistance should be used in the coarse tuning loop. Here, large gate-source voltages are not required, since the noise of CP2 will be minimized by the large external capacitor shown in Fig. 1.

The second CP version is pMOS-based, where the UP current in Fig. 2 is replaced with a DOWN current. This architecture has been used by Osmany et al. (2013) in a low-noise fractional-N PLL synthesizer. In that paper, binary weighted CPs were combined, where the disable functions were realized by multiplexers at the CP inputs and large MOSFET switches in the primary branches of the current mirrors. In

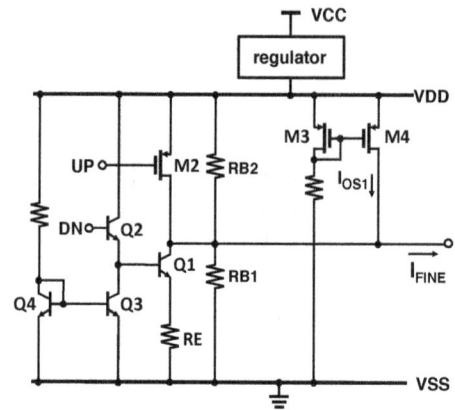

Figure 4. Charge pump CP1 for VCO fine tuning using SiGe HBTs.

this paper, we consider a constant-current CP for simplicity. The inclusion of switches for binary weighted digital current control is easily possible. By using a 4-bit binary weighted offset current source the phase offset can be easily adjusted.

The third CP considered here is shown Fig. 4. Here, the nMOSFET in Fig. 2 was replaced with a SiGe-HBT Q1 including bias resistance RE. Moreover, a level shifter was introduced at the DOWN input to convert the CMOS signal into an HBT compatible signal.

In the next section, the three CP versions are optimized by circuit simulations with respect to linearity and device noise. Subsequently, the resulting figures of merit are compared.

5 Numerical results

5.1 Phase detector linearity

The linearity analysis was performed by a transient analysis using Virtuoso Analog Design Environment. A 100 MHz reference was used at the PFD input, corresponding to a sampling period of $T_s = 10$ ns. In a parametric simulation the phase error at the PFD input was varied between 36 and 72°. This corresponds to a CP duty cycle of 10–20 % and a CP activation time T_{ON} between 1 and 2 ns. For each phase error the charge pump current was averaged over the period of 10 ns. Subsequently, the first and second derivative were numerically calculated by using MATLAB. The resulting linear PD gain K_{PD} and the nonlinearity parameter β were then obtained from Eqs. (1) and (3), respectively. For the latter, a linear fit to the first derivative was calculated to obtain a mean value for the normalized curvature.

The output current pulses are shown in Fig. 5 for different delays at the PFD input for the nMOS-based. The delays where adjusted such that the CP duty cycle varied from 10 to 20 %. The oscillatory behavior at the edges is not critical as long as the plateau is flat. In this case, the area below the curve is a linear function of the delay, and the normalized curvature β is small, as desired.

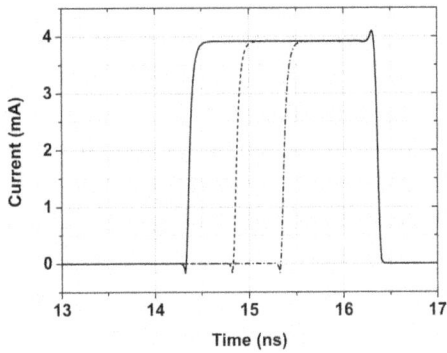

Figure 5. Output current pulses for three different pulse widths $T_{ON} = 1, 1.5$ and 2 ns for nMOS-based CP.

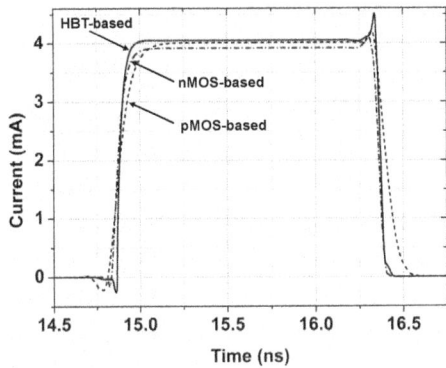

Figure 6. Output current pulses for the three CPs.

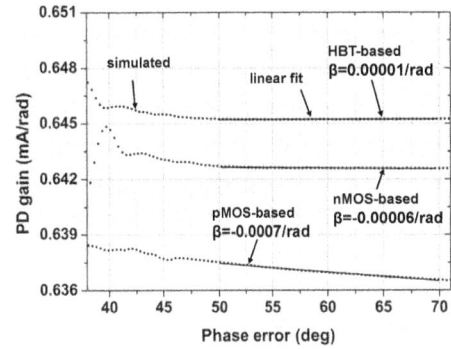

Figure 7. Phase detector gain as a function of phase error at the PD input for the three CPs.

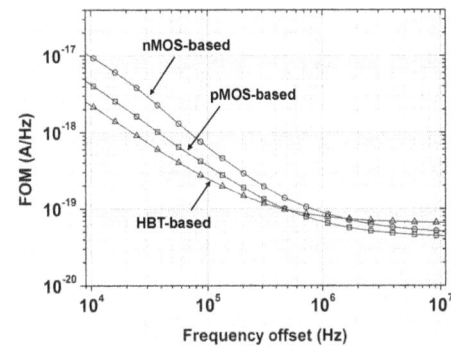

Figure 8. Device noise FOM according to Eq. (5) for the three CPs including DC offset currents and voltage regulator.

A single current pulse for the three investigated CPs is presented in Fig. 6. In order to obtain a high linearity, the turn-on and turn-off times should be as short as possible. It is obvious that the HBT-based CP provides the steepest current slopes. Thus, the highest linearity should be expected for this configuration. The pMOS-based CP shows the worst behaviour due to the low hole mobility.

The PD gain $dI/d\phi$ is shown in Fig. 7, where we have also included the pMOS-based CP for completeness. According to Eq. (3) the derivative of the PD gain needs to be estimated for the calculation of β. The mean slope of the curves depends on the phase region where the fit is calculated. In order to cover the phase excursions in the steady state, a range of $\pm 10°$ around the mean value is adequate as deduced from the calculated phase distribution in a typical fractional-N PLL presented by De Muer and Steyaert (2003). The linear least-squares fits to the simulated gain curves between 50 and 70° result in the estimated values for β at the static phase offset of $\phi_0 \approx 60°$ as given in Fig. 8. Obviously, the HBT-based CP results in the highest PD linearity.

5.2 Device noise

The noise analysis was performed by a periodic steady-state (PSS) analysis followed by a periodic noise analysis us-

ing Virtuoso Analog Design Environment. The device noise analysis includes the low-noise voltage regulator shown in Fig. 3. We used a CP peak current of $I_{CP1} = 4$ mA and an offset current of $I_{OS1} = 0.6$ mA. For the coarse tuning loop these values were reduced by a factor of 10. The ratio $I_{OS1}/I_{CP1} = I_{OS2}/I_{CP2}$ of 0.15 corresponds to a CP duty cycle of 15% and a static phase error of 54° at the PD input.

The FOM is depicted in Fig. 8 as a function of frequency. We observe a plateau in Fig. 8 in the lower MHz range related to white noise sources. In reality, this plateau extents into the kHz range since the $1/f$ noise corner frequency of the CP is proportional to the duty cycle α_{CP}, as explained above. This effect is not correctly reflected by the simulated phase noise spectrum, since long-term correlation effects are not properly included in the simulator. Based on this knowledge, we use an offset of 1 MHz in the circuit simulations as a representative value for the CP phase noise plateau. However, flicker noise from the offset current source is simulated correctly. Therefore, we observe relatively large $1/f$ noise for HBT-based CP even though the f_c of the used bipolar transistors is much lower (lies in 10 kHz range).

At large frequency offsets the phase noise due to device noise is highest for the HBT-based CP. It is composed of the shot noise of the bipolar HBTs in the CP core includ-

Table 1. Performance comparison of charge pumps.

switching device	β ($\phi_0 = 60°$)	FOM (at 1 MHz)	PSRR (at 1 MHz)	DC current (including biasing)
nMOSFET	-0.00006/rad	9×10^{-20} A Hz^{-1}	30 dB	2.9 mA
pMOSFET	-0.0007/rad	7×10^{-20} A Hz^{-1}	30 dB	2.9 mA
SiGe HBT	0.00001/rad	8×10^{-20} A Hz^{-1}	30 dB	4.3 mA

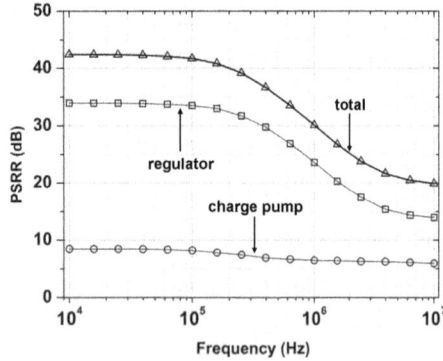

Figure 9. PSRR for the nMOS-based CP.

ing level shifter, the thermal noise of the CMOS offset current, the shot noise of the BGR, and the thermal noise of the voltage regulator. In order to estimate the phase noise plateau due to CP noise we consider a numerical example. We assume a PLL output frequency of 10 GHz corresponding to a feedback divider ratio of $N = 100$, a CP peak current of $I_{CP1} = 4$ mA, and a CP duty cycle of $\alpha_{CP} = 15\%$. Then we obtain from Eqs. (4), (5) and (1) a phase noise level of 1.2×10^{-12} rad^2 Hz^{-1} for the HBT-based CP at an offset frequency of 1 MHz. In decibel, this corresponds to -119 dB rad^2 Hz^{-1}. Despite the large duty cycle, this value is significantly lower than the measured in-band phase noise of state-of-the-art low-noise fractional-N PLLs as published by Osmany et al. (2013).

5.3 Supply noise rejection

The PSRR is depicted in Fig. 9 as a function of the modulation frequency for the nMOS-based CP. Here, we have also shown the PSRR of the regulator only and of the CP only. The poor PSRR of the CP results from the voltage divider at the output. However, in conjunction with the regulator the overall rejection is above 30 dB at frequencies up to 1 MHz. At frequency offsets above 1 MHz the CP noise is of little relevance, since it is effectively filtered by the low-pass filter in a PLL.

5.4 Performance comparison

The results are summarized in Table 1. All numbers in the table include the CP output biasing resistors, the offset cur-

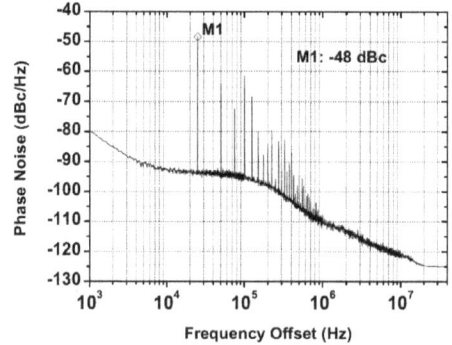

(a)

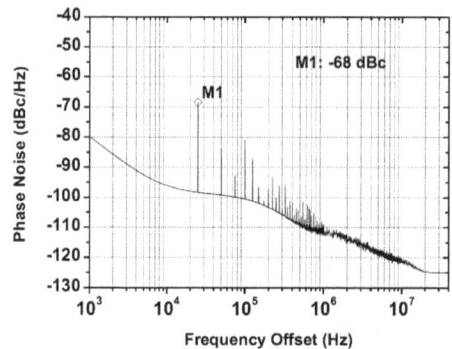

(b)

Figure 10. Simulated output spectrum for **(a)** $\beta = 0.01$ and **(b)** $\beta = 0.001$.

rent and the voltage regulator. As evident from the table, the nMOS-based CP shows the lowest device noise, whereas the HBT-based CP results in the highest PD linearity. In integer-N PLLs linearity is not relevant, and the MOS-based CP is the best solution. For fractional-N PLLs the best choice with respect to phase noise and fractional spur performance is the HBT-based CP for its high linearity. The second best solution is the nMOS-based CP for its better linearity, compared with the pMOS-based CP. Owing to the level shifter at the input the HBT-based CP dissipates slightly more current. We have simulated the PLL phase noise spectrum using the model of Herzel et al. (2010). The quantization noise was modeled as described by De Muer and Steyaert (2003). The simulated spectra for two values of β are shown in Fig. 10. The improvement of the in-band phase noise by the higher PD linearity is only moderate due to other phase noise contributions. By contrast, the in-band spurs are reduced by as much as 20 dB, since the folded quantization noise is proportional to β^2.

6   Conclusions

We have designed and compared three CPs in SiGe BiC-MOS intended for low-noise fractional-N PLLs, where either MOSFETs or SiGe-HBTs were used as switching elements in the steady state. Using large gate-source voltages in conjunction with DC offset currents, linearity and device noise of the CMOS CPs were optimized. The inclusion of SiGe-HBTs for faster current switching is expected to reduce the in-band phase noise of a fractional-N PLLs due to the excellent phase detector linearity.

Acknowledgements. We would like to thank the Co-Editor Jens Anders, the referee Nils Pohl, and the anonymous referee for their numerous comments and questions. We believe that they helped significantly to improve the quality of our manuscript.

References

Arora, H., Klemmer, N., Morizio, J. C., and Wolf, P. D.: Enhanced phase noise modeling of fractional-N frequency synthesizers, IEEE T. Circuits S-I, 52, 379–395, 2005.

Brokaw, A. P.: A simple three-terminal IC bandgap reference, IEEE J. Solid-St. Circ., 9, 388–393, 1974.

Chien, H.-M., Lin, T.-H., Ibrahim, B., Zhang, L., Rofougaran, M., Rofougaran, A., and Kaiser, W. J.: A 4 GHz fractional-N synthesizer for IEEE 802.11a, in: 2004 Symposium on VLSI Circuits Digest of Technical Papers, Honolulu, USA, 46–49, 2004.

De Muer, B. and Steyaert, M. S. J.: On the analysis of $\Delta\Sigma$ fractional-N frequency synthesizers, IEEE T. Circuits S-II, 50, 784–793, 2003.

Hedayati, H. and Bakkaloglu, B.: A 3 GHz wideband $\Sigma\Delta$ fractional-N synthesizer with voltage-mode exponential CP-PFD, in: Proc. of 2009 IEEE Radio Frequency Integrated Circuits Symposium (RFIC 2009), Boston, USA, June 2009, 325–328, 2009.

Herzel, F., Fischer, G., and Gustat, H.: An integrated CMOS RF synthesizer for 802.11a wireless LAN, IEEE J. Solid-St. Circ., 38, 1767–1770, 2003.

Herzel, F., Osmany, S. A., and Scheytt, J. C.: Analytical phase-noise modeling and charge pump optimization for fractional-N PLLs, IEEE T. Circuits S-I, 57, 1914–1924, 2010.

Lacaita, A., Levantino, S., and Samori, C.: Integrated frequency synthesizers for wireless systems, New York: Cambridge Univ. Press, 2007.

Levantino, S., Marzin, G., Samori, C., and Lacaita, A. L.: A wideband fractional-N PLL with suppressed charge pump noise and automatic loop filter calibration, IEEE J. Solid-St. Circ., 48, 2419–2429, 2013.

Osmany, S. A., Herzel, F., and Scheytt, J. C.: Analysis and minimization of substrate spurs in fractional-N frequency synthesizers, Analog Integr. Circ. S., 74, 545–556, 2013.

Pamarti, S., Jansson, L., and Galton, I: A wideband 2.4-GHz delta-sigma fractional-N PLL with 1-Mb/s in-loop modulation, IEEE J. Solid-St. Circ., 39, 49–62, 2004.

Pohl, N., Jaeschke, T., and Aufinger, K.: An ultra-wideband 80 GHz FMCW radar system using a SiGe bipolar transceiver chip stabilized by a fractional-N PLL synthesizer, IEEE T. Microw. Theory, 60, 757–765, 2012.

Riley, T. A. D., Filiol, N. M., Du, Q., and Kostamovaara, J.: Techniques for in-band phase noise reduction in $\Delta\Sigma$ synthesizers, IEEE T. Circuits S-II, 50, 794–803, 2003.

Rücker, H., Heinemann, B., Winkler, W., Barth, R., Borngraber, J., Drews, J., Fischer, G. G., Fox, A., Grabolla, T., Haak, U., Knoll, D., Korndorfer, F., Mai, A., Marschmeyer, S., Schley, P., Schmidt, D., Schmidt, J., Schubert, M. A., Schulz, K., Tillack, B., Wolansky, D., and Yamamoto, Y.: A 0.13 SiGe BiCMOS technology featuring $f_T/f_{\max}$ of 240/330 GHz and gate delays below 3 ps, IEEE J. Solid-St. Circ., 45, 1678–1686, 2010.

Ussmuller, T., Weigel, R., and Eickhoff, R.: Highly integrated fractional-N synthesizer for locatable wireless sensor nodes, in: 2009 IEEE International Conference on Communications, Dresden, Germany, June 2009, 1–5, 2009.

14

Is model-order reduction viable for the broadband finite-element analysis of electrically large antenna arrays?

O. Floch, A. Sommer, O. Farle, and R. Dyczij-Edlinger

Chair of Electromagnetic Theory, Dept. of Physics and Mechatronics, Saarland University, Saarbrücken, Germany

Correspondence to: O. Floch (o.floch@lte.uni-saarland.de)

Abstract. Model-order reduction provides an efficient way of computing frequency sweeps for finite-element models, because the dimension of the reduced-order system depends on the complexity of the frequency response rather than the size of the original model. For electrically large domains, however, the applicability of such methods is unclear because the system response may be very complicated. This paper provides a numerical study of the effects of bandwidth, electrical size, and scan angle on the size and convergence of the ROM, by considering linear antenna arrays. A mathematical model is proposed and validated against numerical experiments.

1 Introduction

Numerical volume discretization methods, such as finite-elements (FE), provide a highly flexible approach to the solution of electromagnetic boundary value problems. The resulting systems of linear equations are typically sparse but of very large scale. They are commonly solved by direct methods, due to their high robustness. However, both computational complexity and memory requirements of such methods are sub-optimal. Iterative solvers, like Krylov sub-space techniques, provide an attractive alternative which, however, depends critically on the availability of efficient preconditioners.

In the frequency domain, the structure of the underlying indefinite vector Helmholtz equation and the effect of dispersion error have long impeded the development of effective preconditions for domains that are electrically large. Recent advances in domain decomposition (DD), as reported in Lee et al. (2005) and Rawat and Lee (2007), have greatly facilitated the solution of the resulting FE systems, especially in

case of antenna arrays. Nevertheless, such methods just provide solutions at single frequency points.

When broad frequency bands are to be analyzed, multi-point methods of model-order reduction (MOR) are more attractive than conventional FE analysis. These techniques employ well-chosen snapshots of the solution manifold to generate an approximation space whose dimension is low and does not depend on the size of the underlying system. State-of-the-art MOR approaches place expansion points adaptively, based on some a posteriori error indicator Chen et al. (2010), Hesthaven et al. (2012).

The dimension of the reduced-order system (ROM) does, however, depend on the complexity of the system response for the specific excitations under consideration. Therefore, it is unclear whether MOR is suitable for the FE-based broadband analysis of large-scale antenna arrays. On the one hand, these structures are electrically large and exhibit high numbers of independent inputs. Moreover, the individual radiators often exhibit resonant behavior. Thus, the resulting frequency response may be expected to be complicated and require a large ROM to be represented well. On the other hand, the considered structures serve the prime purpose of radiating electromagnetic waves. Thus, even though the array may support a great number of beam forms and a wide range of scan angles, the excitations and fields of the individual elements are always characterized by a large amount of phase coherence and slowly changing amplitude distributions. Such system properties are well-suited for MOR.

The first goal of this paper is to numerically investigate the effects of bandwidth, electrical size, and scan angle on the dimension and convergence properties of the ROM. For this purpose, we consider linear arrays of up to 144 wavelengths λ in size. We restrict ourselves to these essentially one-dimensional structures, because they allow to study large-scale effects at minimum computational costs. Nevertheless, for the largest structure the number of FE unknowns exceeds 22 million. Therefore, all FE computations are performed by a general-purpose DD solver which has successfully been applied to truly three-dimensional structures; see Rawat and Lee (2010). Based on the numerical results, we propose a simple mathematical model based on sampling theory to predict the convergence properties of the ROM. The model is validated by comparing its predictions against the results of additional numerical experiments. In view of the general principles behind their core findings, the authors conjecture that they will transfer well to planar and general array configurations.

2 Problem formulation

2.1 Time-harmonic boundary value problem

The electric field is denoted by E, the magnetic field by H, the surface current density by K, the relative magnetic permeability by μ_r, and the relative electric permittivity by ϵ_r. The abbreviations k_0 and η_0 stand for the free-space wave number and characteristic impedance, respectively.

Given a bounded domain $\Omega \subset \mathbf{R}^3$ with boundary $\partial\Omega = \Gamma_P \cup \Gamma_E \cup \Gamma_A$ and unit outward normal vector $\hat{n}$, we consider the time-harmonic boundary value problem (BVP)

$$\nabla \times \mu_r^{-1} \nabla \times E - k_0^2 \epsilon_r E = 0 \quad \text{in } \Omega, \tag{1a}$$

$$\pi_t(E) = 0, \quad \text{on } \Gamma_E, \tag{1b}$$

$$[\mu_r^{-1} \nabla \times E] = -jk_0\eta_0 K, \text{ on } \Gamma_P, \tag{1c}$$

$$\gamma_t(\mu_r^{-1} \nabla \times E) - jk_0\sqrt{\frac{\epsilon_r}{\mu_r}}\pi_t(E) = 0, \quad \text{on } \Gamma_A. \tag{1d}$$

Here, $\gamma_t(E) = n \times E$ denotes the tangential trace and $\pi_t(E) = \hat{n} \times (E \times \hat{n})$ the tangential component trace mapping; see Buffa and Ciarlet (2001). The jump operator $[E]$ is defined by

$$[E] : [E]_{ij} = \gamma_t(E_i) + \gamma_t(E_j). \tag{2}$$

For clarity of presentation, we only consider two sub-domains in the DD formulation and assume that a single material region is cut by the interface Γ: We decompose Ω into two non-overlapping sub-domains such that $\overline{\Omega} = \overline{\Omega_1} \cup \overline{\Omega_2}$, $\Omega_1 \cap \Omega_2 = \varnothing$, $\Gamma = \overline{\Omega_1} \cap \overline{\Omega_2}$, and denote the interface seen from Ω_1 and Ω_2, respectively, by Γ_{12} and Γ_{21}. For the boundaries of the sub-domains, we use the notation $\Gamma_X^m = \partial\Omega_m \cap \Gamma_X$ with $X \in \{E, P, A\}$ and $m \in \{1, 2\}$. The decom-

posed BVP reads

$$\nabla \times \mu_{r,1}^{-1} \nabla \times E_1 - k_0^2\epsilon_{r,1}E_1 = 0 \quad \text{in } \Omega_1, \tag{3a}$$

$$\nabla \times \mu_{r,2}^{-1} \nabla \times E_2 - k_0^2\epsilon_{r,2}E_2 = 0 \quad \text{in } \Omega_2, \tag{3b}$$

$$\pi_t(E_1) = 0 \quad \text{on } \Gamma_E^1, \tag{3c}$$

$$\pi_t(E_2) = 0 \quad \text{on } \Gamma_E^2, \tag{3d}$$

$$[\mu_{r,1}^{-1} \nabla \times E_1] = -jk_0\eta_0 K_1,$$
$$\text{on } \Gamma_P^1 \tag{3e}$$

$$[\mu_{r,2}^{-1} \nabla \times E_2] = -jk_0\eta_0 K_2,$$
$$\text{on } \Gamma_P^2 \tag{3f}$$

$$\gamma_t(\mu_{r,1}^{-1} \nabla \times E_1) - jk_0\sqrt{\frac{\epsilon_{r,1}}{\mu_{r,1}}}\pi_t(E_1) = 0 \quad \text{on } \Gamma_A^1, \tag{3g}$$

$$\gamma_t(\mu_{r,2}^{-1} \nabla \times E_2) - jk_0\sqrt{\frac{\epsilon_{r,2}}{\mu_{r,2}}}\pi_t(E_2) = 0 \quad \text{on } \Gamma_A^2, \tag{3h}$$

$$\mathcal{T}(E_1) - \mathcal{T}(E_2) = 0 \quad \text{on } \Gamma_{12}, \tag{3i}$$

$$\mathcal{T}(E_2) - \mathcal{T}(E_1) = 0 \quad \text{on } \Gamma_{21}. \tag{3j}$$

Equations 3i and 3j enforce the necessary continuity requirements for the electric and magnetic fields on the interface. We here consider a transmission condition $\mathcal{T}$ with two transverse derivatives of second order,

$$\mathcal{T}(E) = \alpha\pi_t(E) + \beta\nabla_t \times \nabla_t \times \pi_t(E)$$
$$+ \gamma\nabla_t\nabla_t \cdot \gamma_t\left(\mu_r^{-1}\nabla \times E\right), \tag{4}$$

with complex coefficients α, β and γ. It was shown in Rawat and Lee (2010) that equivalence of Eqs. (3) and (1) is guaranteed provided that $\Im(\frac{\beta}{\alpha}) \neq 0$ or $\Re(\frac{\beta}{\alpha}) \geq 0$ and $\Im(\gamma) \neq 0$ or $\Re(\gamma) \geq 0$. See Dolean et al. (2015) for a more detailed discussion.

2.1.1 Finite-element formulation

Let $H^{\mathrm{curl}}(\Omega)$, $H^1(\Omega)$, $H^{\frac{1}{2}}(\Gamma)$, and $H^{\frac{1}{2}}(\Gamma)$ denote the Sobolev spaces from Buffa and Ciarlet (2001). We define the subspace H_E^{curl} by

$$H_E^{\mathrm{curl}}(\Omega) := \{v \in H^{\mathrm{curl}}(\Omega) \mid \pi_t(v) = 0 \text{ on } \Gamma_E\} \tag{5}$$

and denote the topological dual spaces of $H^{\frac{1}{2}}(\Gamma_{mn})$ and $H^{\frac{1}{2}}(\Gamma_{mn})$ by $H^{-\frac{1}{2}}(\Gamma_{mn})$ and $H^{-\frac{1}{2}}(\Gamma_{mn})$, respectively. To implement the DD formulation 3, we introduce auxiliary variables $j_m \in H^{-\frac{1}{2}}(\Gamma_{mn})$ and $p_m \in H^{-\frac{1}{2}}(\Gamma_{mn})$ on the interfaces, by

$$j_m = k_0^{-1}\gamma_t\left(\mu_{r,m}^{-1}\nabla \times E_m\right), \quad m \in \{1, 2\}, \tag{6}$$

$$p_m = \left(k_0^2\right)^{-1}\nabla_t \cdot \gamma_t\left(\mu_{r,m}^{-1}\nabla \times E_m\right), \quad m \in \{1, 2\}. \tag{7}$$

The corresponding FE spaces are denoted by $\mathcal{V}_{m,E}^{\mathrm{curl}} \subset H_E^{\mathrm{curl}}(\Omega_m)$, $\mathcal{V}_{mn}^{\Gamma} \subset H^{-\frac{1}{2}}(\Gamma_{mn})$, and $\mathcal{V}_{mn}^{\Gamma} \subset H^{-\frac{1}{2}}(\Gamma_{mn})$. The

FE spaces for the auxiliary variables j_m and p_m are taken to be the range of the tangential component map $\pi_t(\mathcal{V}_{m,E}^{\mathrm{curl}})$ and the two-dimensional grad-conforming space, respectively. The authors' computer implementation employs basis functions of second order. In the discretization process, we apply a Galerkin procedure to Eqs. (3j) and (3b), substitute Eq. (6) for the resulting surface terms, and test the interface condition as well as the definition Eq. (7) by shape functions $u \in \mathcal{V}_{mn}^{\Gamma}$ and $\phi \in \mathcal{V}_{mn}^{\Gamma}$, respectively. The resulting FE-DD system reads

$$\begin{bmatrix} \mathbf{A}_1(k_0,\xi) & \mathbf{C}_{12}(k_0,\xi) \\ \mathbf{C}_{21}(k_0,\xi) & \mathbf{A}_2(k_0,\xi) \end{bmatrix} \begin{bmatrix} x_1(k_0) \\ x_2(k_0) \end{bmatrix} = jk_0\eta_0 \begin{bmatrix} b_1(k_0) \\ b_2(k_0) \end{bmatrix}, \quad (8)$$

wherein $\xi = (\alpha, \beta, \gamma)$, $\mathbf{A}_m \in \mathbf{C}^{N_m \times N_m}$, $\mathbf{C}_{mn} \in \mathbf{C}^{N_m \times N_n}$, and $b_m \in \mathbf{C}^{N_m}$, with $m,n \in \{1,2\}$. The vectors $x_m = e_m$, ${j_m^\Gamma}$, ${p_m^\Gamma}^T$ are of dimension N_m and include FE degrees of freedom for both the electric field E_m and the auxiliary variables j_m and p_m inside the sub-domain Ω_m. The corresponding block matrices are given by

$$\mathbf{A}_m = \begin{bmatrix} \mathbf{A}_m^{e,e} - k_0^2\mathbf{B}_m^{e,e} + k_0\mathbf{R}_m^{e,e} & k_0\mathbf{T}_m^{e,j} & 0 \\ k_0\left(\mathbf{T}_m^{e,j}\right)^T + \frac{k_0\beta}{\alpha}\mathbf{S}_m^{j,e} & \frac{k_0^2}{\alpha}\mathbf{T}_m^{j,j} & \frac{k_0^3\gamma}{\alpha}\mathbf{D}_m^{j,p} \\ 0 & \mathbf{F}_m^{p,j} & k_0\mathbf{T}_m^{p,p} \end{bmatrix}, \quad (9)$$

$$\mathbf{C}_{mn} = \begin{bmatrix} 0 & 0 & 0 \\ -k_0\left(\mathbf{T}_{mn}^{e,j}\right)^T - \frac{k_0\beta}{\alpha}\mathbf{S}_{mn}^{j,e} & \frac{k_0^2}{\alpha}\mathbf{T}_{mn}^{j,j} & \frac{k_0^3\gamma}{\alpha}\mathbf{D}_{mn}^{j,p} \\ 0 & 0 & 0 \end{bmatrix}, \quad (10)$$

$$b_m = \begin{bmatrix} y_m \\ 0 \\ 0 \end{bmatrix}, \quad (11)$$

wherein

$$\left[\mathbf{A}_m^{e,e}\right]_{ij} = \left(\nabla \times v_i, \mu_{r,j}^{-1}\nabla \times v_j\right)_{\Omega_m}, \quad (12a)$$

$$\left[\mathbf{B}_m^{e,e}\right]_{ij} = \left(v_i, \epsilon_{r,j}v_j\right)_{\Omega_m}, \quad (12b)$$

$$\left[\mathbf{R}_m^{e,j}\right]_{ij} = j\langle\pi_t(v_i), \pi_t(v_j)\rangle_{\Gamma_A^m}, \quad (12c)$$

$$\left[\mathbf{T}_{mn}^{e,j}\right]_{ij} = \langle\pi_t(v_i), u_j\rangle_{\Gamma^{mn}}, \quad (12d)$$

$$\left[\mathbf{T}_{mn}^{j,j}\right]_{ij} = \langle u_i, u_j\rangle_{\Gamma^{mn}}, \quad (12e)$$

$$\left[\mathbf{S}_{mn}^{j,e}\right]_{ij} = \langle\nabla \times u_i, \nabla \times \pi_t(v_j)\rangle_{\Gamma^{mn}}, \quad (12f)$$

$$\left[\mathbf{D}_{mm}^{j,p}\right]_{ij} = \langle u_i, \nabla\phi_j\rangle_{\Gamma^{mn}}, \quad (12g)$$

$$\left[\mathbf{T}_{mm}^{p,p}\right]_{ij} = \langle\phi_i, \phi_j\rangle_{\Gamma^{mn}}, \quad (12h)$$

$$\left[\mathbf{F}_{mm}^{p,j}\right]_{ij} = \langle\nabla\phi_i, u_j\rangle_{\Gamma^{mn}}, \quad (12i)$$

$$\left[y_m\right]_i = \langle\pi_t(v_i), \gamma_t(H)\rangle_{\Gamma^H}, \quad (12j)$$

with $v_i, v_j \in \mathcal{V}_{m,E}^{\mathrm{curl}}$, and $u_i, u_j \in \mathcal{V}_{mn}^{\Gamma}$ and $\phi_i, \phi_j \in \mathcal{V}_{mn}^{\Gamma}$. The volume $(\ldots)$ and surface $\langle\ldots\rangle$ bilinear forms are given by

$$(u, v)_\Omega = \int_\Omega u \cdot v\,d\Omega, \quad (13a)$$

$$\langle u, v\rangle_\Gamma = \int_\Gamma u \cdot v\,d\Gamma, \quad (13b)$$

$$\langle\phi, \psi\rangle_\Gamma = \int_\Gamma \phi\psi\,d\Gamma. \quad (13c)$$

3 Reduced-order model

As a prerequisite for projection-based MOR, the system Eq. 8 has to be rewritten in affinely decomposed form with respect to the wavenumber k_0. This is achieved by defining the following wavenumber-independent matrices:

$$\mathbf{A}_m^0 = \begin{bmatrix} \mathbf{A}_m^{e,e} & 0 & 0 \\ 0 & 0 & 0 \\ 0 & \mathbf{F}_m^{p,j} & 0 \end{bmatrix}, \quad \mathbf{A}_m^1 = \begin{bmatrix} \mathbf{R}_m^{e,e} & \mathbf{T}_m^{e,j} & 0 \\ \left(\mathbf{T}_m^{e,j}\right)^T & 0 & 0 \\ 0 & 0 & \mathbf{T}_m^{p,p} \end{bmatrix},$$

$$\mathbf{A}_m^2 = \begin{bmatrix} -\mathbf{B}_m^{e,e} & 0 & 0 \\ 0 & 0 & 0 \\ 0 & 0 & 0 \end{bmatrix}, \quad \mathbf{A}_m^3 = \begin{bmatrix} 0 & 0 & 0 \\ \mathbf{S}_m^{j,e} & 0 & 0 \\ 0 & 0 & 0 \end{bmatrix},$$

$$\mathbf{A}_m^4 = \begin{bmatrix} 0 & 0 & 0 \\ 0 & \mathbf{T}_m^{j,j} & 0 \\ 0 & 0 & 0 \end{bmatrix}, \quad \mathbf{A}_m^5 = \begin{bmatrix} 0 & 0 & 0 \\ 0 & 0 & \mathbf{D}_m^{j,p} \\ 0 & 0 & 0 \end{bmatrix},$$

$$(14)$$

and

$$\mathbf{C}_{mn}^0 = \mathbf{0}, \qquad \mathbf{C}_{mn}^1 = \begin{bmatrix} \mathbf{0} & \mathbf{0} & \mathbf{0} \\ -\left(\mathbf{T}_{mn}^{e,j}\right)^T & \mathbf{0} & \mathbf{0} \\ \mathbf{0} & \mathbf{0} & \mathbf{0} \end{bmatrix},$$

$$\mathbf{C}_{mn}^2 = \mathbf{0}, \qquad \mathbf{C}_{mn}^3 = \begin{bmatrix} \mathbf{0} & \mathbf{0} & \mathbf{0} \\ -\mathbf{S}_{mn}^{j,e} & \mathbf{0} & \mathbf{0} \\ \mathbf{0} & \mathbf{0} & \mathbf{0} \end{bmatrix},$$

$$\mathbf{C}_{mn}^4 = \begin{bmatrix} \mathbf{0} & \mathbf{0} & \mathbf{0} \\ \mathbf{0} & \mathbf{T}_{mn}^{j,j} & \mathbf{0} \\ \mathbf{0} & \mathbf{0} & \mathbf{0} \end{bmatrix}, \quad \mathbf{C}_{mn}^5 = \begin{bmatrix} \mathbf{0} & \mathbf{0} & \mathbf{0} \\ \mathbf{0} & \mathbf{0} & \mathbf{D}_{mn}^{j,p} \\ \mathbf{0} & \mathbf{0} & \mathbf{0} \end{bmatrix},$$

$$\text{(15)}$$

with $m, n \in \{1, 2\}$. Thus 8 takes the form

$$\left(\sum_{i=1}^I \phi_i(k_0)\hat{\mathbf{A}}_i\right)\hat{\mathbf{x}}(k_0) = \left(\sum_{j=1}^J \theta_j(k_0)\hat{\boldsymbol{b}}_j\right), \qquad \text{(16a)}$$

$$\boldsymbol{y}(k_0) = \left(\sum_{j=1}^J \eta_j(k_0)\hat{\boldsymbol{b}}_j^T\right)\hat{\mathbf{x}}(k_0), \qquad \text{(16b)}$$

with wavenumber-dependent functions $\phi_i, \theta_j, \eta_j : \mathbf{R} \to \mathbf{C}$ defined by

$$\phi_0(k_0) = 1, \qquad \phi_1(k_0) = k_0, \qquad \phi_2(k_0) = k_0^2, \qquad \text{(17)}$$

$$\phi_3(k_0) = k_0\frac{\beta}{\alpha}, \qquad \phi_4(k_0) = k_0^2\frac{1}{\alpha}, \qquad \phi_5(k_0) = k_0^3\frac{\gamma}{\alpha}, \qquad \text{(18)}$$

$$\theta_j = jk_0\eta_0, \qquad \eta_j = 1, \qquad \text{(19)}$$

and block matrices $\hat{\mathbf{A}}_i$ and block vectors $\hat{\mathbf{x}}(k_0), \hat{\boldsymbol{b}}_i$ given by

$$\hat{\mathbf{A}}_i = \begin{bmatrix} \mathbf{A}_1^i & \mathbf{C}_{12}^i \\ \mathbf{C}_{21}^i & \mathbf{A}_2^i \end{bmatrix} \in \mathbf{C}^{(N_1+N_2)\times(N_1+N_2)}, \qquad \text{(20a)}$$

$$\hat{\mathbf{x}}(k_0) = \begin{bmatrix} \boldsymbol{x}_1(k_0) \\ \boldsymbol{x}_2(k_0) \end{bmatrix} \in \mathbf{C}^{(N_1+N_2)}, \qquad \text{(20b)}$$

$$\hat{\boldsymbol{b}}_i = \begin{bmatrix} \boldsymbol{b}_1^i \\ \boldsymbol{b}_2^i \end{bmatrix} \in \mathbf{C}^{(N_1+N_2)}. \qquad \text{(20c)}$$

A multi-point reduced-order model (ROM) is built from the FE solutions $\boldsymbol{x}(k_0^i)$ of Eq. 16 at M expansion points $k_0^1, \ldots, k_0^M$, with $M \ll N_1 + N_2$, which are selected adaptively; see Sect. 3.1. The first step is to compute a unitary projection matrix $\mathbf{V} \in \mathbf{C}^{N_1+N_2,M}$ with

$$\text{range}(\mathbf{V}_M) = \text{span}\{\boldsymbol{x}(k_0^i)\}, i = 1 \ldots M. \qquad \text{(21)}$$

Then, Galerkin projection leads to a ROM of the form

$$\left(\sum_{i=1}^I \phi_i(k_0)\widetilde{\mathbf{A}}_i\right)\widetilde{\boldsymbol{x}}(k_0) = \left(\sum_{j=1}^J \theta_j(k_0)\widetilde{\boldsymbol{b}}_j\right), \qquad \text{(22a)}$$

$$\widetilde{\boldsymbol{y}}(k_0) = \left(\sum_{j=1}^J \eta_j(k_0)\widetilde{\boldsymbol{b}}_j\right)\widetilde{\boldsymbol{x}}(k_0) \qquad \text{(22b)}$$

wherein the reduced matrices and vectors are defined as

$$\widetilde{\mathbf{A}}_i = \mathbf{V}_M^*\hat{\mathbf{A}}_i\mathbf{V}_M \in \mathbf{C}^{M\times M}, \qquad \text{(23)}$$

$$\widetilde{\boldsymbol{b}}_i = \mathbf{V}_M^*\hat{\boldsymbol{b}}_i \in \mathbf{C}^M. \qquad \text{(24)}$$

Thus, ROM construction requires the solution of the large-scale FE-DD system Eq. 16 at each expansion point. Since

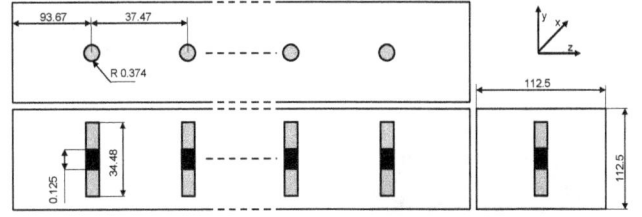

Figure 1. FE model of linear array of dipoles. Black rectangles indicate lumped ports. Dimensions are in mm.

Eq. 16 is solved iteratively, ROM construction time is determined not only by the ROM dimension M but also by the convergence behavior of the linear solver. To improve solver performance, we use an adaptive two-level preconditioner based on the reduced-order system already available at a given adaptive step.

3.1 Adaptivity

As demonstrated in Patera and Rozza (2006–2007), the 2-norm of the residual $\boldsymbol{r}(k_0)$ of Eq. 16a according to the ROM solution $\boldsymbol{x}(k_0)$,

$$\boldsymbol{r}(k_0) = \left(\sum_{j=1}^J \theta_j(k_0)\hat{\boldsymbol{b}}_j\right) - \left(\sum_{i=1}^I \phi_i(k_0)\hat{\mathbf{A}}_i\right)\mathbf{V}_M\,\boldsymbol{x}(k_0), \qquad \text{(25)}$$

can be evaluated very efficiently, using reduced-order quantities only. It was suggested in de la Rubia et al. (2009) to employ the relative residual norm ρ,

$$\rho(k_0) = \frac{\|\boldsymbol{r}(k_0)\|_2}{\left\|\sum_{j=1}^J \theta_j(k_0)\hat{\boldsymbol{b}}_j\right\|_2}, \qquad \text{(26)}$$

as an inexpensive error indicator for guiding the placement of the expansion points k_0^i: At a given stage of ROM generation, ρ is evaluated for a dense sampling $\{k_0^s\}$ of the wavenumber interval under investigation, and the following expansion point is chosen where the relative residual is the largest. ROM generation terminates when ρ has fallen below a user-defined threshold ρ_0 for all k_0^s.

4 Numerical experiments

To simplify mathematical analysis and minimize the costs of numerical tests, we consider a structure that is structurally simple and electrically large in a single direction. Specifically, we choose a linear array of N_A equally spaced, perfectly conducting dipoles in free space. The geometry of the FE domain is shown in Fig. 1. On the outer surface, absorbing boundary conditions are applied. The dipoles are excited at their centers by lumped ports of impedance $Z_P = 73\ \Omega$. To steer the array pattern towards a given angle θ_s, the incident waves at the ports a_n are taken of unit magnitude and

(a) Broadside array; $\phi = 0$ rad.   (b) Broadside array; $\theta = \frac{\pi}{2}$ rad.

(c) End-fire array; $\phi = 0$ rad.   (d) End-fire array; $\phi = \frac{\pi}{2}$ rad.

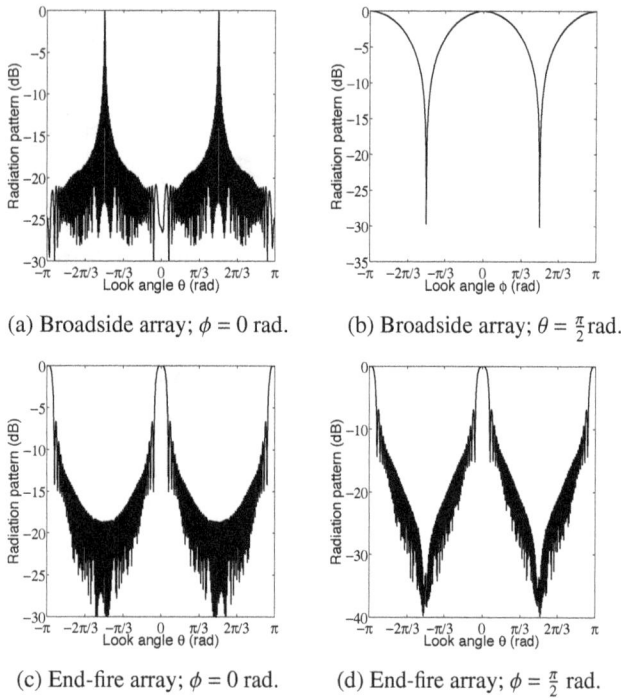

Figure 2. Normalized radiation patterns of $N_A = 192$ dipoles at 4 GHz, where radiators are one-half wavelength apart.

linear phase distribution,

$$a_n = e^{-jk_0 z_n \cos\theta_s}, \tag{27}$$

where z_n is the z coordinate of the dipole axis. Specifically, the limiting cases of broadside ($\theta_s = \frac{\pi}{2}$ rad) and end-fire ($\theta_s = 0$ rad) arrays are considered.

The FE-DD system Eq. (16) is solved by the restarted GMRES(30) iterative method with stopping criterion $\delta = 10^{-6}$; see Saad and Schultz (1986). The ROM termination criterion is $\max\rho(k_0) = 10^{-6}$. Preliminary studies on smaller arrays using standard FE techniques rather than DD have confirmed that the behavior of the ROM is mainly determined by the antenna configuration and almost independent of the FE formulation. Thus no additional data are given.

4.1 Varying number of radiators N_A

Let the center frequency $f_c = 4$ GHz and bandwidth $B = 2$ GHz be given. We investigate the convergence behavior of the adaptive ROM by varying the number of dipoles in the range $N_A = 12\ldots192$ and, correspondingly, the electric size of the array in the range $(6\ldots96)$ λ at 4 GHz.

Figure 2 presents radiation patterns derived from the ROM, for $N_A = 192$. The qualitative behavior is as expected: Since the dipoles are one-half wavelength apart at the chosen frequency, $f = 4$ GHz, the radiation patterns of not only the broadside but also the end-fire array are quarter-symmetric.

Figure 3 shows the convergence behavior of the ROM for different numbers of dipoles. The difference between Fig. 3a

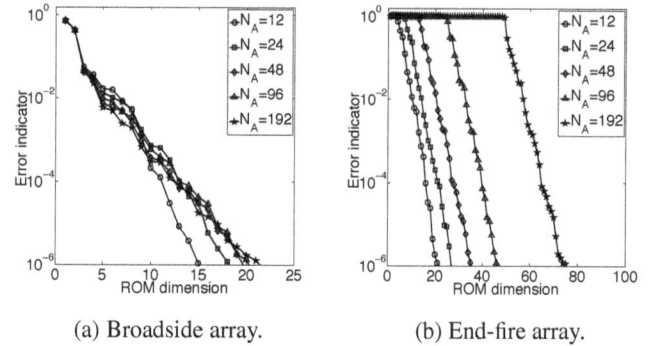

(a) Broadside array.   (b) End-fire array.

Figure 3. Error indicator $\max\rho$ vs. ROM dimension for different dipole counts N_A. Parameters: $f_c = 4$ GHz, $B = 2$ GHz.

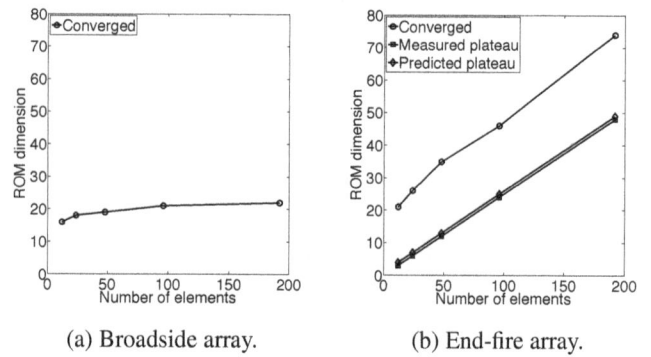

(a) Broadside array.   (b) End-fire array.

Figure 4. ROM dimension versus number of dipoles N_A. Parameters: $f_c = 4$ GHz, $B = 2$ GHz.

and b is striking: the broadside array does not exhibit any pre-asymptotic region and leads to exponential convergence rates right from the start whereas the end-fire configuration leads to a pronounced plateau, followed by exponential convergence. In both cases, convergence rates in the asymptotic region deteriorate slowly with increasing number of dipoles. In consequence, ROM dimension remains almost constant for the broadside array, as illustrated by Fig. 4a. In the end-fire case, in contrast, both plateau width and ROM dimension increase linearly with the number of dipoles; see Fig. 4b. For completeness, the dimensions of the underlying FE-DD systems and corresponding ROMs for different numbers of dipoles are collected in Table 1.

4.2 Varying bandwidth B

For given center frequency $f_c = 4$ GHz and number of dipoles $N_A = 48$, we construct ROMs of varying bandwidth $B = (0.5\ldots4)$ GHz.

Figure 5 illustrates the frequency response of the active reflection coefficient for one dipole at the center and one at the end of the array. It can be seen that broadside and end-fire arrays behave similarly: in both cases, the detuning effects of the periodic environment on the center dipole are clearly visible.

Table 1. Computational data for $f_c = 4\,\text{GHz}$, $B = 2\,\text{GHz}$.

Number of Radiators	Dimension		
	FE-DD	ROM	
	–	End-fire	Broadside
12	$1.8 \cdot 10^6$	21	16
24	$3.2 \cdot 10^6$	27	18
48	$5.9 \cdot 10^6$	35	19
96	$11.4 \cdot 10^6$	46	21
192	$22.5 \cdot 10^6$	75	22

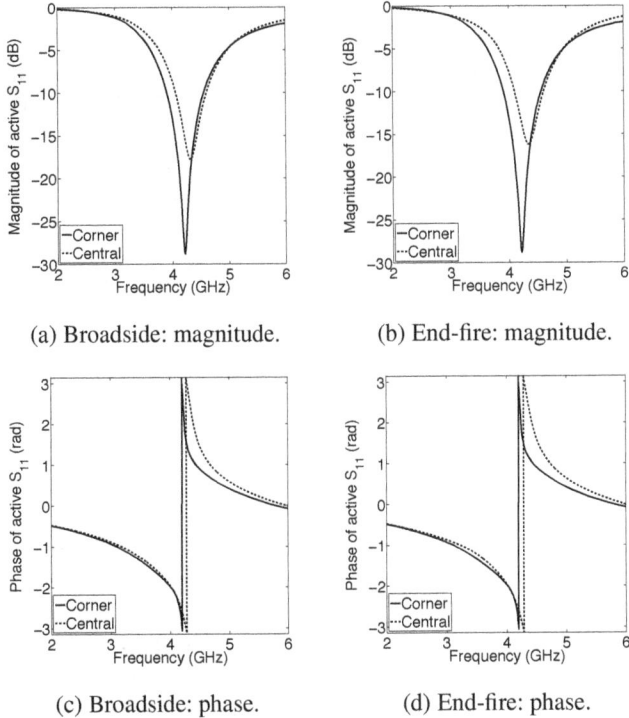

(a) Broadside array.   (b) End-fire array.

Figure 6. Error indicator $\max \rho$ versus ROM dimension for different bandwidths B. Parameters: $N_A = 48$, $f_c = 4\,\text{GHz}$.

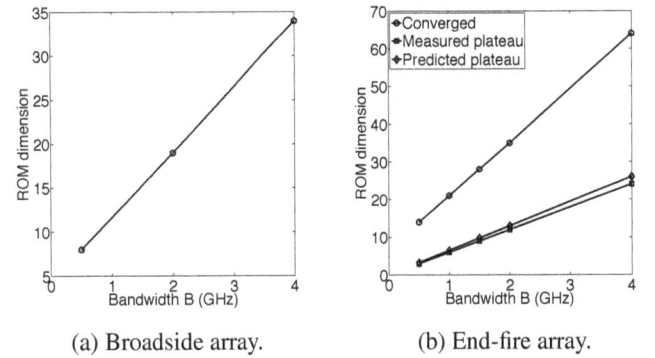

(a) Broadside array.   (b) End-fire array.

Figure 7. ROM dimension versus bandwidth B. Parameters: $N_A = 48$, $f_c = 4\,\text{GHz}$.

(a) Broadside: magnitude.   (b) End-fire: magnitude.

(c) Broadside: phase.   (d) End-fire: phase.

Figure 5. Frequency response of active reflection coefficient for selected dipoles. Array size: $N_A = 48$ elements.

Figure 6 presents the convergence behavior of the adaptive ROM for different choices of bandwidth: the broadband array leads to a very short pre-asymptotic region at most, whereas the end-fire configuration exhibits a pronounced plateau. In the asymptotic region, both cases result in exponential convergence which deteriorates with increasing bandwidth. Figure 7 illustrates the behavior of the ROM dimension as a function of bandwidth for both broadside and end-fire configurations.

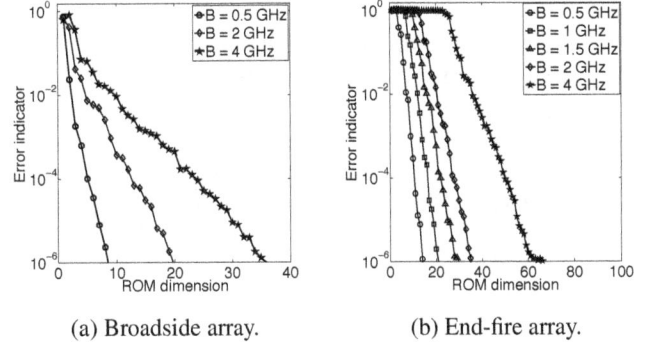

5 Proposed explanation of convergence properties

A first hint at the origin of the peculiar convergence behavior is given by the distributions of the electric near-fields on the side face of the FE domain plotted in Fig. 8: Fig. 8a and b illustrate that, for the broadside array, the field distribution does not change much over the frequency range (2–4) GHz, because the excitations are in phase, and the domain is electrically small perpendicular to the z direction. In case of the end-fire configuration, however, the fields vary strongly with frequency, in accordance with Eq. (27); see Fig. 8c and d.

Changes in spatial distribution are crucial because the projection-based MOR of Sect. 3 is an approximation method that employs the electric field at the expansion wavenumbers, corresponding to the FE solutions $x(k_0^i)$, as basis functions in the spatial domain.

As long as the structure is electrically large in the z direction only, and provided that the electric behavior of a single radiator does not change fundamentally over the considered frequency range, one may decompose the fields into localized contributions that do not vary much with frequency and wave-like components $w(z, k_0)$ in z direction, in accordance with the steering angle θ_s:

$$w(z, k_0) = e^{-jk_0 z \cos\theta_s} = e^{-j2\pi f \frac{z}{c_0} \cos\theta_s}. \tag{28}$$

Herein, c_0 is the vacuum speed of light. It is therefore a necessary requirement for convergence of the ROM that the projection space of Eq. (21) must resolve all $w(z, k_0)$ that get excited by Eq. (27) over the considered frequency range. The authors suppose that it is the non-fulfillment of such condition that causes the plateau in the convergence behavior. Our mathematical model is as follows: we express the frequency as

$$f = f_c + \Delta f \qquad (29)$$

with center frequency f_c and the equivalent baseband frequency Δf. By introducing the spatial frequency

$$f_z = \Delta f \frac{\cos\theta_s}{c_0}, \qquad (30)$$

the waves (Eq. 28) take the form

$$w(z, f_z) = e^{-j2\pi f_c \frac{z}{c_0} \cos\theta_s} e^{-j2\pi f_z z} \qquad (31)$$

with

$$f_z \in \left[-\frac{B\cos\theta_s}{2c_0}, \frac{B\cos\theta_s}{2c_0} \right]. \qquad (32)$$

We next assume an equidistant spatial sampling in z direction and apply the discrete Fourier transformation. According to Oppenheim and Schafer (1989, p. 698), the length L of the field domain leads to a spatial frequency resolution of

$$\delta f_z = \frac{1}{L}. \qquad (33)$$

Thus, the minimum number of frequencies M_{min} required to sample the interval (Eq. 32) is given by

$$M_{min} = \frac{BL\cos\theta_s}{c_0}. \qquad (34)$$

5.1 Varying number of radiators

Equation (34) suggests that the required number of expansion frequencies M_{min} depends on the length L of the computational domain weighted by the cosine of the steering angle, $L\cos\theta_s$, rather than the number of radiators. However, it is common use to embed the array in an air buffer of constant size: In the examples of Sect. 4.1, L was always taken as

$$L = (N_A + 4)d, \qquad (35)$$

where d is the distance between radiators. Hence

$$M_{min} = \frac{B(N_A + 4)d \cos\theta_s}{c_0}. \qquad (36)$$

Equation 36 predicts that plateaus do not occur in the case of broadside arrays, $\cos\theta_s = 0$, whatever the number of radiators N_a. This agrees with the experimental findings of Fig. 3a. For the end-fire case, $\cos\theta_s = 1$, the results of Eq. (36) have been included in Fig. 6a. It can be seen that the theoretical results are slightly conservative. However, they track the actual width of the plateau very well, over the entire range of 12–192 dipoles.

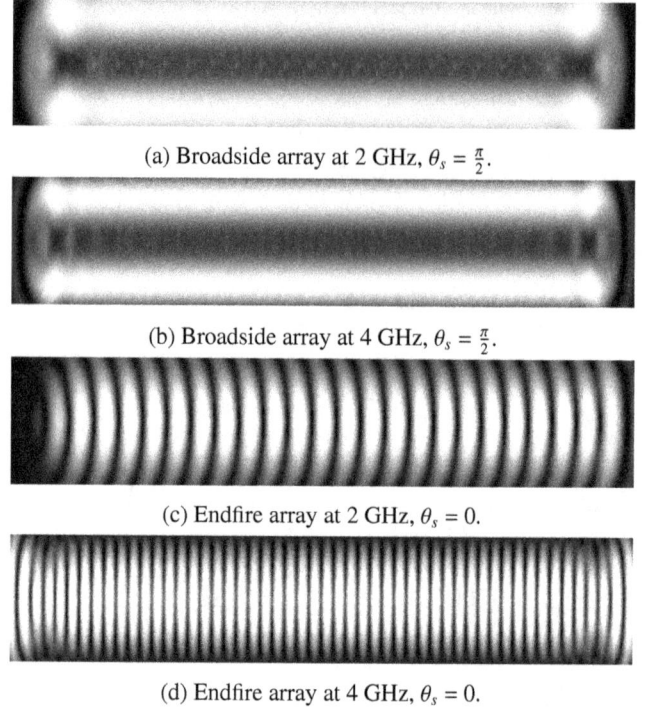

(a) Broadside array at 2 GHz, $\theta_s = \frac{\pi}{2}$.

(b) Broadside array at 4 GHz, $\theta_s = \frac{\pi}{2}$.

(c) Endfire array at 2 GHz, $\theta_s = 0$.

(d) Endfire array at 4 GHz, $\theta_s = 0$.

Figure 8. Magnitude of electric field on outer boundary of FE domain. Array size: $N_A = 48$ dipoles.

5.2 Varying bandwidth

Again, Eq. (36) predicts no plateau for broadside arrays, which agrees well with the experimental findings of Fig. 6a. The authors conjecture that the short pre-asymptotic region of 2 iterations present in Fig. 6a for $B = 4$ GHz is caused by the fact that, at the highest operating frequency, the transversal dimensions of the FE model correspond to 2.25λ which is no longer electrically small. For the the end-fire case, $\cos\theta_s = 1$, the theoretical results of Eq. (36) have been included in Fig. 7b. Again, they are somewhat conservative but track the actual width of the plateau very well, for all considered bandwidths.

6 Further validations

6.1 Effectiveness of equidistant expansion frequencies

The proposed model predicts that adaptive MOR construction based on some error indicator is not necessary in the pre-asymptotic region. Instead, it will suffice to take M_{min} frequencies spaced equally over B.

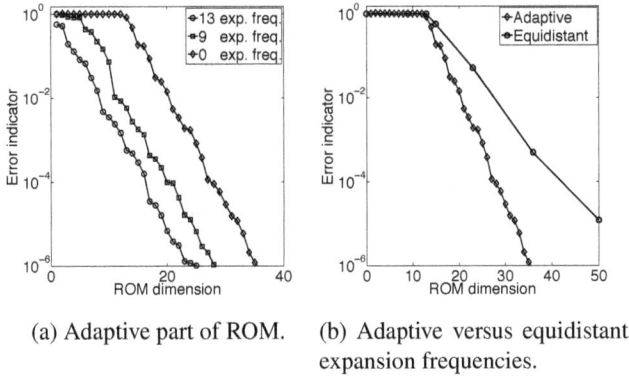

(a) Adaptive part of ROM.

(b) Adaptive versus equidistant expansion frequencies.

Figure 9. End-fire array ($\theta_s = 0$) of $N_A = 48$ dipoles, with $f_c = 4\,\text{GHz}$ and $B = 2\,\text{GHz}$: Error indicator $\rho(k_0)$ versus ROM dimension. Parameter: number of pre-computed equidistant expansion frequencies.

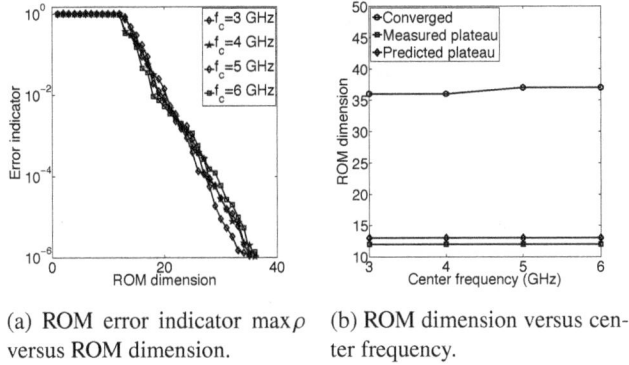

(a) ROM error indicator $\max\rho$ versus ROM dimension.

(b) ROM dimension versus center frequency.

Figure 10. End-fire array ($\theta_s = 0$) of $N_A = 48$ dipoles: dependence of adaptive ROM on center frequency f_c for constant bandwidth of $B = 2\,\text{GHz}$.

To test this hypothesis, we consider an end-fire array of 48 dipoles, center frequency 4 GHz, and bandwidth 4 GHz. According to Eq. (36), 13 equidistant expansion frequencies are required to prevent a plateau from occurring. We initialize the ROM basis with the solutions at n_{eq} equidistant frequency points before starting the adaptive MOR process. Figure 9a shows convergence curves for different choices of n_{eq}. For $n_{\text{eq}} < M_{\text{min}}$, a plateau occurs, and its width is approximately $N_{\text{min}} - n_{\text{eq}}$. At $n_{\text{eq}} = M_{\text{min}}$, the onset of exponential convergence is immediate, which confirms that the corresponding ROM space contains all critical information.

To demonstrate the advantages of adaptive MOR in the asymptotic region, Fig. 9b gives a comparison to a series of ROMs that always use equidistant expansion points: the convergence rate of the adaptive method is significantly higher.

6.2 Varying center frequency

The proposed model also implies that the width of the plateau will be independent of the center frequency f_c.

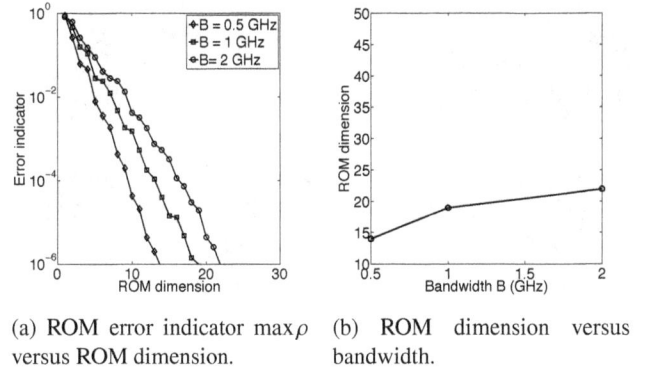

(a) ROM error indicator $\max\rho$ versus ROM dimension.

(b) ROM dimension versus bandwidth.

Figure 11. Dependence of adaptive ROM on bandwidth B for single excited radiator at center of 48 dipoles; $f_c = 4\,\text{GHz}$.

In our numerical test, we investigate an end-fire array of 48 dipoles and bandwidth 2 GHz. Equation 36 predicts a constant plateau width of 13 iterations. Figure 10a gives convergence curves for different center frequencies in the range 3–6 GHz, and Fig. 10a presents the corresponding ROM dimensions. It can be seen that the plateau width is constant, at 12 iterations.

6.3 Spatial distribution of excitations

Whenever the levels of sidelobes are of importance, excitations of tapered amplitude are employed. In the examples above, however, it was assumed that the excitations of the dipoles were of constant magnitude and linear phase shift. The authors believe that, despite its simplicity, this wave form is well chosen because it strongly excites wave-like fields along the z direction and allows to vary the spatial periodicity in the z direction from infinity to the actual wavelength at the highest operating frequency; see Eq. (28).

To complement our numerical tests, we investigate the frequency response as a function of bandwidth for a spatially impulse-like excitation, a single driven dipole at the center of an array of 48 dipoles, with $f_c = 4\,\text{GHz}$. Figure 11 illustrates the convergence behavior for different bandwidths in the range 0.5–2 GHz: there are no plateaus, and iteration counts are moderate. The nearfield plots of Fig. 12 suggest the following qualitative explanation: the dipoles adjacent to the driven element direct the fields in the z direction. However, since their ports provide nearly matched terminations, they also absorb a great amount of energy. In consequence, the fields are strongly damped, so that the amplitudes of many wave-like modes along the z direction become very weak. Since the MOR process is driven by the system response, such modes are disregarded.

(a) Frequency: 2 GHz.

(b) Frequency: 4 GHz.

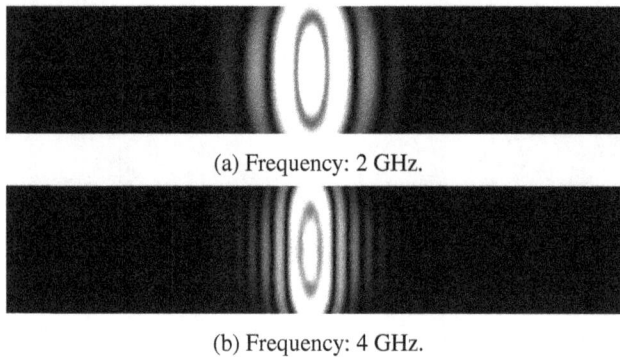

Figure 12. Magnitude of electric field on boundary of FE domain for single excited radiator at center of 48 dipoles.

7 Conclusions and outlook

The numerical studies presented in this paper have shown that projection-based MOR converges even for linear arrays of large electrical size (144λ at $6\,\mathrm{GHz}$). Depending on excitation, the convergence curves may exhibit a pronounced pre-asymptotic region which depends on steering angle, electric length of the model domain, and bandwidth; the worst case occurs in the end-fire direction, $|\theta_\mathrm{s}| = \frac{\pi}{2}$ rad.

The proposed mathematical model has proved to predict the width of the plateau very well. It also allows to replace the adaptive, error-driven selection of MOR expansion frequencies, which is intrinsically sequential, in the pre-asymptotic region by a well-defined number of equally spaced points. Not only does this remove the computational burden of error estimation; it is also perfectly suited for parallelization. Note that, in practice, the impact of the plateau on overall performance may be much higher than in the examples given in this paper: The MOR termination criterion of 10^{-6} was chosen very low to demonstrate numerical robustness and lack of stagnation. In many real-world applications, residual norms in the order of $10^{-2}\ldots10^{-3}$ will suffice. Thus the asymptotic region will be much shorter whereas the plateau width will stay the same.

In the case of planar arrays, the authors conjecture from Eq. (34) that the plateau width will be in the order of $(BL|\cos\theta_\mathrm{s}|/c_0)^2$. More detailed studies are the subject of ongoing research.

References

Buffa, A. and Ciarlet, P.: On traces for functional spaces related to Maxwell's equations Part I: An integration by parts formula in Lipschitz polyhedra, Mathemat. Method. Appl. Sci., 24, 9–30, 2001.

Chen, Y., Hesthaven, J., Maday, Y., and Rodríguez, J.: Certified Reduced Basis Methods and Output Bounds for the Harmonic Maxwell's Equations, SIAM J. Scientif. Comput., 32, 970–996, 2010.

de la Rubia, V., Razafison, U., and Maday, Y.: Reliable Fast Frequency Sweep for Microwave Devices via the Reduced-Basis Method, IEEE Transactions on Microwave Theory and Techniques, 57, 2923–2937, 2009.

Dolean, V., Gander, M. J., Lanteri, S., Lee, J.-F., and Peng, Z.: Effective transmission conditions for domain decomposition methods applied to the time-harmonic curl-curl Maxwell's equations, J. Computat. Phys., 280, 232–47, 2015.

Hesthaven, J. S., Stamm, B., and Zhang, S.: Certified Reduced Basis Method for the Electric Field Integral Equation, SIAM J. Scientif. Comput., 34, A1777–A1799, 2012.

Lee, S.-C., Vouvakis, M. N., and Lee, J.-F.: A non-overlapping domain decomposition method with non-matching grids for modeling large finite antenna arrays, J. Computat. Phys., 203, 1–21, 2005.

Oppenheim, A. and Schafer, R.: Discrete-Time Signal Processing, Prentice Hall, international Edn., 1989.

Patera, A. T. and Rozza, G.: Reduced Basis Approximation and A Posteriori Error Estimation for Parametrized Partial Differential Equations, to appear in (tentative rubric) MIT Pappalardo Graduate Monographs in Mechanical Engineering, Version 1.0, 2006–2007.

Rawat, V. and Lee, J.: Nonoverlapping Domain Decomposition with Second Order Transmission Condition for the Time-Harmonic Maxwell's Equations, SIAM J. Scient. Comput., 32, 3584–3603, 2010.

Rawat, V. and Lee, J.-F.: Treatment of cement variables in the domain decomposition method for Maxwell's equations, in: Antennas and Propagation Society International Symposium, 2007 IEEE, 5937–5940, 2007.

Saad, Y. and Schultz, M.: GMRES: A Generalized Minimal Residual Algorithm for Solving Nonsymmetric Linear Systems, SIAM J. Scient. Stat. Comput., 7, 856–869, 1986.

15

An active UHF RFID localization system for fawn saving

M. Eberhardt, M. Lehner, A. Ascher, M. Allwang, and E. M. Biebl

Fachgebiet Höchstfrequenztechnik, Technische Universität München, Munich, Germany

Correspondence to: M. Eberhardt (michael.eberhardt@tum.de)

Abstract. We present a localization concept for active UHF RFID transponders which enables mowing machine drivers to detect and localize marked fawns. The whole system design and experimental results with transponders located near the ground in random orientations in a meadow area are shown. The communication flow between reader and transponders is realized as a dynamic master-slave concept. Multiple marked fawns will be localized by processing detected transponders sequentially. With an eight-channel-receiver with integrated calibration method one can estimate the direction-of-arrival by measuring the phases of the transponder signals up to a range of 50 m in all directions. For further troubleshooting array manifolds have been measured. An additional hand-held receiver with a two-channel receiver allows a guided approaching search without endangering the fawn by the mowing machine.

1 Introduction

In the spring, when the vegetation increases, the high season of forage crop for farmers begins. The time pressure is high and the technology enables very large mowing performance. At the same time it is the main period for fawns to be born (Rieck, 1955). The does look for calm and safe places and prefer meadows nearby a border of the wood. The high grass affords an excellent protection against natural threats. In addition, the fawns have a native instinct in the first weeks of their lives (Jarnemo, 2002). They press down flat on the ground if any danger is approaching and keep calm even when a mowing-machine runs over them. Many fawns are killed every year in this way (Jarnemo, 2002).

In Germany, the farmers are obliged by law to avoid injuring animals. To date, the fields will be searched on foot for fawns, which means a considerable expense. Many researches have been made for a fawn detection technology.

Developed systems in the recent years based on infrared are shown in Tank et al. (1992) and Israel (2011) or based on radar in Fackelmeier and Biebl (2009) and Reichthalhammer (2012). These technologies facilitate the search substantially. Nevertheless the search performance is still insufficient compared to peak mowing performance caused by many farmers mowing at the same time. Another drawback is possible false-positive alarms, which are very user-unfriendly because they slow the searching process in addition. Thus, there is a need for a system which allows to separate the search for fawns from the mowing.

To overcome this problem, a new approach is a four-step procedure to separate the searching- from the mowing-process. The four steps are Finding, Marking, Recovering and Saving. In this way any searching technology, whether based on infrared or radar, can be used to find the fawns several days before the mowing. After finding the fawns the first time, one can mark them with a RFID-transponder. Marked fawns can be reliably detected during the mowing without false-positive alarms and localized to secure them out of the danger area. This approach expands the existing searching technologies to a very flexible fawn saving concept.

In this paper we present a first RFID system prototype for the mentioned fawn saving concept and a direction finding system to localize the RFID transponders. The direction finding system consists of a vehicle mounted and a hand-held receiver for estimating the direction-of-arrival of active RFID transponders in the UHF band. The mowing machine mounted system measurements have been performed under real world conditions, in which transponders were located near the ground, in random orientations and short distances up to 50 m. The hand-held receiver has been tested under simple conditions.

2 Preliminary

The intent is to develop a RFID localization system which is able to gather information about the whole working area around the mowing machine. A mowing machine driver shall be aware if there is a marked fawn and if there is one, where it is located. Transponders, which will be detected beside the mowing-machine's current lane, could be registered for the next turn and used as backup information.

To simplify the use case and the measurements, we assumed a mowing machine not moving, in a wide, open and flat meadow area surrounded by a couple of fawns on random positions and random orientations. If a doe bears more than one fawn, than they can lay very close to each other.

In practice, there is only one useful position on mowing machines for a sensor node; the vehicle's roof. With one sensor node the only way to determine a relative position of a transponder is estimating the distance and the direction-of-arrival (DOA). In this paper the focus is set only on DOA and a two step procedure is presented to find a transponder only by bearing information under assistance of a hand-held receiver. Due to the possible short distance between several fawns, a classical DOA device would need a very high resolution.

A detected transponder can be localized in two steps only by getting bearing information. First step is a first estimation of the DOA by the sensor node of the mowing machine. With a known direction to a marked fawn, the driver can stop the machine and start the second step, an approach to the fawn guided through hand-held receiver. In waist-high grass a person can not hold a precise walking direction over a distance longer than 10 m. This reveals the need for a device that corrects in real-time the walking direction. The corrected bearing information given by the hand-held receiver enables a person to approach up to a needed distance of 0.5 m.

The increasing speed of mowing-machines up to $20 \, \text{km} \, \text{h}^{-1}$ and a receiving antenna fixed in the center of the machine lead to a minimum range of at least 15 m in driving direction. Today's deckle width are up to 9 m. The transmit power regulations for RFID applications are limited to 2 W ERP in many countries. Therefore, a passive UHF RFID-transponder with a matched dipole antenna reaches a maximum read range of 11 m under good conditions (Finkenzeller, 2008). Due to a given antenna mismatch caused by the fawn body, the position near above the ground and random orientation of dipole antenna, a passive RFID system can be excluded. By using an active RFID transponder, the possible range for a certain detection is expected to be much larger and the antenna mismatch on the transponder side is a second-tier problem.

An active RFID transponder requires a special communication concept. The transponder's maximum lifetime is limited on its battery capacity. A needed lifetime of at least eight weeks will not be realizable in a small transponder with small battery, which is in receive mode for eight weeks. Therefore the transponder has to be in a sleep mode, wakes up frequently, sends a short beacon, waits a short duration for a response and goes back to sleep mode. With a limited capacity the lifetime depends mainly on beacon-sending-interval and receiving-duration. On the other hand, the detection duration depends on the sending-interval, too.

3 Communication concept

The bottleneck is the very small transponder size, which has to be able to be clipped to a fawn. In future, it is planned to design small transponders with integrated communication ICs, small chip antennas and small batteries. Almost all of the time the reader interface on the mowing machine will be out of range of the marked fawns. Thus, the transponder has to be nearly all of its lifetime in a sleep mode to save enough power for a maximum service life (power mode 2, PM2). If a transponder and the reader are in range, they have to detect each other as fast as possible because of the fast mowing speed.

For this reason the communication concept was designed based on two aspects; the maximum possible lifetime and the maximum response duration, which is needed for detecting the transponder. A long transponder lifetime due to limited battery capacity is important because of the varying birth period of the fawns and the weather which changes the mowing period. The maximum duration for detecting the transponder depends on the mowing speed and the range of the RFID-system. With a safety distance of 20 m, a range of 50 m, a braking distance of 10 m and the moving speed of $20 \, \text{km} \, \text{h}^{-1}$, it leads to a maximum detection duration of 3.6 s.

Figure 1 shows the first approach for the transponder behavior. After assembly, the transponder is in the "Initialization State" which is a deep sleep mode (power mode 3, PM3) with lowest possible current consumption until it is clipped on a fawn and it is initialized for usage. This can not be done by a timer but only by an external action like a button or a magnetic switch. From this point, the transponder is in the "Basic State" and it reacts as a master and will frequently sent a beacon with its ID. It is not possible to fall back to "Initialization State". If the reader is not in range, which is true for most of its lifetime, then the transponder will fall back to a regular sleep mode (PM2) to save battery power. If the reader gets in range, it can react and set the transponder by command in the "Second State". Now the transponder is the slave and the reader is the master for this registered transponder.

By doing this, possible collisions between several transponders can be avoided and all further communications and interactions will be controlled by the reader. Multiple registered transponders can be sequentially processed and to each the DOA determined. Later, the results will show that every DOA-algorithm is working well in this scenario. By determining the DOA to each transponder sequentially

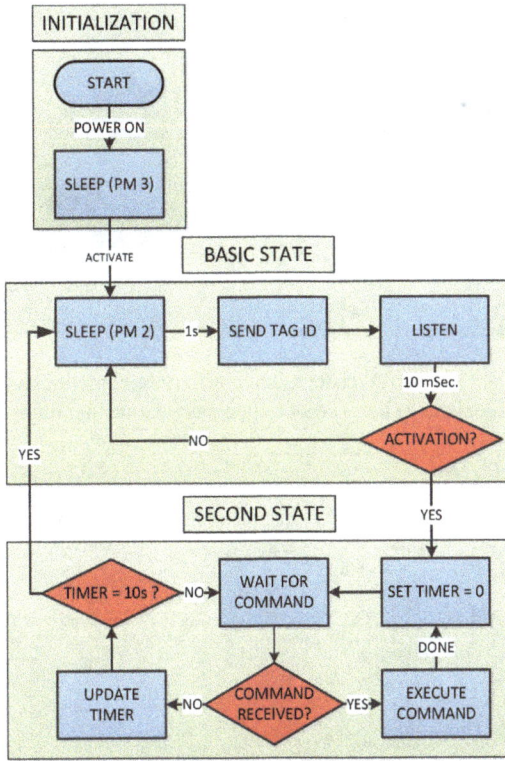

Figure 1. Transponder firmware states and work flow.

the limited resolution of simple algorithms like beamforming and correlative interferometry is sufficient and computation power can be saved.

At the moment, there is no further state for transponder recycling. But it is conceivable to implement a further process for detecting low battery capacity, which sets the transponder in a "Recycle State". In the "Recycle State", the beacon interval could be much longer. This function may be needed to be able to find the transponders, which are no longer attached to a fawn. Due to missing experience a standard return procedure for all used transponders is not projectable at the moment.

4 Theory of operation for DOA estimation

There are a lot of different algorithms to estimate the DOA of an impinging electromagnetic wave. Zekavat and Buehrer (2011) splits them into delay-and-sum algorithms and those with high resolution like MUSIC and ESPRIT. All of them need an array manifold which is defined as

$$\mathbf{A} = [\boldsymbol{a}(\theta_1)\boldsymbol{a}(\theta_2)...\boldsymbol{a}(\theta_n)], \tag{1}$$

where

$$\boldsymbol{a}(\theta_n) = \begin{bmatrix} 1 \\ e^{j\Delta\Phi_2} \\ e^{j\Delta\Phi_3} \\ ... \\ e^{j\Delta\Phi_M} \end{bmatrix} \tag{2}$$

is the direction-dependent steering vector and θ is the azimuthal DOA. The terms $\Delta\Phi_n$ are the phase-differences between the antenna n and the first antenna element in an array with M elements. How many columns (steering-vectors) in the manifold are needed depends on the used algorithm.

Here we choose for first experimental results some known algorithms, for example the Bartlett–Beamformer, which is defined as

$$\mathbf{P}(\theta) = \boldsymbol{a}(\theta)^{\mathrm{H}} \cdot \mathbf{R}_{\mathrm{XX}} \cdot \boldsymbol{a}(\theta), \tag{3}$$

where $\mathbf{R}_{\mathrm{XX}}$ is the signal covariance matrix of the M received signals. A more simpler way is to correlate all steering vectors with one measured steering vector. The phase-differences can be extracted in the first column of the signal covariance matrix or from the complex spectrum of each channel.

$$\mathbf{P} = (\boldsymbol{a}_{\mathrm{measured}}^{\mathrm{H}} \cdot \mathbf{A})^{\mathrm{T}} \tag{4}$$

This method is sometimes called correlative interferometry and is very similar to the Bartlett–Beamformer. The well known MUSIC algorithm with high resolution is defined as

$$\mathbf{P}(\theta) = \frac{1}{\boldsymbol{a}(\theta)^{\mathrm{H}}\mathbf{E}_{\mathrm{N}}\mathbf{E}_{\mathrm{N}}^{\mathrm{H}}\boldsymbol{a}(\theta)}, \tag{5}$$

where $\mathbf{E}_{\mathrm{N}}$ is a matrix with noise eigenvectors (Schmidt, 1986). All algorithms have in common to take samples, calculating the signal covariance matrix and estimating angle-dependent power with the array manifold.

Typical error sources are disrupted wave propagation, unknown parameters about the target signal, co-channel interference, noise caused by the receiver, measurement errors and modeling errors (Tuncer and Friedlander, 2009). The wave propagation conditions in the open meadow area without obstacles are very good, so only influences from the ground and possible diffractions by the mowing-machine are expected. The transponders work cooperative, so all parameters about the signals are known. Measurement errors can be avoided very well by calibrating the receiver with a reference signal. The architecture, as shown in Fig. 2, is realized with high frequency switches to select either the signal received by the antennas or a reference signal generated by a third local oscillator. The main problem is the modeling errors due the antenna array.

Typical modeling errors are due to mutual coupling between antenna elements, coupling between antenna elements

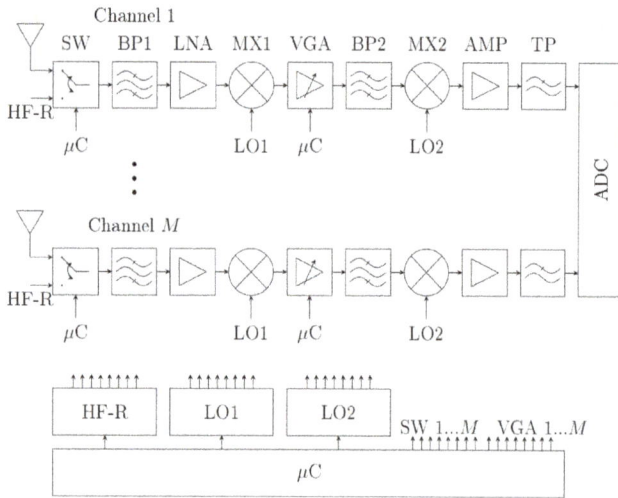

Figure 2. Receiver architecture schema of 8 channel receiver and 2 channel receiver with integrated calibration device.

Figure 3. Used RFID-reader module with power amplifier and connectors. Red mark shows needed elements for a smaller transponder.

Figure 4. 8 channel receiver with integrated calibration device for mowing machine mounted DOA-system.

and mechanical support structures and reflection and diffraction caused by the objects near the antenna array (Tuncer and Friedlander, 2009). The mowing machines are up to 3.5 m high and the limit in Germany for vehicles is 4 m. For this reason we tried to mount the antenna array very close to the vehicle's roof. The expectation is that this will lead to the biggest error source. Furthermore, there will be non-line-of-sight connections to some measurement points depending on transponder distance and roof size.

A possible approach to compensate the expected errors could be done by measuring many reference points with known directions (Pierre and Kaveh, 1995). For every reference point a carrier is sent by an oscillator connected to a dipole antenna and the DOA-system samples the received signals on each channel. Then we build for every reference direction a measured steering vector and build the array manifold.

For reference measurements it is important to know a minimum distance between receiving antennas from the direction-finding system and the source antenna. A minimum distance can be calculated by Tuncer and Friedlander (2009):

$$d \geq \frac{\pi A^2}{4\delta_{\text{tol}}\lambda},\tag{6}$$

where A is the aperture diameter, λ the shortest operated wavelength and δ_{tol} the tolerable phase deviation at the antenna array's edge elements. With $\delta_{\text{tol}} = 2.5°$ and $A = 0.5$ m leads to $d \geq 12.99$ m.

5 Hardware setup

This section gives a short overview of all relevant hardware realizations. There are the RFID communication system, the eight-channel-receiver and a uniform circular antenna array

for the mowing machine mounted DOA system and a hand-held receiver prototype for the guided approach.

For the communication we choose a fully integrated communication chip, the CC1110 from Texas Instruments, because of the very small current consumption and different programmable power modes. A very powerful reader device which in the beginning can be used as transponder as well, is shown in Fig. 3. In the red circle, the components are marked, which are only needed for a small transponder in the future. The additional components in this realization are a front-end amplifier and several connections like USB, UART and SPI. So, it is conceivable to be able to design a very small transponder, which can be clipped to a fawn's ear.

The detection range of the RFID-system was determined experimentally. A reader module with additional amplifier has been mounted on a mowing machines roof in 3 m height and a transponder without amplifier was placed right on the ground. The transponder's sending power was 10 dBm and a certain detection range of at least 300 m could be reached. Via a SMA connector a trigger pulse for incoming transponder signals can be generated. The analog-digital-converters

Figure 5. Uniform circular array prototype with 7 dipole elements, mounted on mowing machines roof. Two additional antennas for RFID-reader communication and received signal strength measurement.

Figure 6. Prototype of hand-held receiver with 2 channel receiver and patch antenna array.

(ADC) can trigger on this pulse and only incoming signals will be sampled and very efficient timings are possible.

The proposed eight-channel-receiver was constructed and is shown in Fig. 4. The hardware was optimized for receiving narrowband signals between 865 and 868 MHz and convert it to an intermediate frequency between 1 and 8 MHz. Using a third local oscillator and a network with power dividers, a precise reference signal can be fed in over the switches to calibrate all channels and compensate the phase errors caused by the receiver hardware. Reference signals themselves have a maximum phase error of $1.5°$, which was measured by a network analyzer. The known phase errors of the reference signals in each channel are factored in the whole phase error correction. A baseband conversion with IQ-modulator is not necessary. The complex analytical signal can be computed with the Hilbert transform and several ADCs for Q channels can be saved. The ADC is realized by a TI development kit with the AD5292 which was connected to a PC.

An antenna array prototype was realized without paying attention on high precise symmetry and is shown in Fig. 5. It is a uniform circular array with seven folded dipole elements. The apertures diameter is 40 cm which matches a relation of $\frac{d}{\lambda} = 0.5$ between neighbor elements. The smallest distance between two elements in propagation direction matches $\frac{d}{\lambda} = 0.21$.

The hand-held receiver prototype was built up with a two-channel-receiver which has the same architecture shown in Fig. 2. The ADC was built with the AD9231 by Analog Devices and is realized as extension on an Arduino Due board. The hand-held receiver only needs to cover a sector in walking direction, so patch antennas by Kathrein as linear array with two elements are used. The whole prototype is shown in Fig. 6.

6 Measurements and experimental results

Experimental tests with the mowing machine mounted device have been made in an open meadow environment. The hand-held receiver was initially tested in an outdoor environment under simple conditions. The setups and results will be shown in this section.

6.1 Mowing machine mounted device

An initial stand-alone test with the eight-channel-receiver and the uniform circular array (UCA) was done in an outdoor environment. A transmitter oscillator with a dipole antenna and the UCA from the direction-finding system were arranged in 1 m height and 15 m distance. Within a maximum tolerance of $\pm 10°$ every measurement was a correct DOA estimation with a mean error of $3°$ without paying attention on high precise symmetry in the UCA and mutual coupling compensation between antenna elements.

For use case conditions, the presented antenna array was mounted close to a mowing machines roof. The total distance between a dipole element and the roof was 4 cm, the vehicle roof height was 3.2 m. For the calibration, we took a regular oscillator with the carrier frequency of 868 MHz and a dipole antenna vertically orientated. It was mounted in 1 m height in $10°$-steps and a distance of 15 m around the mowing machine. For every direction the signal was captured with 4096 samples and the steering vector calculated. Later the manifold was algorithm-dependent interpolated to be able to reach a theoretical accuracy of $0.1°$.

In the test scenario a transponder was positioned in various directions with a horizontal distance to the antenna array of 15 m and 30 m. The dipole antenna height above ground was 7 cm and aligned in various orientations on one measurement position. In some exemplary measurements, a maximum distance of 50 m was used. The detection of transponder by the

Table 1. Correctly estimated positions with tolerance and used array-manifold, mowing machine mounted system.

	$p(5°)$ [%]	$p(10°)$ [%]	$p(15°)$ [%]
Ideal manifold	25.77	50.92	64.42
Free manifold	32.52	57.67	76.67
Mounted manifold	49.69	74.23	83.44

Table 2. Mean error of correctly estimated positions and used array-manifold, mowing machine mounted system.

	$p(5°)$[°]	$p(10°)$[°]	$p(15°)$[°]
Ideal manifold	2.1	5.6	7.1
Free manifold	2.9	5.0	6.8
Mounted manifold	2.0	4.0	4.8

Figure 7. Measurement setup with handheld-receiver prototype.

reader worked without problems, so signals have been captured by the direction-finding system every time. Due to the short distances, the received power was high enough with signal-to-noise-ratios of at least 25 dB. A total of 163 measurements were made. Each measurement was tested with every algorithm mentioned in Sect. 4. The number of samples were varied via software between 2 and 1024.

Before discussing the results, it is important to know that every algorithm estimated nearly the same result in every measurement position. In some cases a direction estimation difference of 0.5° occurred, which is negligible. This shows that a simple algorithm is adequate for this problem and the liability to errors is due to a corrupted array manifold. Because of this, only results in dependency of the used manifold are presented. With at least 16 samples the results are identical. Now the captured samples are evaluated with different manifolds.

Table 1 shows the percentage of correctly estimated directions as a function of the tolerance. At first we tried to estimate the DOA with the ideal manifold for a UCA, which was just computed. For example, with a tolerance of $\pm 10°$ 50.92 % of the 163 measurements were correctly estimated. Table 2 shows the mean direction error additional for the correctly estimated positions.

Then two different calibrations are proceeded. The first one was a calibration, where the UCA was mounted on a tripod in an open meadow area. From this calibration we calculated a manifold, which is in Table 1 and 2 signed as "Free manifold". A second calibration was done with the UCA mounted close on the mowing machine's roof, which is signed as "Mounted manifold".

It is obvious that the ideal manifold provides the worst results and with the calibration with the mounted UCA the best. With the ideal manifold no mutual coupling or distortions by the vehicle and its support structure are considered. But a calibration process with the mounted UCA in the use case is

an improvement of 19 %, compared to using an ideal manifold. A tolerance of 15° might be huge, but it is for an initial walking direction in an open outdoor environment sufficient. In addition, these are results from a static scenario without movement of the mowing machine and could be improved just by several observations in a dynamic scenario, e.g. with tracking.

6.2 Hand-held receiver

The hand-held receiver can be used to guide a person to a needed distance of 0.5 m only by DOA estimation. Equation (6) shows that with smaller target distances the phase deviation compared to plane waves is increasing. With a distance of 0.5 m the phase deviation is $\delta_{tol} \geq 23.4°$, which implicates higher DOA errors. On this account, measurements are made without influences of the ground or additional elevation.

The hand-held receiver was positioned in an outdoor environment in 80 cm height. The transponder with a vertical oriented dipole antenna was also arranged in 80 cm height. Between -30 and $30°$ in heading direction with $10°$-steps and a distance d of 10, 5, 2, 1 and 0.5 m, several measurements have been made, as shown in Fig. 7. In every of these 35 different positions, 5 measurements are made without rearranging the transmit antenna.

Within a tolerance of 10°, every measurement point was estimated correctly. 8.9° was the maximum direction error. The standard deviation of direction errors within a certain direction was 1.17° minimum and 2.40° maximum, as it is shown in Table 3. There are mean direction errors of minimum 0.45° in direction $-20°$ and maximum 4.77° in direction 0°. As a function of the distance, the standard deviation of estimated direction goes from 2.2 to 3.61°. With a distance of 0.5 m, the standard deviation was only 3.61°, as it shows Table 4.

Table 3. Mean and standard deviation of estimated directions, direction-dependent, handheld-receiver.

θ [°]	$\mathbf{E}\{\hat{\theta}\}$ [°]	σ [°]
−30	−28.72	2.09
−20	−20.45	2.30
−10	−12.3	1.23
0	−4.77	1.17
10.0	6.08	1.90
20.0	17.04	2.40
30.0	25.01	2.35

Table 4. Mean error and standard deviation of estimated directions, distance-dependent, handheld-receiver.

d [m]	0.5	1	2	5	10
$\mathbf{E}\{|\hat{\theta}-\theta|\}$ [°]	3.54	3.74	3.66	2.79	2.75
σ [°]	3.61	2.49	3.12	2.20	2.75

The total mean direction error of all 35 measurement positions is −2.58° and standard deviation is 2.89°. These are the results without calibration process for the patch antenna array. Only the receiver was calibrated after 10 measurements for correct phase measurement. With a tolerance of 10°, a user would be able to find a fawn with the hand-held receiver, so further considerations are meaningful.

7 Conclusion and outlook

In this paper we presented an active RFID localization system, which enables a mowing machine driver to detect and localize RFID-marked fawns. The active RFID-system works in the UHF band at 868 MHz. A mowing machine mounted direction-of-arrival system was realized with an eight-channel-receiver and a seven-element uniform circular array to estimate the direction-of-arrival of the transponder signals in every azimuthal direction. The additional hand-held receiver with a two-channel-receiver and a two-element linear patch array shall enable the user to find marked fawns by a guided approach close to a distance of 0.5 m.

The communication concept, as shown in Sect. 3, is a simple solution for an active RFID system, in which transponder and reader are out of range the most of the lifetime. In dependency of sleep interval time the trade-off between the maximum lifetime and the maximum duration for transponder acquisition can be adjusted. A sequential eradication of many transponders in range avoids the need of a high angular resolution.

The mowing machine mounted device was tested under real world conditions and with the antenna array mounted only with a distance of 4 cm to the vehicle's roof. It could be shown that a calibration procedure including the mowing ma-

chine leads to a big performance improvement and within a direction tolerance of 15°, 83.44 % of all measurement points could be estimated correctly. Furthermore, all these results were independent of all tested DOA algorithms.

The hand-held device has been tested under simple conditions and showed a very good performance. Within a 10° direction tolerance, every measurement point could be estimated correctly. In a short distance of 0.5 m, the mean error was 3.54° without calibration procedure.

Future work will be a more detailed investigation on the possibilities for mounting a uniform circular array close to a vehicle's roof. One possibility could be determining the mutual coupling factors between antenna elements by modern calibration algorithms and separating these influences from the distortions caused by the support structure. For the hand-held device real world conditions will be tested.

Acknowledgements. The project is supported by funds of the Federal Ministry of Food, Agriculture and Consumer Protection (BMELV) based on a decision of the Parliament of the Federal Republic of Germany via the Federal Office for Agriculture and Food (BLE) under the innovation support programme. We would like to thank Roland Kröner and Stefan Edstaller for many hardware realizations.

References

Fackelmeier, A. and Biebl, E.: A multistatic radar array for detecting wild animals during pasture mowing, in: European Radar Conference, Rome, Italy, 30 September–2 October 2009, 477–480, 2009.

Finkenzeller, K.: RFID-Handbuch – Grundlagen und praktische Anwendungen von Transpondern, kontaktlosen Chipkarten und NFC, Hanser, 5. aktualisierte und erweiterte Auflage, Carl Hanser Verlag, München, Germany, 2008.

Israel, M.: A Uav-Based Roe Deer Fawn Detection System, Proceedings of the International Conference on Unmanned Aerical Vehicle in Geomatics (UAV-g), Zürich, Switzerland, 14–16 September 2001, XXXVIII-1/C22, 1–5, doi:10.5194/isprsarchives-XXXVIII-1-C22-51-2011, 2011.

Jarnemo, A.: Roe deer Capreolus carpreolus fawns and mowing – mortality rates and countermeasures, Wildlife Biol., 8, 211–218, 2002.

Pierre, J. and Kaveh, M.: Experimental evaluation of high-resolution direction-finding algorithms using a calibrated sensor array testbed, Digit. Signal Process., 5, 243–254, 1995.

Reichthalhammer, T.: Ein Radar mit synthetischer Apertur für den Nahbereich, Verlag Dr. Hut, München, Germany, 2012.

Rieck, W.: Die Setzzeit bei Reh-, Rot- und Damwild in Mitteleuropa, Zeitschrift für Jagdwissenschaft, 1, 69–75, 1955.

Schmidt, R. O.: Multiple emitter location and signal parameter estimation, Antennas and Propagation, IEEE Transactions, 34, 276–280, 1986.

Tank, V., Haschberger, P., Dietl, H., and Lutz, W.: Infrarotoptis-
    cher Wildsensor – eine Entwicklung zur Detektion von Wild in
    Wiesen und zur Wildrettung bei der Frühjahrsmahd, Zeitschrift
    für Jagdwissenschaft, 38, 252–261, 1992.
Tuncer, E. and Friedlander, B.: Classical and Modern Direction-of-
    Arrival Estimation, Academic Press, New York, USA, 2009.

Zekavat, R. and Buehrer, R. M.: Handbook of position location: the-
    ory, practice and advances, John Wiley & Sons, Hoboken, New
    Jersey, USA, 2011.

16

High-voltage circuits for power management on 65 nm CMOS

S. Pashmineh and D. Killat

Brandenburg University of Technology, Department of Microelectronics, Cottbus, Germany

Correspondence to: S. Pashmineh (pashmine@b-tu.de) and D. Killat (killat@b-tu.de)

Abstract. This paper presents two high-voltage circuits used in power management, a switching driver for buck converter with optimized on-resistance and a low dropout (LDO) voltage regulator with 2-stacked pMOS pass devices. The circuit design is based on stacked MOSFETs, thus the circuits are technology independent.

High-voltage drivers with stacked devices suffer from slow switching characteristics. In this paper, a new concept to adjust gate voltages of stacked transistors is introduced for reduction of on-resistance. According to the theory, a circuit is proposed that drives 2 stacked transistors of a driver. Simulation results show a reduction of the on-resistance between 27 and 86 % and a reduction of rise and fall times between 16 and 83 % with a load capacitance of 150 pF at various supply voltages, compared to previous work. The concept can be applied to each high-voltage driver that is based on a number (N) of stacked transistors.

The high voltage compatibility of the low drop-out voltage regulator (LDO) is established by a 2-stacked pMOS transistors as pass device controlled by two regulators: an error amplifier and a 2nd amplifier adjusting the division of the voltages between the two pass transistors. A high GBW and good DC accuracy in line and load regulation is achieved by using 3-stage error amplifiers. To improve stability, two feedback loops are utilized.

In this paper, the 2.5 V I/O transistors of the TSMC 65 nm CMOS technology are used for the circuit design.

1 Introduction

Power management plays an increasing role in electronic systems for consumers, sensors and automotive electronics. However, the standard transistors of nanometer CMOS technologies are only capable to handle low voltages within technology limits and are therefore not compatible with voltages of standard device interfaces and batteries. Thus one common method to design high-voltage circuits is to use high-voltage transistors, which are technology dependent (Bandyopadhyay et al., 2011). In contrast, high-voltage circuits based on stacked low-voltage CMOS transistors are more efficient because of their full compatibility with scaled technologies (Serneels and Steyaert, 2008; Nam et al., 2012; Bradburn and Hess, 2010).

High voltage circuits using stacked devices are a problem, when fast switching or high currents are required. Therefore this paper summarizes the design of two important high voltage circuits for power management, a driver for switching applications (Pashmineh et al., 2013b, c) and a low-dropout voltage regulator (Pashmineh et al., 2013a), both based on stacked transistors. The voltage between terminals of each transistor have been kept within the technology limit.

This work is organized as follows: Sect. 2 describes the structure and operation of a high-voltage driver based on stacked standard CMOS transistors. For reducing the on-resistance of drivers, a theory to calculate gate voltages of stacked transistors to drive the maximum drain current is presented. According to the theory a circuit design methodology is described to generate these voltages. Section 3 introduces the structure and operation of an LDO based on 2-stacked pMOS pass transistors. The circuit design of two regulators, which control the pass transistors, thus regulate the output and the voltages between the two pass devices, is described. Section 4 presents the simulation results of the proposed LDO and a 2-stacked CMOS driver in 65 nm TSMC technology. The results demonstrate a significantly low dropout voltage of the LDO and considerably improved rise and fall times of the driver. Finally, conclusions are given.

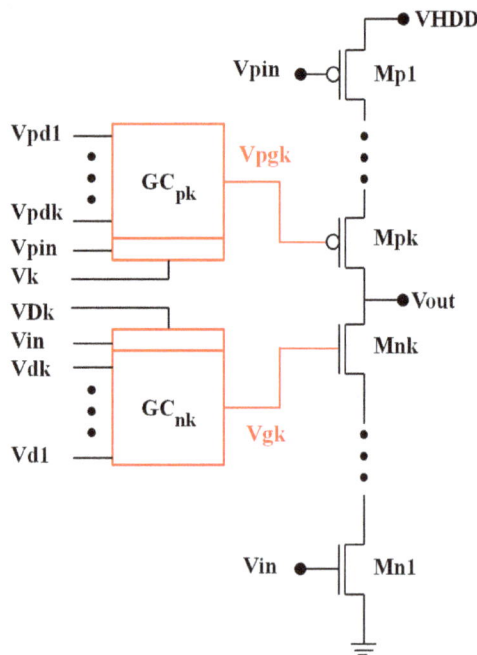

Figure 1. A high-voltage N-stacked CMOS driver circuit.

2 Drivers

Drivers are one of the most important circuit blocks used in power management to switch converters and amplifiers.

In this paper, the proposed high-voltage drivers are based on stacked low-voltage standard CMOS transistors and are technology independent. Their disadvantage however, is that depending on the number of stacked transistors, switching speed may not satisfy requirements because of raised on-resistance of the pull-up and pull-down driver transistors, resulting in slower charge and discharge characteristics of capacitive output nodes.

This work focuses on reduction of the on-resistance of high-voltage drivers. In the following sections a new circuit topology for high-voltage drivers with a minimum on-resistance will be introduced.

2.1 System description

A high-voltage driver based on stacked low-voltage standard CMOS transistors is shown in Fig. 1.

The number of stacked transistors depends on the supply voltage, because the voltage between the terminals of each standard transistor has to be equal to or less than the nominal operating voltage Vn. With a supply voltage of VHDD, which is in the range of $(N-1) \times$ Vn to $N \times$ Vn, the high-voltage driver was designed using N stacked CMOS transistors (N pMOS transistors in the pull-up and N nMOS transistors in the pull-down path).

The driver is controlled by two input signals. The first is Vin, which varies between the ground and the nominal oper-

ating voltage Vn and switches the first nMOS transistor Mn1. The second input signal Vpin, which is level shifted from Vin and varies between VHDD–Vn and VHDD, switches the first pMOS transistor Mp1.

The main challenge in designing high-voltage drivers is the generation of gate voltages of cascode transistors of the stack (Mn2...Mnk, Mp2...Mpk), which need to fulfil two requirements.

First, the transistor in the stack of the driver output must be switched in a way that the voltage between the terminals of each transistor is kept within the technological limits. Second, the driver should pull-up and pull-down with the maximum possible current by setting the appropriate gate voltages for each of the N-stacked transistors. Such a driver realizes a minimum on-resistance.

Therefore a theory describing the optimal gate drive voltages of the N-stacked transistors must be developed.

According to the theory, external circuits (GCnk and GCpk) have been designed to generate these voltages (Fig. 1). This will be fully described in the following sections.

2.2 Circuit theory

In this paper, the gate voltages of N-stacked transistors of a high-voltage driver, except for the first CMOS transistors which are switched by the input signals, have been calculated using a computer algebra system for each output voltage.

First, the gates of the N-stacked nMOS transistors of the pull-down path are considered.

The calculation has been performed for both closed (1) and open (2) scenarios:

1. Operation in on-condition for a maximum drain current at an input signal of 2.5 V, which switches the driver on. The gate voltages of nMOS transistors have been calculated to switch the corresponding nMOS transistors on, enabling a maximum drain current and a minimized on-resistance of the pull-down path. As a result the driver's output can be discharged to the ground.

2. Operation in off-condition at an input signal of 0 V, which switches the first nMOS transistor off. The calculated gate voltages of this condition turn off the respective nMOS transistors. As a consequence, the output can be charged to the high-voltage VHDD.

In both cases, the voltages between each transistor's terminals were kept within the technologically required range.

In the next sections, both conditions are described in further detail.

2.2.1 Operation in on-condition

In this case, the input signal is logical high, which is 2.5 V in this work, turning transistor Mn1 on. The gate voltages of the other transistors have been calculated to switch the

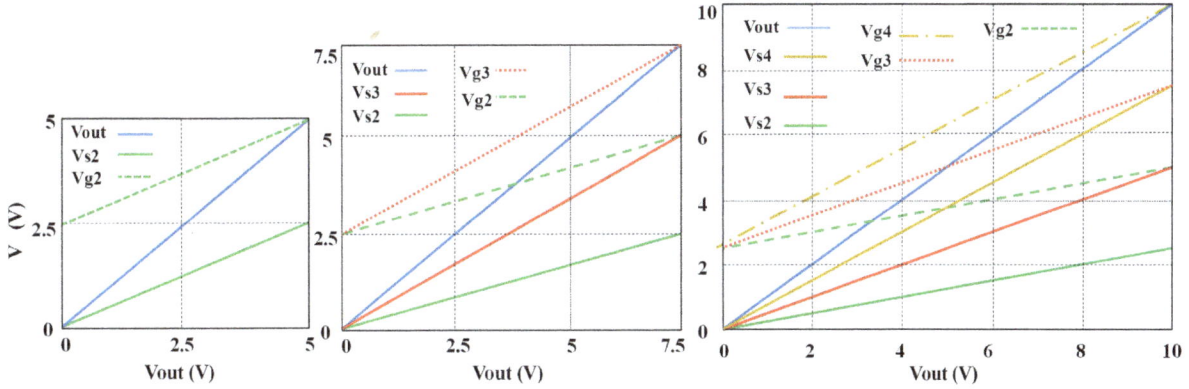

Figure 2. Node voltages characteristics of a **(a)** 2- **(b)** 3- **(c)** 4-NMOS driver for a maximum drain current (on-condition).

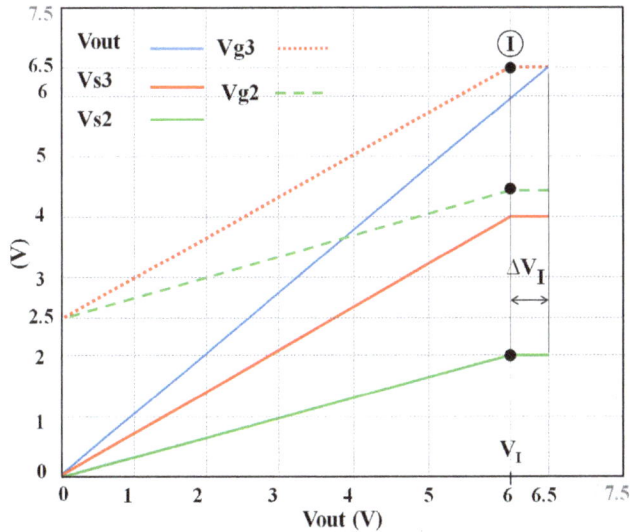

Figure 3. Node Voltages of 3-stacked NMOS driver (on-condition, VHDD = 6.5 V.)

shown in the following equations:

$$Vsk = \frac{(k-1) \times Vout}{N}, \quad k \geq 2 \tag{1}$$

$$Vgk = \frac{(k-1) \times Vout}{N} + 2.5 \text{ V}, \quad k \geq 2 \tag{2}$$

$$Vdk = \frac{k \times Vout}{N}, k \geq 2 \tag{3}$$

N stands for the total number of nMOS transistors; Vsk, Vgk and Vdk denotes source, gate and drain node voltages of the kth nMOS transistor respectively, as shown in Fig. 1.

According to the above equations, the relations between node and output voltages of a 3-stacked nMOS driver can be expressed with the following conditions:

$$Vs2 = \frac{Vout}{3}, \quad Vs3 = \frac{2 \times Vout}{3} \tag{4}$$

$$Vg2 = \frac{Vout}{3} + 2.5 \text{ V}, \quad Vg3 = \frac{2 \times Vout}{3} + 2.5 \text{ V} \tag{5}$$

$$Vd2 = \frac{2 \times Vout}{3}, \quad Vd3 = \frac{3 \times Vout}{3} = Vout \tag{6}$$

Vs2 and Vs3 are the source, Vg2 and Vg3 the gate and Vd2 and Vd3 the drain node voltages of the 2nd and 3rd nMOS transistor respectively. The above relations of node voltages correspond with the calculation results of a 3-stacked nMOS driver in Fig. 2b.

The gate voltages are calculated for a driver with a high supply voltage, which is N-times greater than the nominal operating voltage Vn.

From the calculation results of the required gate voltages of an N-stacked nMOS driver with a supply voltage different from $N \times$ Vn, the following expression can be obtained:

$$\begin{cases} VgN = V_{HDD} \\ Vgk = k \times Vn - \Delta VI, \quad Vout > V_I, \ 2 \leq k < N \\ \\ Vgk = \frac{(k-1) \times Vout}{N} + 2.5 \ V. \quad Vout \leq VI \end{cases}$$

corresponding transistors on and also to drive a maximum drain current. The driver output is discharged from the high-supply voltage VHDD to the ground.

The characteristics of the calculated gate voltages of 2-, 3- and 4-stacked nMOS devices are mapped over the driver output voltages for maximum drain currents as depicted in Fig. 2a, b and c. The driver has a high supply voltage, which is N times greater than the nominal operating voltage Vn. Depending on the number of driver stacked nMOS, the output node Vout is discharged from 5, 7.5 and 10 to 0 V.

The calculated results prove that the source node voltage of each nMOS transistor is proportional to the driver's output voltage. The gate terminal has an offset to the source voltage, which is equal to the nominal operating voltage (2.5 V). Due to this, the gate, drain and source voltage of each nMOS transistor can be described as functions of the output voltage, as

Table 1. ΔVI of a 2, 3 and 4-stacked driver with various supply voltages.

2-stacked driver		3-stacked driver		4-stacked driver	
VHDD [V]	ΔVI [V]	VHDD [V]	ΔVI [V]	VHDD [V]	ΔVI [V]
5.0	0.0	7.5	0.0	10.0	0.0
4.5	0.5	7.0	0.25	9.5	0.125
4.0	1.0	6.5	0.5	9.0	0.25
3.5	1.5	6.0	0.75	8.5	0.375
3.0	2.0	5.5	1.0	8.0	0.5

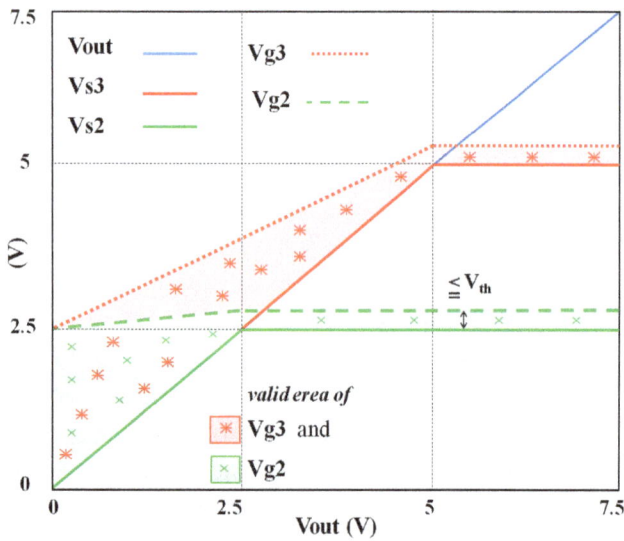

Figure 4. Node Voltages of 3-stacked NMOS driver (off-condition).

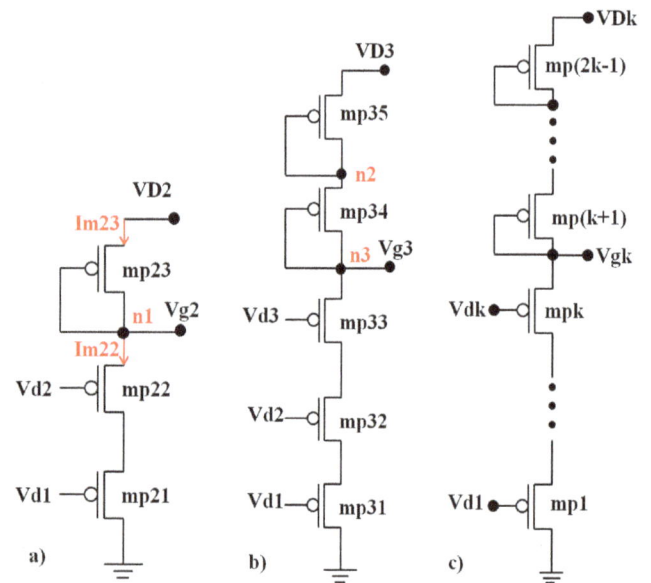

Figure 5. (a) Circuits to generate gate voltages Vg2, **(b)** Vg3 and **(c)** Vgk.

$$(7)$$

VI is the voltage of the point I, as can be seen in Fig. 3. At the beginning of the discharge of the output and pull-down nodes, the gate voltage of each nMOS transistor is constant down to point (I). When the discharge of the driver output falls below this point, the source and gate voltages follow the rule according Eqs. (1) and (2).

Table 1 shows the results of ΔVI, the difference between the supply voltage and VI. By increasing the number of stacked nMOS-transistors, ΔVI decreases. This means that the gate voltages follow the rule (1) for higher N-stacked nMOS.

With an input signal of 0 V, which switches the first nMOS transistor off, the input signal of the pull-up path Vpin is equal to VHDD–Vn, which switches the first pMOS transistor on.

The gate voltages of the other pMOS transistors have been calculated for switching the corresponding transistor on and also driving a maximum drain current in the pull-up path to charge the output from the ground to the high-supply voltage VHDD.

The calculated results of gate and source voltages of the pMOS transistors, which are related to the output voltage, can be described with the following functions:

$$\text{Vpsk} = \frac{(k_p - 1) \times \text{Vout}}{N} + (N + 1 - k_p) \times 2.5\,\text{V},$$

$$k_p \geq 2 \tag{8}$$

$$\text{Vpgk} = \text{Vpsk} - 2.5\,\text{V} \Rightarrow \text{Vgk} = \frac{(k_p - 1) \times \text{Vout}}{N}$$

$$+ (N - k_p) \times 2.5\,\text{V}, \quad k_p \geq 2. \tag{9}$$

Number k_p denotes the kth pMOS-transistor of the high-voltage driver.

2.2.2 Operation in off-condition

In off-condition the input signal Vin is 0 V, which switches the first nMOS transistor off. The gate voltages of the other nMOS transistors need to be adjusted in such a way that the corresponding transistors can be switched off as quickly as possible in order to avoid shot-through currents in the push-

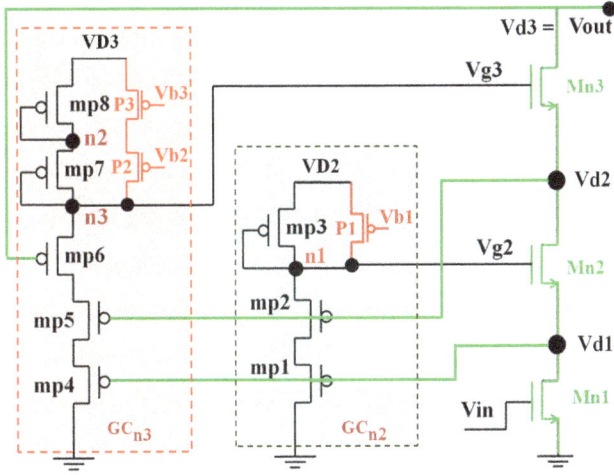

Figure 6. Circuits to generate gate voltages of a 3-stacked NMOS driver.

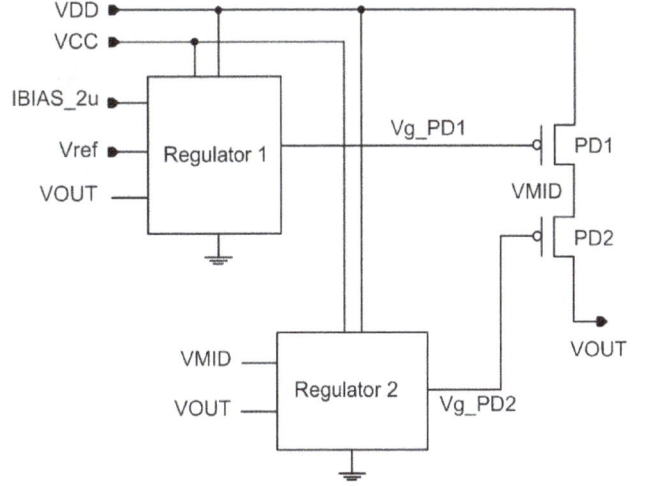

Figure 7. Principe of a LDO with two cascaded pass devices PD1 and PD2.

pull driver and to maintain the voltage between the nodes of each transistor within the technology limit. As a result, the output load will be charged with the highest possible rate.

To meet the above conditions, a stacked transistor must be switched off, if the source node voltage rises to the limit:

$$\text{Vsk\_limit} = (k-1) \times \frac{\text{VHDD}}{N}. \tag{10}$$

In the off-state of each transistor, the gate-source voltage must be equal to or less than the threshold voltage.

Figure 4 shows the calculated gate and source voltages of a 3-stacked nMOS driver with a supply voltage of 7.5 V. The second nMOS transistor switches off at an output voltage of 2.5 V and the third at 5 V. Finally the driver output node is charged to 7.5 V.

2.3 Circuit design

A circuit design methodology for generation of voltages according the theoretical results is described in this section.

The circuit that generates the gate voltages of the 2nd nMOS transistor Vg2 is depicted in Fig. 5a. As can be seen, the circuit is supplied by VD2 and contains 3 pMOS transistors (mp21, mp22 and mp23) in series. The transistor mp23 is gate-drain connected and the gate nodes of the other transistors are determined by the voltages of the driver nodes Vd1 and Vd2 (drain voltages of the driver transistors Mn1 and Mn2 in the pull-down path). The dimensions of mp21 have been set for operation of transistor mp22 in saturation region during the on-condition. The dimensions of the transistors mp22 and mp23 are the same. Therefore, node n1 between mp22 and mp23 (Fig. 5a) supplies the required gate voltage Vg2. According to the on- and off-conditions, the supply voltage of this circuit (VD2) switches between 5 V–ΔVI and 2.5 V respectively. The value of ΔVI depends on

the high-supply voltage of the driver and can be read from Table 1.

The expression in Eq. (10) describes the required voltage of Vg2 during the on-condition. It is derived from two equations with equal drain currents of mp22 and mp23 in the saturation region.

$$\left. \begin{array}{l} I_{\text{mp23}} = \frac{\beta_p}{2} \times (\text{VD2} - \text{Vg2} - \text{Vthp})^2 \\ I_{\text{mp22}} = \frac{\beta_p}{2} \times (\text{Vg2} - \text{Vd2} - \text{Vthp})^2 \end{array} \right\} \begin{array}{l} (3): \text{Vd2} = \frac{2 \times \text{Vout}}{N} \\ \Rightarrow \end{array}$$

$$\text{Vg2} = \frac{\text{Vout}}{N} + \frac{\text{VD2}}{2}. \tag{11}$$

In the on-condition, when the high supply voltage of driver VHDD is N times greater than the nominal voltage Vn, the supply voltage (VD2) of the gate-control circuit GCn2 switches to 5 V. The generated voltage Vg2 from Eq. (10) is equal to the calculated gate voltage of the second stacked nMOS transistor, as in Eq. (2).

The current Imp23 of the transistor mp23 begins to flow when the gate-source voltage of mp23 exceeds its threshold voltage. In this case, the desired voltage of 5 V at node n1 is limited to 5 V–Vth. To solve this problem, a pMOS transistor (such as P1 in Fig. 6) has been connected in parallel to mp3. The gate of this transistor (P1) is biased by Vb1. When the nominal voltage is not an exact fraction of the high supply voltage VHDD, this parallel pMOS transistor (P1) enables the generated voltage at the node n1 to approach the conditions in Eq. (7).

Figure 5b shows a circuit generating the gate voltage (Vg3) of the third nMOS transistor. This gate control circuit GCn3 comprises 5 pMOS transistors (mp31-mp35) in series, with a supply voltage VD3, which switches between 7.5 V–ΔVI and 5 V, respectively, according to the on- or off-conditions. Both transistors mp34 and mp35 are gate-

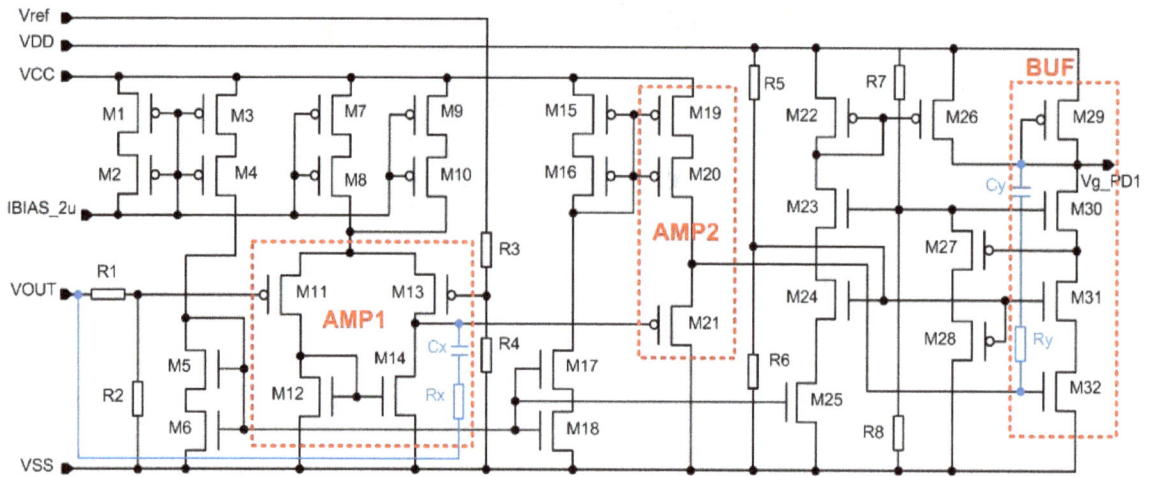

Figure 8. Circuit of the regulator 1.

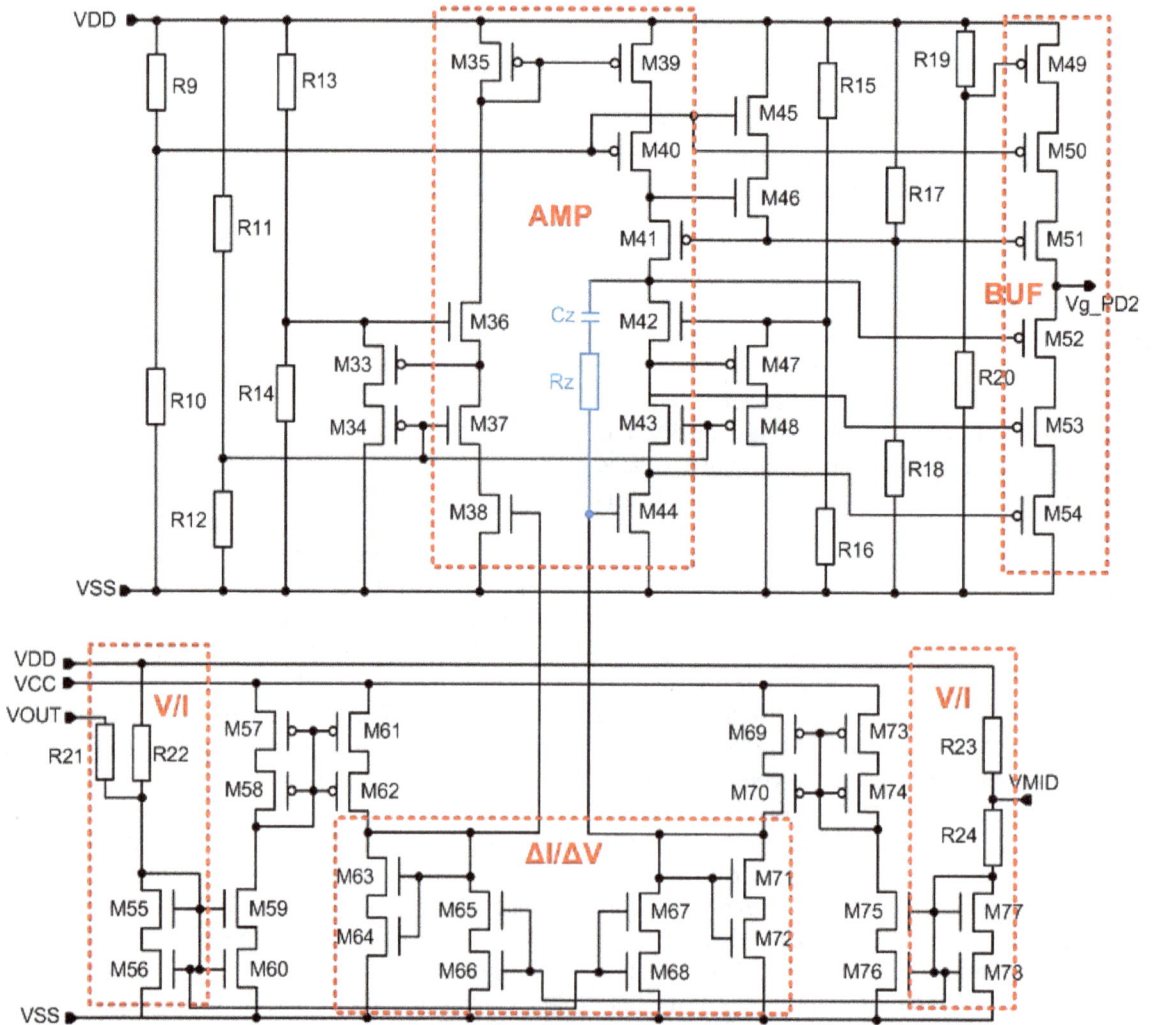

Figure 9. Circuit of the regulator 2.

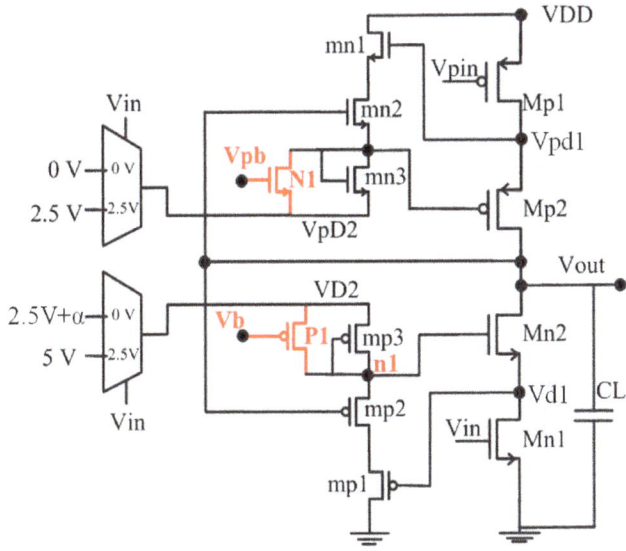

Figure 10. 2-stacked CMOS HV-driver with gate-control circuits.

drain connected and the gate nodes of the other transistors are controlled by the node voltages of the driver Vd1, Vd2 and Vd3 (drain voltages of the transistors Mn1, Mn2 and Mn3).

In on-condition, the voltage of the node 3 (n3) can be calculated from the drain current equations of mp33, mp34 and mp35, which are operated in saturation region:

$$\left. \begin{array}{ll} \text{n2}: & V2 = \frac{VD3}{2} + \frac{V3}{2} \\ \\ \text{n3}: & V3 = Vg3 = \frac{Vd3}{2} + \frac{V3}{2} \end{array} \right\} \overset{(3):\ Vd3 = \frac{3 \times V_{Vout}}{N}}{\Longrightarrow}$$

$$Vg3 = \frac{2 \times Vout}{N} + \frac{VD3}{3}. \quad (12)$$

When the nominal voltage Vn is a fraction of the high-supply voltage VHDD, the supply voltage VD3 is switched to 7.5 V in on-condition. The generated voltage Vg3 (Eq. 12) follows Eq. (2). By connecting 2 pMOS transistors in series (P2 and P3 in Fig. 6), parallel to the gate-drain connected pMOS (mp34 and mp35), the generated voltage Vg3 can approach condition of Eq. (7), when VHDD is unequal to $N \times$ Vn.

With a similar procedure a circuit generating the gate voltage of the kth stacked nMOS transistor of a high-voltage driver consisting of two groups of pMOS transistors can be described: $k - 1$ gate-drain-connected pMOS and k pMOS transistors must be connected in series. The gate of the k pMOS transistors are controlled by the nodes of the stacked driver. The node between both groups generates the desired gate voltage Vgk (Fig. 5c). The supply voltage VDk switches between $k \times$ Vn$-\Delta$VI and $(k-1) \times$ Vn V according to the on- or off-condition. Current can only flow when the gate-source voltage of each transistor exceeds its threshold voltage, thus limiting the required Vgk.

This problem can be solved by connecting pMOS transistors in series biased by reference voltages, in parallel to the gate-drain connected pMOS transistors. When the driver's supply voltage VHDD is not equal to $N \times$ Vn, the generated voltage Vgk can approach Eq. (6) using these extra pMOSs.

The circuits generating gate voltages of pMOS transistors in the pull-up driver are made up of the complement form of the described circuits. Thus nMOS transistors are used instead of pMOSs (mpk), which are used in the proposed 2-stack CMOS high-voltage driver in 65 nm technology (Fig. 10).

3 Low drop-out voltage regulator (LDO)

3.1 System description

In this work, a low drop-out voltage (LDO) is designed, as can be seen in Fig. 7. It is supplied with 5 V and based on standard low-voltage transistors in 65nm TSMC technology with a nominal voltage of 2.5 V (Dearn et al., 2005; D'Souza et al., 2011; Kuttner et al., 2011). Two-stacked pMOSs (PD1 and PD2) are used as pass transistors. The circuit contains two regulators (Regulators 1 and 2) connected to the gates of pass transistors, respectively. Regulator 1 controls the output voltage (VOUT) according to the reference voltage (Vref) by controlling the first pass transistor PD1. Regulator 2 controls the gate of the second pass transistor (PD2) the partitioning of the high voltage between supply and output between the pass devices PD1 and PD2. In the following sections, the design of both regulators will be described in more detail.

3.2 Regulator 1

Figure 8 shows Regulator 1 composed of a 3-stage amplifier. It is supplied by low voltage VCC of 2.5 V and high voltage VDD of 5 V while comparing the output Vout with a reference voltage Vref. The first stage is a single-ended differential amplifier (AMP1) with a pMOS current mirror as active load. Furthermore, the differential amplifier is supplied by the (nominal) low voltage of 2.5 V. To avoid an overvoltage between transistor terminals, both high-voltage input signals (Vout and Vref) are reduced to lower voltages by voltage dividers.

The second stage, a common source amplifier (AMP2), is also supplied with a low voltage VCC. It drives the 3rd stage, consisting of a common source amplifier and a MOS diode load operating as a buffer (BUF). It provides both level shifting and low impedance drive to the pass device PD1. This stage utilizes stacked transistors. The first is a main transistor of the CS amplifier, and the 2nd and 3rd transistor shield high voltage. The 4th transistor is a pMOS diode connected to PD1 and controls the drain current of this pass transistor of the LDO.

Table 2. Comparison results between this work (A) and model B.

VDD [V]	Model	t_{LH} [ns]	Δt [ns]	t_{HL} [ns]	Δt [ns]	$Rn_{on}\Omega$	$Rp_{on}\Omega$
5	B	57		99		687	385
	this work A	42	15	75	24	484	234
	IMPROV.	26.3 %		24 %		30 %	39 %
4.5	B [5]	69.8		90.1		640	471
	this work A	41.2	28.5	71.5	18.6	464	220
	IMPROV.	41 %		21 %		28 %	53 %
4	B [5]	100		82.3		581	680
	this work A	39.6	60.8	66.8	15.5	423	204
	IMPROV.	60.6 %		18.8 %		27.2 %	70 %
3.5	B [5]	220		75		515	1367
	this work A	37.9	182	63.3	11.7	384	195
	IMPROV.	82.7 %		15.6 %		25.4 %	85.7 %

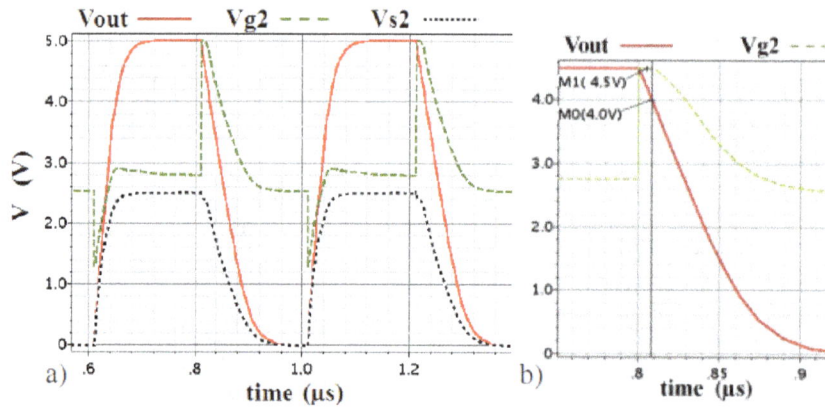

Figure 11. Simulation results of a 2-stacked CMOS driver **(a)** VHDD = 5 V **(b)** VHDD = 4.5 V.

Table 3. Transistor dimensions width/length ($\mu m\,nm^{-1}$), resistor (Ω) and capacitor (F) values of the Regulator 1.

M1, M2, M3, M4, M7, M8, M9, M10, M15, M16; M19, M20	M5, M6, M17, M18, M22, M23, M24, M25
3/500	1/500
M12, M14, M28	M11, M13, M21
6/500	9/280
M30, M31	M29
4/500	5/500
M32	M26, M27
32/280	2/500
R1, R2, R3, R4, R7	R5, R8
499 K	1.5 M
Rx	Ry
91.2 K	1 M
Cx	Cy
50 p	57 p

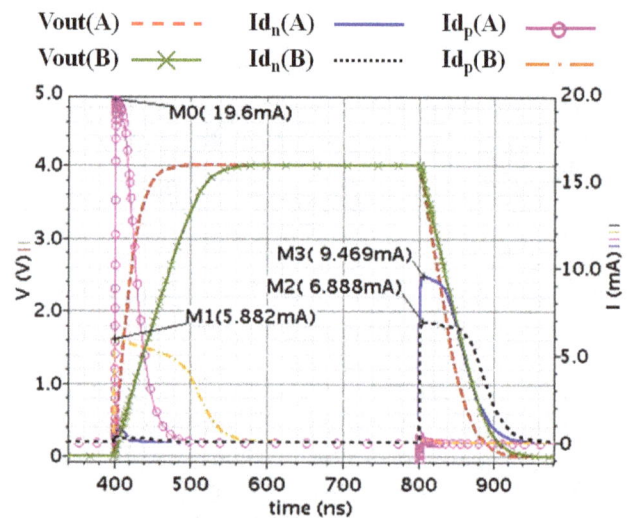

Figure 12. Simulation results of this work in comparison with B.

Table 4. Transistor dimensions width/length $(\mu m\,nm^{-1})$, resistor (Ω) and capacitor (F) values of the Regulator 2 and pass transistors.

M35, M39, M38, M44, M55, M56, M65, M66, M67, M68, M77, M78	M33, M46, M47, M57, M58, M61, M62, M63, M64, M69, M70, M71, M72, M73, M74
3/500	2/500
M37, M40, M43	M36, M42, M52
4/500	4/280
M49	M34, M48, M45
5/500	6/500
M50, M51, M53, M4	M41
8/500	8/280
M59, M60, M75, M76	R10, R11, R14, R16, R17, R20
2.4/500	1.5 M
R9, R12	R13, R15, R18
857 K	652.3 K
R19	Rz
326 K	99.59
Cz	PD1, PD2
485.4f	320/280

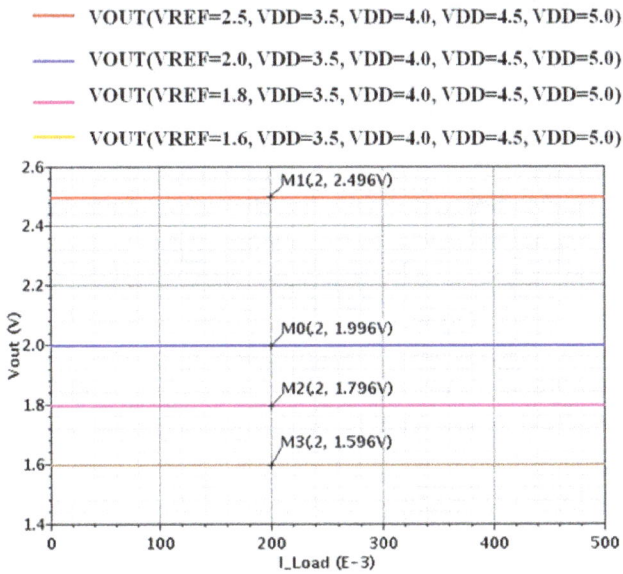

— VOUT(VREF=2.5, VDD=3.5, VDD=4.0, VDD=4.5, VDD=5.0)

— VOUT(VREF=2.0, VDD=3.5, VDD=4.0, VDD=4.5, VDD=5.0)

— VOUT(VREF=1.8, VDD=3.5, VDD=4.0, VDD=4.5, VDD=5.0)

— VOUT(VREF=1.6, VDD=3.5, VDD=4.0, VDD=4.5, VDD=5.0)

Figure 13. DC simulation of the LDO.

3.3 Regulator 2

Figure 9 shows Regulator 2. This circuit regulates the 2nd-stacked pass transistor PD2 and the actual voltage between PD1 and PD2 (VMID) by comparing this voltage with VMID_REF. VMID_REF is the voltage generated by a voltage divider between the high voltage input VDD and the output VOUT.

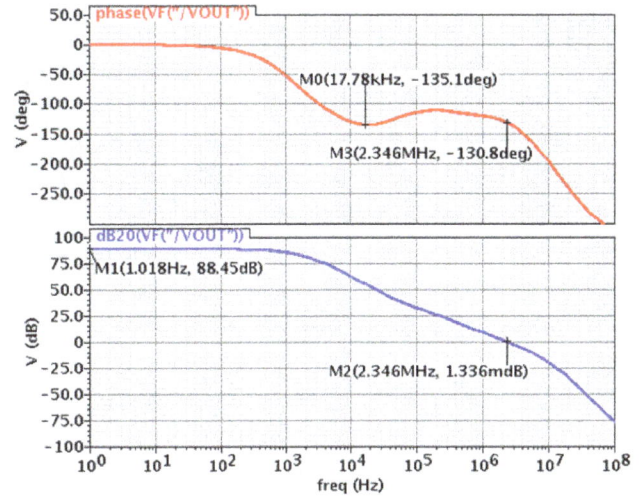

Figure 14. AC characteristics of the open loop (REG1, P1, load).

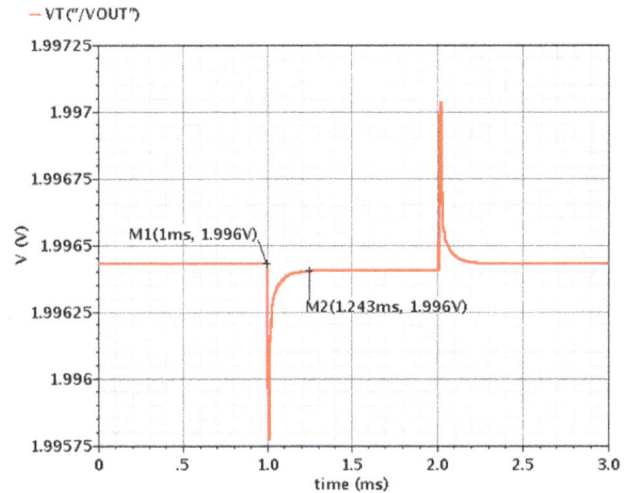

Figure 15. Line regulation (VOUT) with 500 mV input voltage step.

Regulator 2 consists of two parts: a voltage to current converter input with high-to-low voltage level shift function, a differential current to voltage converter and a 3-stage error amplifier. To simplify the design of this high-voltage circuit, the high input voltages VMID and VMID_REF are converted into currents by the voltage to current converter.

The currents are then subtracted from each other in a differential current to voltage converter $\Delta I/\Delta V$, whereas the output voltage is referenced on a MOS diode voltage. This differential voltage controls the pseudo-differential amplifier AMP in the second part. This AMP drives the 2nd stage with nearly rail-to-rail output range. The second stage is intended to operate as a buffer (BUF) and drives the pass transistor PD2 to regulate VMID depending on the difference with VMID_REF.

Transistor dimensions, capacitor and resistor values of the designed circuits are given in Tables 3 and 4.

— VT("/VMID_REF-VMID")

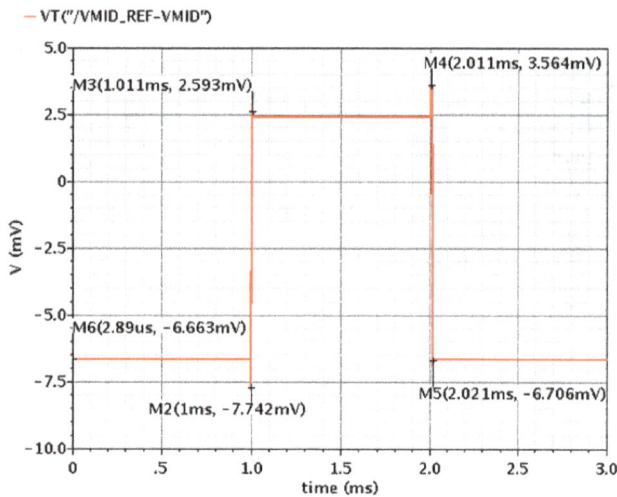

Figure 16. Line regulation (ΔVMID) with 500 mV input voltage step.

— VT("/VOUT")

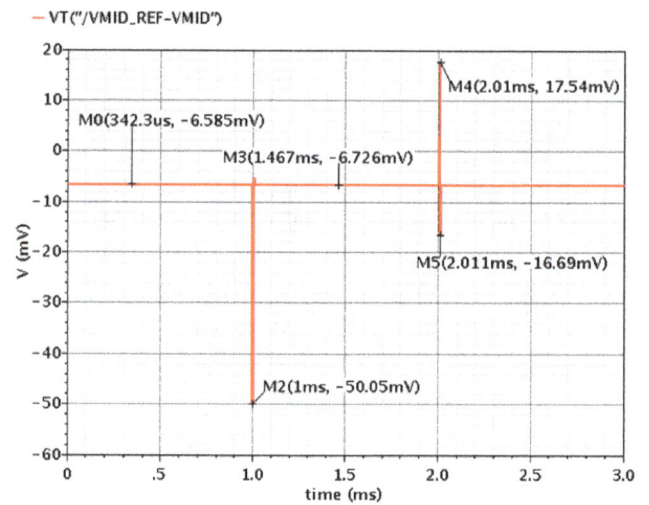

Figure 17. Load regulation (VOUT) with 250 mA load current step.

— VT("/VMID_REF-VMID")

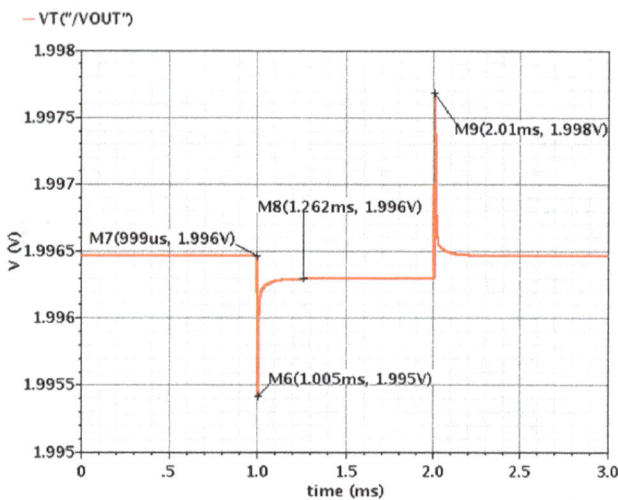

Figure 18. Load regulation (ΔVMID) with 250 mA load current step.

4 Simulation results

In this section, the simulation results of both proposed high-voltage circuits (the high-voltage driver based on 2-stacked CMOS and the LDO with 2-stacked pass transistors), are presented and described.

4.1 Simulation results of the proposed 2-stacked CMOS high voltage-driver

Figure 10 shows the proposed 2-stack CMOS high-voltage driver in 65 nm technology. Vin is the input signal of the pull-down path and Vpin, level-shifted from Vin, switches the pull-up path. The simulation results of this circuit, supplied with 4.5 and 5 V are shown in Fig. 11a and b, respectively. Figure 11a shows that at logic low input the gate volt-

age Vg2 switches transistor Mn2 off by reducing the gate-source voltage below Vth. The output voltage VOUT has been charged to 5 V and the source node Vs2 to 2.5 V, which is half of the output voltage. When the input signal is 2.5 V, the output is discharged from 5 V and Vs2 from 2.5 to 0 V. The simulation results in Fig. 11a and b show that the voltage Vg2 follows the rule according to Eqs. (2) and (7) respectively.

Figure 12 shows the output and the drain current of this work (A) in comparison to previous work B (Serneels and Steyaert, 2008) with a supply voltage of 4 V. In the previous work, the gate voltages of the second nMOS and pMOS transistors of a 2-stacked CMOS driver are fixed to the high level of the input signal. The principle of the work B is applied on a 2-stack CMOS driver in 65 nm technology with a nominal voltage of the I/O devices of 2.5 V. The comparison between the results of the rise/fall time and on-resistance of both works with different supply voltages (3.5, 4, 4.5 and 5 V) are given in Table 2. The initial pull-down and pull-up on-resistances of this work are respectively 27–30 and 39–86 % less than B. The rise and fall times of the output voltage of this paper are improved by approximately 24–83 and 16–20 % with a load capacitance of 150 pF. This indicates that the driver is able to switch faster.

The voltage between each transistor's terminals was kept within the nominal technology limit.

4.2 Simulation results of the proposed LDO

Figure 13 shows the simulation results of the proposed LDO output vs. the load current for different values of reference voltages VREF. The output VOUT is maintained constant according to VREF. As the supply voltage drops from 5 to 3.5 V, the output remains constant and follows the reference voltage VREF.

Figure 14 shows the frequency response of the Regulator 1. The AC characteristics are obtained from the open loop gain and phase at the operating point VDD = 5 V, VOUT = 2 V and a load current ILOAD = 100 mA. The frequency of the dominant pole is set with a load capacitance of 4.7 μF and a load resistance of 20 Ω at 1.69 kHz. The stability of Regulator 1 is achieved by using pole splitting. The second pole, which is set by AMP1, is shifted to a higher frequency at fp2 = 371 kHz with a zero at 41.8 kHz. With compensation, a unity gain bandwidth of 2.7 MHz is achieved. At unity-gain frequency the phase approaches −131°, indicating stability of Regulator 1.

The simulation results of the line regulation response of the proposed LDO are depicted in Figs. 15 and 16. The high-supply voltage VDD is varied from 4.5 to 5 V with rise and fall slew rates of 50 mV μs^{-1}, while the load current remains constant at 100 mA. The steady-state output voltage is ΔVOUT = 27 μV and the value of the line transient response is about 0.66 mV. The difference between the optimum mid-voltage VMID_REF and VMID is 6.6 mV, and the line regulation response of this difference is 9 mV.

The load regulation response is obtained by the load current ILOAD varied from 5 to 250 mA with rise and fall slew rates of 50 mA μs^{-1}. The simulation results for VOUT and the difference between VMID_REF and VMID are shown in Figs. 17 and 18 respectively.

The load regulation transient response and the steady-state output ΔVOUT are 1 and 0.17 mV respectively. The load regulation of the steady-state of Δ(VMID_REF−VMID) is 0.14 mV and its load transient response is 5 and 34 mV respectively for the rising and falling edges of the input signal.

5 Conclusions

In this paper, two high-voltage circuits for power management, a high-voltage driver and an LDO, are presented. The circuits are based on stacked transistors and are compatible with scaled technologies.

For high-voltage drivers, a theory to calculate and design circuits generating gate voltages of stacked transistors to drive a maximum drain current is introduced. By optimally adjusted gate voltages, the driver output provides a minimum on-resistance.

The theory is applied to a 2-stack CMOS driver in 65 nm with a nominal voltage of 2.5 V. Considering the simulation results the gate and source voltages follow the theoretical optimum characteristics. The rise and fall times of the output are improved considerably, which indicate a reduced on-resistance driver. The principle can be applied to N-stack driver transistors as well.

As a second high-voltage circuit for power management, a high-voltage LDO voltage regulator is presented. The circuit is based on the same technology using stacked 2.5 V transistors. An error amplifier controls the main pass transistor and regulates the output voltage. The error amplifier uses 3 stages and has 2 feedback loops, achieving high DC accuracy, as well as good AC and transient characteristics. The second of the stacked pass transistors is controlled by a separate amplifier with lower bandwidth, allowing seamless operation from power down to high load currents. The amplifier equalizes voltage drop across both pass transistors. Therefore, transistor lifetime can be extended and the overvoltage between transistor terminals is avoided. The LDO with stacked devices is suitable for the integration of power management on standard CMOS technologies.

Acknowledgements. The authors would like to thank Stefan Bramburger for his contribution to design and layout of the LDO and Kay-Uwe Schulz for supporting the measurement setup and the printed circuit board. This work is funded by the German National Science Foundation (DFG).

References

Bandyopadhyay, S., Ramadass, Y. K., and Chandrakasan, A. P.: 20 μA to 100 mA DC–DC Converter With 2.8-4.2 V Battery Supply for Portable Applications in 45 nm CMOS, Solid-State Circuits, IEEE Journal of Solid-State Circuits, 46, 2807–2820, 2011.

Bradburn, S. R. and Hess, H. L.: An integrated high-voltage buck converter realized with a low-voltage cmos process, Circuits and Systems (MWSCAS), 2010 53rd IEEE International Midwest Symposium, 1021–1024, 2010.

Dearn, D., Stuart, M. J., and Pannwitz, A.: LDO regulator with wide output load range and fast internal loop, US6856124, 15 February 2005.

D'Souza, A. J., Singh, R., Prabhu, J. R., Chowdary, G., Seedher, A., Somayajula, S., Nalam, N. R., Cimaz, L., Le Coq, S., Kallam, P., Sundar, S., Shanfeng Cheng, Tumati, S., and Huang, W. C.: A fully integrated power-management solution for a 65nm CMOS cellular handset chip, Solid-State Circuits Conference Digest of Technical Papers (ISSCC 2011), 382–384, 2011.

Kuttner, F., Habibovic, H., Hartig, T., Fulde, M., Babin, G., Santner, A., Bogner, P., Kropf, C., Riesslegger, H., and Hodel, U.: A digitally controlled DC-DC converter for SoC in 28nm CMOS, Solid-State Circuits Conference Digest of Technical Papers (ISSCC 2011), 384–385, 2011.

Nam, H., Ahn, Y., and Roh, J: 5-V Buck Converter Using 3.3-V Standard CMOS Process With Adaptive Power Transistor Driver Increasing Efficiency and Maximum Load Capacity, Power Electronics, IEEE Transactions on Power Electronics, 27, 463–471, 2012.

Pashmineh, S., Bramburger, S., Hongcheng Xu, Ortmanns, M., and Killat, D.: An LDO using stacked transistors on 65 nm CMOS,

Circuit Theory and Design (ECCTD), Dresden, September 2013, 1–4, 2013a.

Pashmineh, S., Hongcheng Xu, Ortmanns, M., and Killat, D.: Design of high speed high-voltage drivers based on stacked standard CMOS for various supply voltages, Circuits and Systems (MWSCAS), 2013 IEEE 56th International Midwest Symposium, Columbus (Ohio, USA), August 2013, 529–532, 2013b.

Pashmineh, S., Hongcheng, Xu, and Killat, D.: Technique for reducing on-resistance of high-voltage drivers based on stacked standard CMOS, PhD Research in Microelectronics and Electronics (PRIME), June 2013, 185–188, 2013c.

Serneels, B. and Steyaert, M.: Design of High Voltage xDSL Line Drivers in Standard CMOS, Springer, 2008.

17

Energy conserving coupling through small apertures in an infinite perfect conducting screen

J. Petzold, S. Tkachenko, and R. Vick

Chair of Electromagnetic Compatibility, Otto-von-Guericke-University, Magdeburg, Germany

Correspondence to: J. Petzold (joerg.petzold@ovgu.de)

Abstract. Apertures in shielding enclosures are an important issue for determining shielding efficiencies. Various mathematical procedures and theories were employed to describe the coupling between the regions connected via an aperture in a well conducting plane. Bethe's theory describes the coupling via the equivalent problem of field excited dipole moments at the location of the aperture. This approach neglects the reaction of the dipole moments on the exciting field and therefore violates energy conservation. This work emphasizes an analytical approach for coupling between halfspaces through small apertures, inspired by the so called *method of small antenna*, which allows an understandable generalization of Bethe's theory.

1 Introduction

In earlier works, Tkachenko et al. (2012) and Rambousky et al. (2013) used the method of small antenna (MSA) to analytically calculate the current induced in small near resonance antennas by external fields in various environments (Tkachenko et al., 2012; Rambousky et al., 2013). The MSA uses the formalism of Green's functions which describe the scatterer and the boundary conditions imposed by the environment. In the vicinity of the scatterer the Green's functions are split in two parts, one singular part and one regular part. The singular part is identified with the singular part of the Green's function for free space, for which asymptotic solutions of the coupling problems are known. The regular part is seen to contain all information of the environment and its integral can be evaluated since the singularity is extracted into the first part. In this way it is possible to construct the induced sources in a specific environment from the source induced in free space and some regularized function representing the

specific boundary conditions. This is also called the method of regularization (Nosich, 1999). The scattering of electrically small apertures in the infinite screen can be considered under the approximation of quasi static fields at the aperture, since the fields will not vary much along the aperture. The solution was first published by Bethe (1944), which gives the simple model of magnetic and electric dipole moments, representing the aperture (Bethe, 1944). The assumption of quasi static conditions neglects the influence of radiation of the moments on themselves, which leads to the violation of power conservation (Collin, 1982). So the problem of coupling between two regions separated by an infinite perfect conducting plane via an aperture will be attended by investigating an equivalent problem, which is the induction of electric and magnetic dipole moments located where the aperture is in the original problem. The structure of the present work is to apply the MSA to the dipoles which represent the aperture. This gives an expression for the renormalized moments, which then depend on radiation boundary conditions and fulfill power conservation.

2 Classical theory

2.1 Boundary conditions at the aperture

At the aperture the fields must fulfill certain boundary conditions. It is useful to split the total fields existing in the considered volume such that

$$
E = \begin{cases} E^0 + E^{\mathrm{I}} & \text{for} \quad x < 0 \\ E^{\mathrm{II}} & \text{for} \quad x > 0, \end{cases} \tag{1a}
$$

$$
H = \begin{cases} H^0 + H^{\mathrm{I}} & \text{for} \quad x < 0 \\ H^{\mathrm{II}} & \text{for} \quad x > 0, \end{cases} \tag{1b}
$$

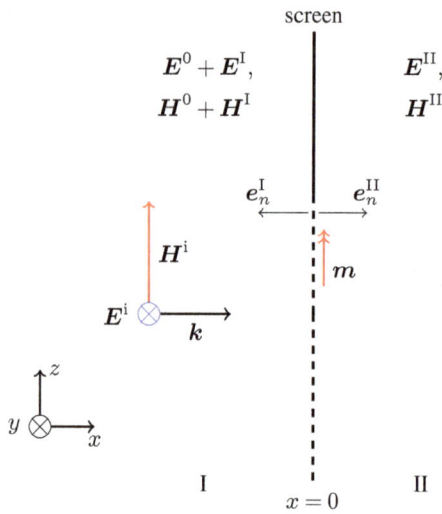

Figure 1. Geometry of the problem with incoming plane wave and induced magnetic moment m.

to write the boundary conditions in a convenient way. The fields H^0 and E^0 are the shortcut fields, which only exist in the left half space, if the aperture is closed. The other fields are the scattered fields, generated by the presence of the aperture. The tangential electric component and normal magnetic component of the shortcut fields must vanish everywhere at $x = 0$. In addition the total fields at the aperture must be continuous. This gives

$$E_\perp^0 + E_\perp^I = E_\perp^{II}, \tag{2a}$$

$$E_\|^I = E_\|^{II}, \tag{2b}$$

$$H_\perp^I = H_\perp^{II}, \tag{2c}$$

$$H_\|^0 + H_\|^I = H_\|^{II}. \tag{2d}$$

The equations Eq. (2) are the boundary conditions at the aperture. Note that they do not depend on size or shape of the aperture (Bethe, 1944).

2.2 Equivalent sources

The conditions described by Eq. (2) may be fulfilled by placing equivalent sources at the closed aperture. These sources depend on the geometry of the aperture, the incoming fields and the radiation conditions in both regions and radiate the scattered fields in the respective regions. To find these sources is the first step to solve the coupling problem. To keep the following considerations more simple a polarization of the incident fields is chosen as shown in Fig. 1. Then only Eq. (2d) is needed as it will be illustrated later.

The shortcut fields are the superposition of the incoming and reflected fields in region I only whereas the scattered fields are radiated by the equivalent sources, so with basic

electromagnetic theory it can be written for region I

$$H^{tot} = \underbrace{H^i + H^r}_{H^0} + \underbrace{\text{rot}A_e^I + \frac{1}{j\omega\mu_0}\text{grad div}A_m^I - j\omega\varepsilon_0 A_m^I}_{H^I}, \tag{3}$$

and for region II

$$H^{tot} = \underbrace{\text{rot}A_e^{II} + \frac{1}{j\omega\mu_0}\text{grad div}A_m^{II} - j\omega\varepsilon_0 A_m^{II}}_{H^{II}}, \tag{4}$$

with

$$A_e^I(r) = -A_e^{II}(r) = \iint_S \overline{\overline{G}}_h^{A_e}(r, r')J_e^S(r')dr', \tag{5}$$

$$A_m^I(r) = -A_m^{II}(r) = \iint_S \overline{\overline{G}}_h^{A_m}(r, r')J_m^S(r')dr'. \tag{6}$$

The vector potentials A_e and A_m are associated with the equivalent surface currents J_e^S as electric source and J_m^S as magnetic source. S is the surface of the aperture. For region II the sign of the vector potentials is negative due to the change of direction of the surface normal vector in the respective region as shown in Fig. 1. $\overline{\overline{G}}_h^{A_e}$ and $\overline{\overline{G}}_h^{A_m}$ are the dyadic Green's functions of the half space which can be found by the mirror procedure and the Green's function of free space. This leads to the conclusion that electric sources do not contribute to the fields, if there is no normal electric field component, because the tangential components of $\overline{\overline{G}}_h^{A_e}$ are zero at the screen. To keep the considerations as simple as possible, a specific configuration as shown in Fig. 1 is chosen. In this configuration only magnetic sources need to be considered.

Interpreting $\overline{\overline{G}}^{A_m}$ as the dyadic vector potential of a dyadic point source, one may write

$$\overline{\overline{G}}^{H_m} = \frac{1}{j\omega\mu_0}\left(\text{grad div}\overline{\overline{G}}^{A_m} + k^2\overline{\overline{G}}^{A_m}\right), \tag{7}$$

where $\overline{\overline{G}}^{H_m}$ is the dyadic Green's Function for the magnetic field of a magnetic source. Using Eq. (7) on Eq. (3) and inserting in Eq. (2d) the more compact form of an aperture integral equation

$$-H_\|^0 = 2\left(\iint_S \overline{\overline{G}}_h^{H_m}J_m^S dr'\right)_\|\tag{8}$$

is achieved.

To evaluate the integrals in Eq. (8) it is necessary to deal with the high order singularities of the Green's functions, which prevent a direct analytical calculation. But by combining the method of analytical regularization and Bethe's theory of diffraction by small apertures an asymptotic analytical solution is possible (Bethe, 1944; Nosich, 1999).

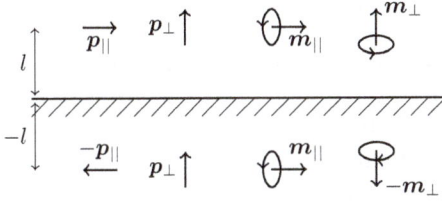

Figure 2. Mirror procedure for electric and magnetic sources.

3 Generalized theory

3.1 Analytical regularization of the Green's function

The basic idea of analytical regularization is the following. The Green's function representing the interaction between source and field while fulfilling the respective boundary conditions is split into two parts, as

$$G = G_s + G_r. \tag{9}$$

The singular part G_s represents the interaction of field and source near the scatterer, for which quasi-static solutions exist if the scatterer is electrically small. In case of a small aperture the singular part is Bethe's solution for the quasi-static scattering. The regular part G_r represents the radiation conditions far away from the scatterer, which is in present work the radiation in half space in region I and II. Since the singularity is extracted into the singular part, the integral of the regular Green's function can be evaluated. The problem is now to write $\overline{\overline{G}}_h^{H_m}$ in the form of Eq. (9).

For the half space the Green's function can be found by applying the mirror procedure. The dyadic Green's function for half space is the superposition of the dyadic Green's function for free space of the original and mirrored source.

$$\overline{\overline{G}}_h^{A_m} = \frac{1}{4\pi}\left(\frac{e^{-jk|r-r'|}}{|r-r'|} + \frac{e^{-jk|r-r''|}}{|r-r''|}\right)\overline{\overline{I}} \tag{10}$$

The single primed vector r' is the position of the original point source, while r'' is the position of the mirrored point source. In the setup of Fig. 1 this means

$$r' = (x', y', z')$$

and

$$r'' = (-x', y', z'),$$

which allows together with Fig. 2 to see clearly, that the normal components of $\overline{\overline{G}}_h^{H_m}$ at the screen cancel out each other, while the tangential components at the screen are simply

$$\left(\overline{\overline{G}}_h^{A_m}\right)_{||} = 2\left(\overline{\overline{G}}_f^{A_m}\right)_{||}. \tag{11}$$

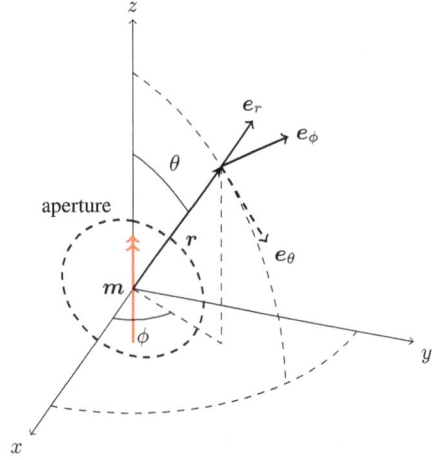

Figure 3. Coordinate systems and aperture.

For the Green's function of free space near a small scatterer regularization can be done by expanding the exponential term into a Taylor series with $k|r-r'| \ll 1$ as the expansion variable around zero, which leads to

$$\frac{e^{-jk|r-r'|}}{|r-r'|} = \frac{1}{|r-r'|} - jk - \frac{k^2}{2}|r-r'| + \frac{jk^3}{6}|r-r'|^2 + \dots \tag{12}$$

By inserting Eq. (12) into Eq. (11) and applying Eq. (7) one arrives at the zz-component of the dyadic Green's function for free space as

$$\left(\overline{\overline{G}}_f^{H_m}\right)_{zz} \approx \frac{1}{4\pi j\omega\mu_0}\left[\underbrace{\frac{2}{|z-z'|^3}}_{\text{singular}} + \underbrace{\frac{k^2}{|z-z'|} - \frac{2}{3}jk^3}_{\text{regular}}\right], \tag{13}$$

which is the only component needed if the incoming magnetic field has only a z-component as in the case of the chosen geometry (see Fig. 1). The regular and singular parts can now be clearly distinguished as indicated in Eq. (13).

3.2 Renormalized Polarizability

To calculate the induced renormalized magnetic moment Eqs. (13), (11) and (8) are used to derive the expression

$$-\frac{H_z^i}{2} = \iint_S \left(\overline{\overline{G}}_{f,s}^{H_m}\right)_{zz} J_{m,z}^S \, dr' + \left(\overline{\overline{G}}_{f,r}^{H_m}\right)_{zz} \iint_S J_{m,z}^S \, dr'. \tag{14}$$

The indices s and r relate to the singular and regular parts of the Green's function. On the left hand side of Eq. (14) the boundary condition for the shortcut magnetic field was

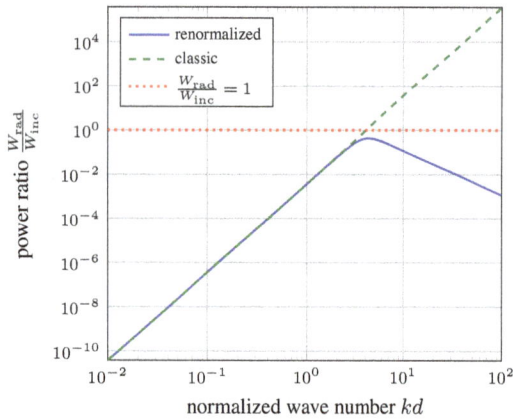

Figure 4. Power ratio over normalized wave number; on logarithmic scale a ratio above 0, as seen for the classic approach, indicates violation of power conservation.

used. The first integral on the right hand side represents the quasi-static solution by the classical Bethe theory and is the singular part of the magnetic field on the left hand side of Eq. (14). Then the integral can be expressed as

$$\iint_S \left(\overline{\overline{G}}^{H_m}_{f,s}\right)_{zz} J^S_{m,z}\, \mathrm{d}r' = -\frac{H^i_{z,s}}{2} = \frac{m_z}{2\overline{\overline{\alpha}}_{zz}} \tag{15}$$

with

$$m_z = -\overline{\overline{\alpha}}_{zz} H^i_{z,s} \tag{16}$$

as given by the classical Bethe Theory. The second integral is per definition the magnetic dipole moment

$$\iint_S J^S_{m,z}\, \mathrm{d}r' = j\omega\mu_0 m_z. \tag{17}$$

In Eq. (15) $\overline{\overline{\alpha}}$ is the dyadic magnetic polarizability, which depends on the shape and size of the aperture. For further information on the polarizability see (van Bladel, 2007, p. 489) and (Tesche et al., 1997, p. 210). With Eqs. (15) and (17) inserted in Eq. (14) the magnetic moment may be written as

$$m_z = -\underbrace{\left[\overline{\overline{\alpha}}^{-1}_{zz} + 2j\omega\mu_0\left(\overline{\overline{G}}^{H_m}_{f,r}\right)_{zz}\right]^{-1}}_{\overline{\overline{\overline{\alpha}}}_{zz}} H^i_z. \tag{18}$$

Here $\overline{\overline{\overline{\alpha}}}_{zz}$ is the zz-component of the renormalized polarizability, which now depends also on the boundary conditions for radiation, which are represented by the first term in the brackets. For other environments such as apertures between cavities and free space one may follow a similar approach. Note that when renormalization is absent Eq. (18) coincides with the classical equation.

4 Calculation Example

To demonstrate the improved coupling theory a specific example will be calculated. The focus lies on power calculation and flow. To investigate power conservation, the radiated power of circular aperture is considered. In this case the classical magnetic polarizability is

$$\overline{\overline{\alpha}}_{zz} = \frac{d^3}{6}, \tag{19}$$

with d as the diameter of the aperture (Tesche et al., 1997, p. 210). For a circular aperture one gets with Eq. (19) and Eq. (18) for the renormalized polarizability

$$\overline{\overline{\overline{\alpha}}}_{zz} = \left(\frac{6}{d^3} - \frac{jk^3}{3\pi}\right)^{-1}. \tag{20}$$

Far from the aperture the scattered fields are the fields radiated by the magnetic dipole moment in half space and with regard to Fig. 3 can be expressed as

$$\frac{jk^2 m_z}{2\pi} e^{-jkr_0} e_\theta \frac{j\sin\theta}{r_0} = \begin{cases} H^I & \text{if } 0 < \theta \le \pi \\ H^{II} & \text{if } \pi < \theta \le 2\pi \end{cases}, \tag{21a}$$

and

$$-\frac{jk^2 m_z \eta_0}{2\pi} e^{-jkr_0} e_\phi \frac{j\sin\theta}{r_0} = \begin{cases} E^I & \text{if } 0 < \theta \le \pi \\ E^{II} & \text{if } \pi < \theta \le 2\pi \end{cases}. \tag{21b}$$

Here η_0 is the wave impedance of free space. Note that Eq. (21) are the far fields of a magnetic dipole moment in free space multiplied by 2 since in this case the moment radiates in half space. With Eqs. (21) and (20) the total radiated power in the far field is

$$\begin{aligned} W_{\mathrm{rad}} &= \int_0^\pi \int_0^{2\pi} P r_0^2 \sin\theta\, \mathrm{d}\theta\, \mathrm{d}\phi \\ &= \frac{\eta_0 k^4}{3\pi} m_z m_z^* \\ &= \frac{\eta_0 k^4 |H^i_z|^2}{12\pi\left(\frac{9}{d^6} + \frac{k^6}{36\pi^2}\right)}. \end{aligned} \tag{22}$$

If the incoming wave is a plane wave too, the ratio of incoming and radiated power can be calculated as

$$W_{\mathrm{inc}} = \eta_0 |H^i_z|^2 S, \tag{23}$$

$$\frac{W_{\mathrm{rad}}}{W_{\mathrm{inc}}} = \frac{(kd)^4}{27\pi^2 + (kd)^6}. \tag{24}$$

The relation Eq. (24) is plotted in Fig. 4.

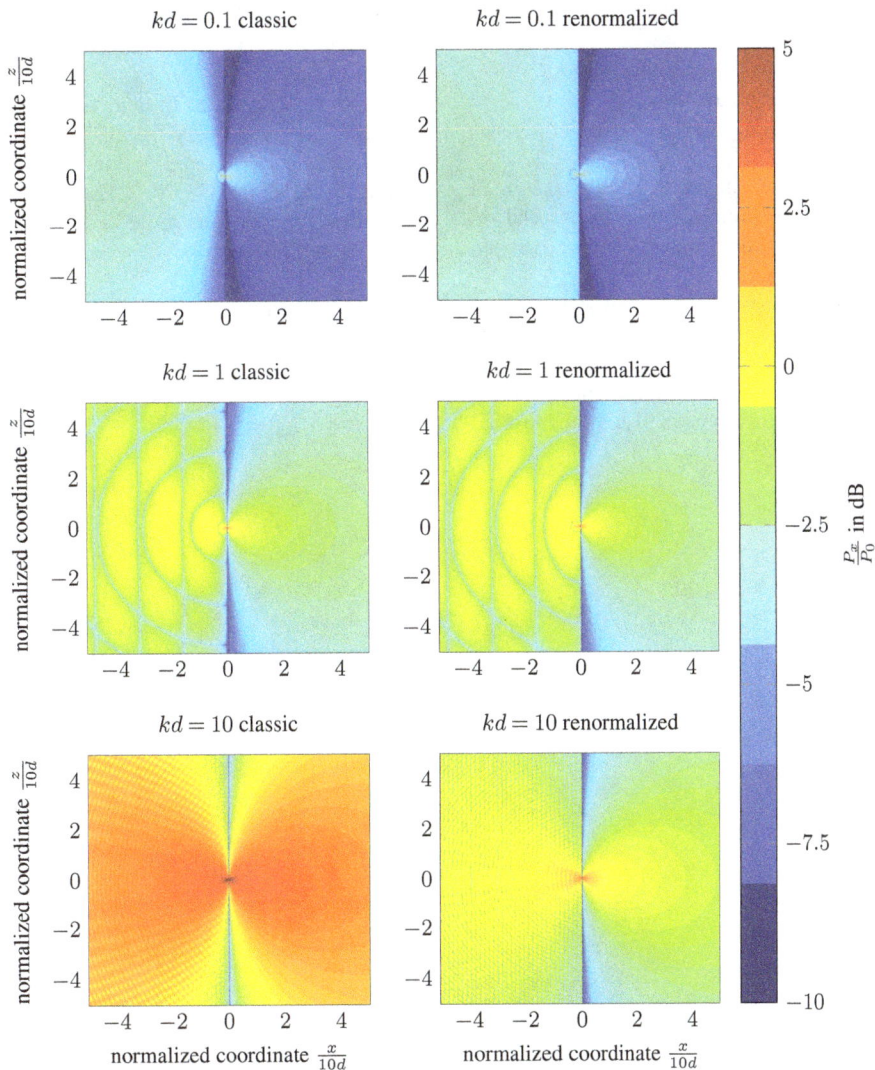

Figure 5. x component of the normalized real power density in the vicinity of the aperture for different normalized wavenumbers. the calculations visualize differences between renormalized and non renormalized moments even for lower frequencies ($kd \leq 1$).

The difference is also evident in direct analytical calculation of the power density around the aperture. Figure 5 shows the x component of the normalized real part of the total power density calculated with the renormalized and non renormalized aperture moments for different normalized wave numbers. One can see significant change in field distributions even for low frequencies. Note that the calculations for $kd > \frac{\pi}{2}$ are not correct due to neglected multipole moments. One should expect higher angle dependence at this frequencies. This will also be part of future investigations.

case of aperture coupling, i.e. an aperture in infinite plane, was used. Calculations of the radiated power of the renormalized moments show that in this way power conservation is established for the model. One has to note, that in the case of radiation in half-space, power conservation is violated for high frequencies, where the dipole model isn't valid at all. But for radiation in cavities and waveguides, the radiation resistance, which is represented by the regular part of the Green's function, can be dominant even for lower frequencies. An application of the presented formalism gives the possibility of forward and backward aperture scattering in different environments. This will be attended in future works.

5 Conclusions

A understandable generalization of a simple model for aperture coupling was derived which features a physical interpretation of aperture radiation. As an example the simplest

References

Bethe, H.: Theory of Diffraction by Small Holes, Phys. Rev., 66, 163–182, doi:10.1103/PhysRev.66.163, 1944.

Collin, R. E.: Small aperture coupling between dissimilar regions, Electromagnetics, 2, 1–24, 1982.

Nosich, A. I.: The method of analytical regularization in wave-scattering and eigenvalue problems: foundations and review of solutions, IEEE Antenn. Propag. M, 41, 34–49, doi:10.1109/74.775246, 1999.

Rambousky, R., Tkachenko, S., and Nitsch, J.: Calculation of currents induced in a long transmission line placed symmetrically inside a rectangular cavity, in: 2013 IEEE International Symposium on Electromagnetic Compatibility – EMC 2013, Denver, CO, USA, 5–9 August 2013, pp. 796–801, doi:10.1109/ISEMC.2013.6670519, 2013.

Tesche, F. M., Ianoz, M., and Karlsson, T.: EMC analysis methods and computational models, John Wiley & Sons, New York, USA, 1997.

Tkachenko, S., Nitsch, J., and Al-Hamid, M.: High-Frequency Electromagnetic Field Coupling to Small Antennae in a Rectangular Resonator, Int. J. Antenn. Propag., 1–6, doi:10.1155/2012/897074, 2012.

van Bladel, J.: Electromagnetic fields, 2nd Ed., Wiley-IEEE Press, 1176 pp., 2007.

18

Definition and test of the electromagnetic immunity of UAS for first responders

C. Adami, S. Chmel, M. Jöster, T. Pusch, and M. Suhrke

Fraunhofer Institute for Technological Trend Analysis (INT), Euskirchen, Germany

Correspondence to: M. Jöster (michael.joester@int.fraunhofer.de)

Abstract. Recent technological developments considerably lowered the barrier for unmanned aerial systems (UAS) to be employed in a variety of usage scenarios, comprising live video transmission from otherwise inaccessible vantage points. As an example, in the French-German ANCHORS project several UAS guided by swarm intelligence provide aerial views and environmental data of a disaster site while deploying an ad-hoc communication network for first responders. Since being able to operate in harsh environmental conditions is a key feature, the immunity of the UAS against radio frequency (RF) exposure has been studied. Conventional Electromagnetic Compatibility (EMC) applied to commercial and industrial electronics is not sufficient since UAS are airborne and can as such move beyond the bounds within which RF exposure is usually limited by regulatory measures. Therefore, the EMC requirements have been complemented by a set of specific RF test frequencies and parameters where strong sources are expected to interfere in the example project test case of an inland port environment. While no essential malfunctions could be observed up to field strengths of $30\,\mathrm{V\,m^{-1}}$, a sophisticated, more exhaustive approach for testing against potential sources of interference in key scenarios of UAS usage should be derived from our present findings.

1 Introduction

Advances in technology allow for sophisticated, unmanned robots and vehicles to explore environments otherwise inaccessible to human investigation due to hazardous conditions like high temperatures, noxious chemicals or ionizing radiation exceeding safety thresholds. In case of large scale disasters, obstructions on ground level may point to aerial reconnaissance as a viable option. The use of single UAS equipped with sensor systems has already seen some application (Pölläna et al., 2009). In order to quickly and continuously chart a disaster site, operating several robotic systems in parallel is necessary. Coordination schemes for multiple UAS in joint operation have been investigated (Simi et al., 2013). There are attempts at controlling clouds of air contaminants by UAS networks (White et al., 2008). Projects like SENEKA involve unmanned ground systems (UGS) in conjunction with UAS (Kuntze et al., 2012). The same is true for the ANCHORS (UAV-Assisted Ad Hoc Networks for Crisis Management and Hostile Environment Sensing) project where a special focus was laid on detection of ionizing radiation (Berky et al., 2014). Within this project running 2012–2015, a system is developed to combine UAS into an autonomous swarm of sensors while at the same time providing a dynamic communication infrastructure of digital radio network cells.

When relying on such systems in order to improve situational awareness of first responders, robustness against any disturbance is crucial. Depending on the deployment scenario, the UAS will have to withstand electromagnetic interference (EMI) by strong RF sources. In previous studies, UAS operation in close proximity to radio broadcasting station has been investigated (Torrero et al., 2013). Other studies have been focussing on actually avoiding strong radar sources by flight path planning algorithms (Chen et al., 2014; Duan et al., 2014).

Radar sources may be prevalent when considering logistics, such as cargo processing in airports and ports. As an example, the ANCHORS project test case conceives a strong industrial source of ionising radiation averaging in the port of Dortmund, Germany. In this case, port radar and ship radar transmitters would have to be taken into account.

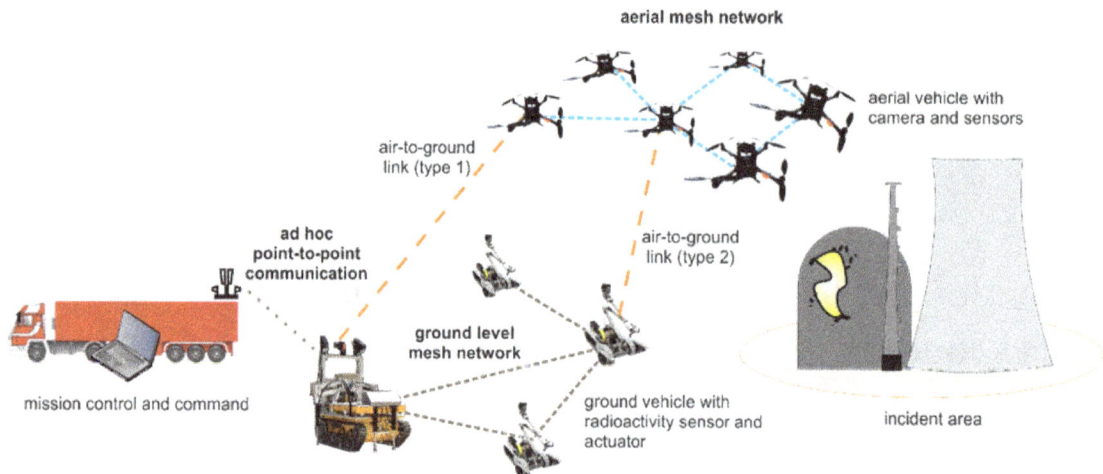

Figure 1. ANCHORS: All actors of the autonomous systems including network connections during an incident (Source: ANCHORS Consortium).

In addition to flight stability and operability, the wireless data link to base stations represents an essential part of the system functionality. Potential disruptions have been subject to research as well (Zhang et al., 2013; Guo et al., 2013).

In our present study, we address the issue of electromagnetic compatibility (EMC) of one of the ANCHORS UAS. We expand on a classic EMC test procedure (DIN Deutsches Institut für Normung e.V., 2006) by selecting additional RF frequencies and test parameters based on the test case scenario analysis. We will present details of the ANCHORS system concept in order to assess potential vulnerabilities, as well as details about the employed test methods and the diagnostics setup. In our tests ranging up to field strengths of $30\,\mathrm{V\,m^{-1}}$, no essential malfunctions could be observed. Nevertheless, we deem further research to be essential. We do point out some starting points in our conclusion, expanding on the scenario-based approach.

2 Description of the ANCHORS system concept

The ANCHORS concept foresees several UAS deployed by a UGS base station in the immediate vicinity of a disaster site. In Fig. 1, the UAS coalescing to a self-regulated swarm are depicted. They communicate with each other via radio, allowing for a flexible role management system where the swarm dynamically accommodates for vacated network nodes. These occur during power replenishment or decontamination actions the UAS perform automatically by autonomous landing on a mobile UGS base station.

All swarm members can be variably equipped with cameras or radiation sensors and send the collected data via radio network to the incident command.

In addition, the UAS are airborne relay stations for radio network cells in ad-hoc communications. As shown in Fig. 2, groups of response personnel are provided with a local net-

work cell which combines with other relay stations, local incident command and possibly remote headquarters to an all-encompassing network. The wireless standards "Long Term Evolution" (LTE) and "Private Mobile Radio" (PMR) have been settled upon for communications.

3 Analysis of the electromagnetic environment of the swarm system

3.1 Common aspects of typical usage scenarios

The whole system of mobile and stationary swarm members as shown in Fig. 2 might be influenced by the local electromagnetic environment of the operational area. In a first step, some assumptions common to typical environments can be made. Because of the high coverage of mobile services in many countries, stationary base stations are part of this electromagnetic environment with a high probability.

Wireless communication and IT devices with relatively low transmitter power are common in residential zones. In industrial zones, we expect a considerable variety of business-specific transmitters like local wireless communication and IT networks with low transmitter power, broadcast stations with medium and high power, and stationary as well as mobile radar facilities in harbors, on ships, at airports and on aircrafts with high and very high pulsed power. Even electromagnetic interferences by unintentional RF transmissions are possible in industrial areas, as generated for example by power inverters.

Moreover, usual protection measures for persons against high field values generated by stationary transmitters by keeping them at distance by structural measures like fences do not work for UAS automatically.

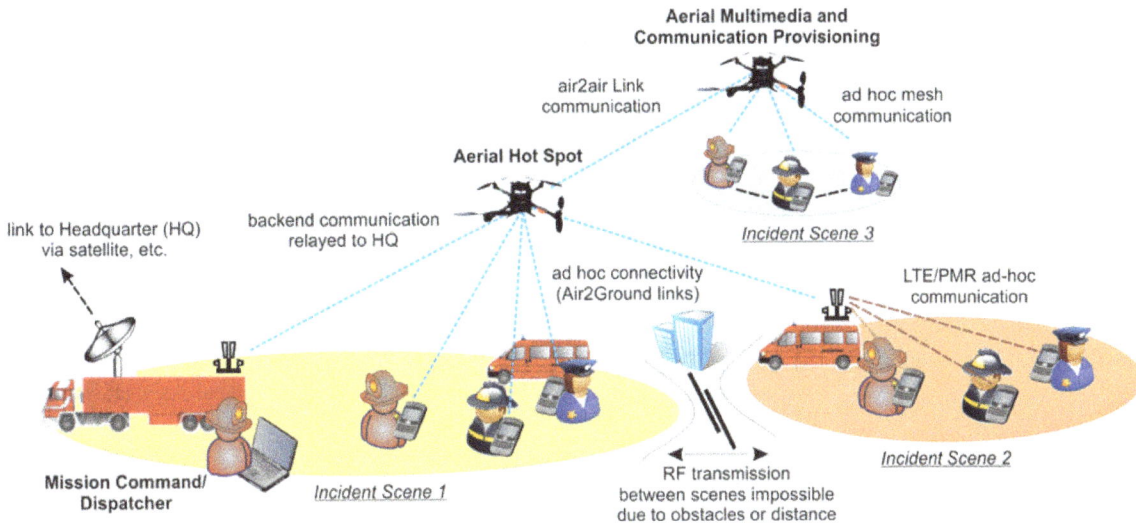

Figure 2. ANCHORS: UAVs as relay stations in an ad-hoc communications radio network (Source: ANCHORS consortium).

3.2 The ANCHORS test case environment

In the following, we will refer to the test case scenario of the ANCHORS project as an example for a potential application case for UAS. We have compiled a comprehensive list of RF sources possibly interfering with system functionality. This includes a suggestion for a geographical analysis of RF exposure based upon which further measures could be developed, like no-fly areas to be integrated into the system software.

As a part of the ANCHORS project, a large scale incident scenario in the harbor of Dortmund has been developed. Two radioactive sources used for on-site material inspection have sprung a leak during an accident and radioactive material is released in the environment.

With regard to the specified location, a mobile network base station has been identified in the direct incident area of the scenario. Figure 3a shows this location on a map of the harbor.

In Germany, information about such base stations is available in a public database (Federal Network Agency, http://emf3.bundesnetzagentur.de/karte/Default.aspx) and can be used for location-based RF exposure assessment. A safety distance for each sector antenna is given, at which the electric field strength meets the limits for human exposure given in the related German electromagnetic fields regulations (BMUB, 2013). As the transmitter frequencies are not accessible in the database, whereas the field strength limit is frequency dependent, the electric field limit value at the indicated safety distance has been calculated back to 1 m in front of the antenna for all possible network service frequencies. As a worst case estimate, the highest value of field strength per antenna has been taken to calculate the distance needed to keep electric fields below $E = 10\,\mathrm{V\,m^{-1}}$. The area within this safety perimeter has been shaded in red in Fig. 3b.

The field strength of $E_{CE} = 10\,\mathrm{V\,m^{-1}}$ represents the EMC immunity required by EU regulations for electronic equipment used in industrial environments (BMUB, 2013), valid for LTE800 and GSM900 mobile networks. For services using frequencies in the range 1.4 up to 2 GHz, like GSM1800 and LTE1900, the tested severity level is just $E_{CE} = 3\,\mathrm{V\,m^{-1}}$. Above 2 GHz where services like UMTS and LTE2600 are located the tested level is lowered to $E_{CE} = 1\,\mathrm{V\,m^{-1}}$. The EMC immunity performance of typical commercial UAS beyond these test levels is most probably unknown for lack of any additional conformity requirements taking the extended spatial mobility into account.

When maintaining a rather cautious stance regarding system safety, flight paths infringing the red-marked area in Fig. 4b have to be avoided in case the network service GSM900 is installed at the mobile base station. The required distance is up to 55 m. In most cases, even more network services are provided by a single base station, so the safety distance will be much higher with 183 and 549 m relating to the two lower test levels of 3 and $1\,\mathrm{V\,m^{-1}}$ in the related frequency bands, respectively.

When considering the electromagnetic environment of the ANCHORS scenario, it can be assumed that there are radar facilities present, mobile ones on ships and a stationary one in the port area. Additionally, a scenario-independent source of electromagnetic energy is inherent in the ANCHORS concept, the ad-hoc network capability. To realize an airborne network relay, a transmitter will be mounted directly underneath each UAS. To complete the list of electromagnetic sources in the environment, there are transmitters for wireless communication and UAS/UGS remote control located in the safe boundary area where the first responders act on ground.

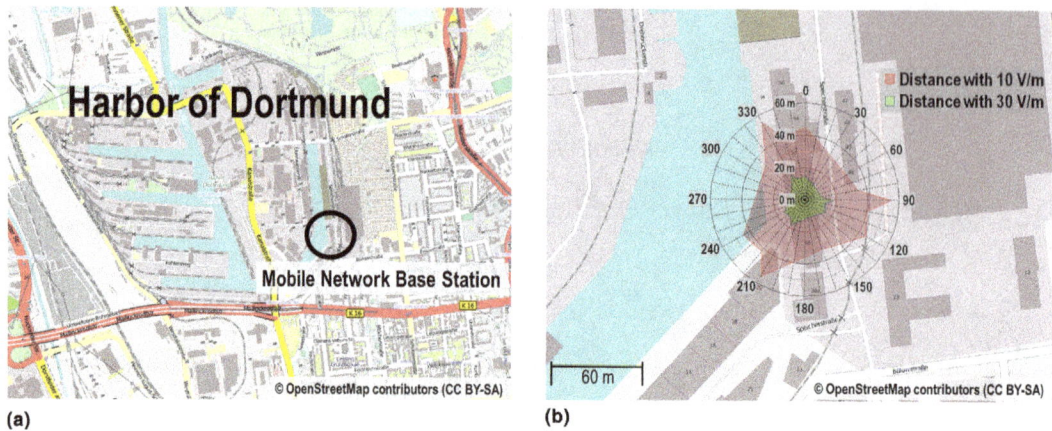

Figure 3. (a) Location of a mobile network service base station within the direct incident area of the scenario developed in the ANCHORS project. **(b)** Calculated areas of $E \geq 10\,\mathrm{V\,m^{-1}}$ (red color) and $E \geq 30\,\mathrm{V\,m^{-1}}$ (green color) around the mobile network base station (source of maps: http://openstreetmap.org).

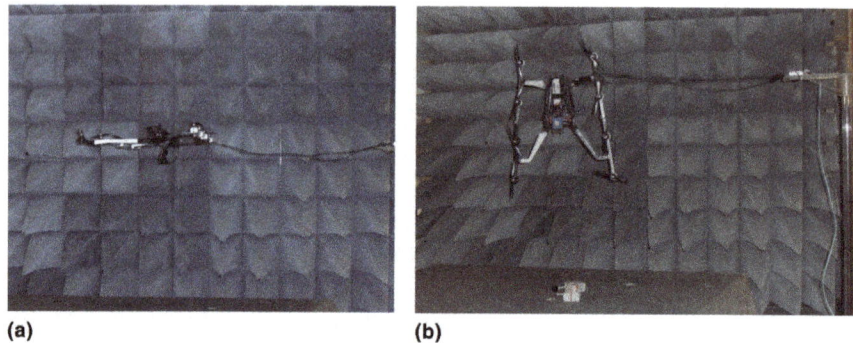

Figure 4. (a) Horizontal and **(b)** vertical test positions of the UAS in the TEM waveguide.

4 Elaboration of parameters for laboratory testing

To address a common susceptibility risk of the entire AN-CHORS system, a basic normative immunity test has been defined according DIN EN 61000-6-2 (DIN Deutsches Institut für Normung e.V., 2006), covering the range of most technically used frequencies. This test standard covers the frequency range from 80 to 2700 MHz with electrical field strength test values from $E = 10\,\mathrm{V\,m^{-1}}$ down to $1\,\mathrm{V\,m^{-1}}$, as detailed in the previous section. When testing with these parameters, no relevant function of the Device Under Test (DUT) is allowed to degrade. As the key result of the scenario-related electromagnetic environment analysis this basic immunity requirement has been expanded by a set of frequencies where the DUT is expected to be particularly vulnerable. Table 1 gives an overview of all tested frequencies.

All additional frequencies have been tested with $E = 30\,\mathrm{V\,m^{-1}}$, the highest defined test level of the test standard (DIN Deutsches Institut für Normung e.V., 2006). Figure 4b shows a green area where the field strength exceeds $30\,\mathrm{V\,m^{-1}}$ for the mobile network services provided

by this base station. The maximum distance a UAS immune to this field strength level shall observe with regard to the base station reduces from 55 m in case of the previously mentioned $10\,\mathrm{V\,m^{-1}}$ limit to 18 m.

To simulate the digital modulation, the GSM time slot pulse with $570\,\mu s$ length and 4.6 ms pulse repetition time has been taken as basis for all communication frequencies. Radar signals are simulated by pulses with $1\,\mu s$ length and 1 ms pulse repetition time. An open TEM waveguide has been used for normative immunity testing according DIN EN 61000-4-20.

5 Test setup and diagnostics

The EMC immunity test standard our study refers to (DIN Deutsches Institut für Normung e.V., 2006), classifies temporary and permanent degradation of functionality in steps of usability of the DUT. Four classes are defined, ranging from "a" for no degradation during RF exposure to "d" for permanent malfunction after exposure. The functional degradation a UAS can cope with is obviously limited, as many functions have a direct influence on navigation and flight sta-

Table 1. Specified test frequencies for the ANCHORS project as derived from scenario.

Frequency	Service	Immunity test value
80–1000 MHz	Basic EMC immunity requirement	$10\,\mathrm{V\,m^{-1}}$
1400–2000 MHz	Basic EMC immunity requirement	$3\,\mathrm{V\,m^{-1}}$
2000–2700 MHz	Basic EMC immunity requirement	$1\,\mathrm{V\,m^{-1}}$
400 MHz	LTE/PMR communication within ANCHORS, on-board transmitter	$30\,\mathrm{V\,m^{-1}}$
2400 MHz	Remote control, other services on 2.4 GHz ISM (Industrial, Scientific, and Medical) band	$30\,\mathrm{V\,m^{-1}}$
5200 MHz, 5800 MHz	UAS remote control downlink channel, other services on 5 GHz ISM band	$30\,\mathrm{V\,m^{-1}}$
810 MHz, 2660 MHz, 1840 MHz	GSM/LTE stationary base stations, ANCHORS LTE/PMR communication with on-board transmitter	$30\,\mathrm{V\,m^{-1}}$
3020 MHz, 9375 MHz	Stationary and mobile naval radar facilities in S- and X-band	$30\,\mathrm{V\,m^{-1}}$

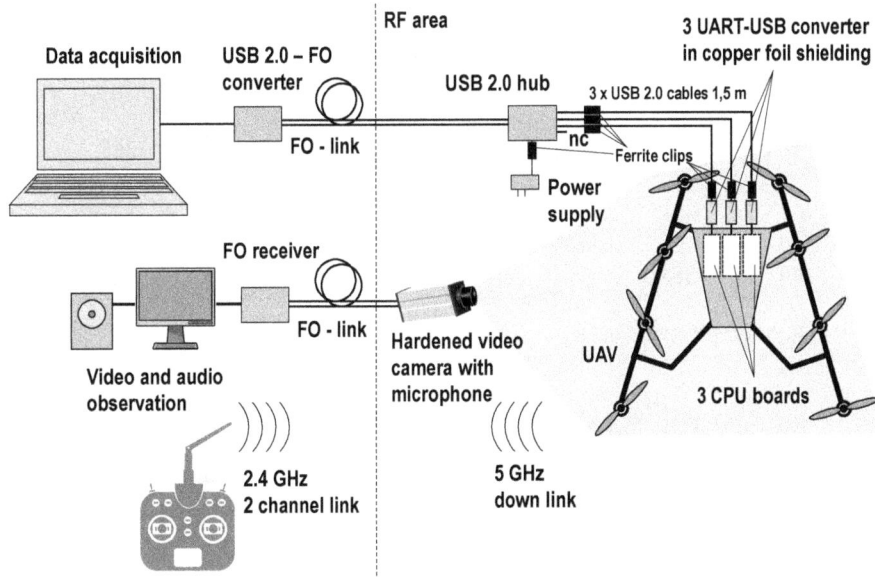

Figure 5. Overview sketch of the monitoring test setup.

bilization. Therefore the flight relevant functions have been identified and assigned to the normative performance class "a". In case of the ANCHORS UAS, some basic functions are designed redundantly in the system. While the respective functional unit as a whole is required to comply with performance class "a" criteria, substructures may degrade to performance class "b", designating a temporary malfunction during RF exposure. As an example, a certain number of the four motor drivers per side may fail and the UAS will nevertheless remain in a stable flight condition.

The functionality classification being complete, the functions identified are described with electrical parameters and tolerances, which have to be monitored during testing. The UAS has been tested in horizontal and vertical position within the open TEM waveguide as it is shown in Fig. 4. The data stream allowing the monitoring of the relevant functions

passed a fiber optic link to a monitoring PC outside of the RF test hall.

Figure 5 gives an overview of the monitoring test setup. By a hardened video camera with microphone the revolution speed of the rotors could be observed.

The tests have been performed with the substitution method described in the test standard. After calibrating the empty TEM waveguide to the desired electric field value at each test frequency, the stored forward power values are then provided again during a test run with the DUT in place. The power adjustment tolerance foresees an appropriate slight over-testing. Figure 6 gives an overview of the RF test site used for the immunity tests.

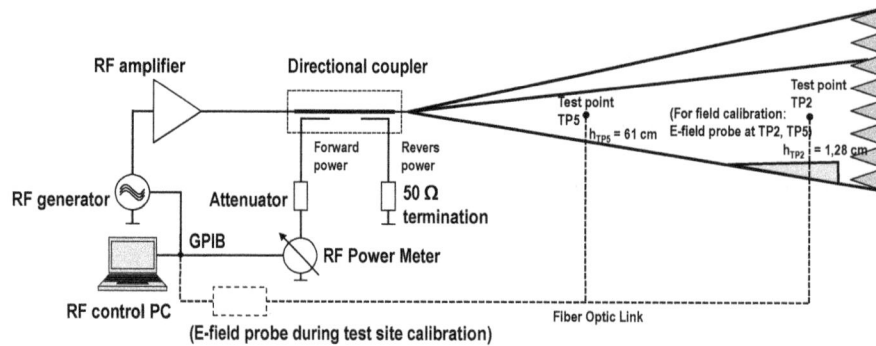

Figure 6. Overview sketch of the RF test setup based on the open TEM waveguide.

Table 2. Test result summary.

Frequency	Test parameter*	Observed effects	Performance class required/reached	Evaluation
80 MHz–1 GHz	$10\,\mathrm{V\,m^{-1}}$, AM	none	a/a	pass
1.4–2 GHz	$3\,\mathrm{V\,m^{-1}}$, AM	2.4 GHz remote control link disturbed between 1887–2002 MHz	b/b	pass
2–2.7 GHz	$10\,\mathrm{V\,m^{-1}}$, AM	2.4 GHz remote control link disturbed between 2040–2616 MHz	b/b	pass
400 MHz	$30\,\mathrm{V\,m^{-1}}$, GSM	none	a/a	pass
2400 MHz	$30\,\mathrm{V\,m^{-1}}$, GSM	none	a/a	pass
5200 MHz, 5800 Mhz	$30\,\mathrm{V\,m^{-1}}$, GSM	none	a/a	pass
810 MHz, 1840 MHz, 2660 MHz	$30\,\mathrm{V\,m^{-1}}$, GSM	none	a/a	pass
3020 MHz, 9375 MHz	$30\,\mathrm{V\,m^{-1}}$, Pulse	none	a/a	pass

* AM = Amplitude modulation 80 %, 1 kHz; GSM = Pulse modulation 570 µs/4.6 ms; Pulse = Pulse modulation 1 µs/1 ms.

6 Test results

As Table 2 reflects, the UAS fulfilled all requirements fixed in Sect. 4.

As predicted and considered in the test requirements, one of two redundant remote control links in the 2.4 GHz band quit during exposure to test pulses at frequencies within this ISM band. Surprisingly the same issue occurred with test frequencies between 1880 and 2400 MHz, and 2500 and 2620 MHz, this might be related to an issue with band filtering in the 2.4 GHz receiver module. As the remote link is still operational with the remaining link channel, the required functional state is fulfilled. By an oversight related to the RF test setup, the frequency range of 1400 up to 2700 MHz has been slightly over-tested with a minimum factor of two, but the DUT kept full functionality. Some additional tests have been performed in order to probe the DUT at even higher field strengths. The electronic design of the UAS shows a good immunity margin up to $80\,\mathrm{V\,m^{-1}}$, as the UAS still fulfilled the functional performance requirements.

7 Conclusion

Battery technologies with high energy density and high computing power in small and light weight units made UAS to very appealing devices to be used for aerial reconnaissance in crisis situations, where the direct incident area is not accessible by humans anymore. But during a mission, they can enter areas of high electromagnetic fields, typically controlled against access by structural measures taking no aerial approach at low altitudes into account. These electromagnetic fields are larger than the EMC immunity standard levels consumer and industrial electronics are designed for, derived from the electromagnetic environment in normal daily usage situations. Therefore, the issue of RF immunity to high level exposure has to be addressed by users of UAS in critical missions where the reliability of the devices has to be high.

In this work, a baseline electromagnetic environment has been estimated for residential and industrial areas. With a very high probability, a mobile network base station will be part of it. In order to focus on usage of UAS in large scale incidents, a scenario at a real location in Germany has been defined within the ANCHORS project. Analysis of the scenario led to a definition of basic and additional test frequencies. As a project task, the RF immunity of the UAS has been evaluated applying the EMC immunity test standard for industrial electronic devices. The UAS is immune to the set of selected frequencies and showed an immunity margin up to a minimum factor of two compared to the normative test levels.

Further hardening measures to increase the immunity especially at mobile network service frequencies might be undertaken, but will possibly reduce the amount of payload, flight duration and mission range of the UAS by increasing weight. As a hybrid solution, immunization measures could be combined with a map indicating keep-out areas around stationary transmitters like mobile network base stations, broadcast stations and radar stations. Thus, the UAS could navigate around areas with high RF field strengths.

In ANCHORS, a key capacity in development is the swarm capability. As the devices interact with each other, the whole swarm can be seen as a controlled system. The behavior of each member in association with others in the swarm can be conceived as control loops, reacting on expected disturbances like wind, but to as of yet unexpected electromagnetic disturbances, too. Suppressed communication can break the swarm as well as corrupted or missing navigation and position information of single swarm members.

Acknowledgements. The project ANCHORS is sponsored by the German Federal Ministry of Education and Research (BMBF) and the French National Research Agency (ANR).

The authors thank the company Ascending Technologies for the close cooperation during setup preparation and tests.

References

Berky, W., Chmel, S., Friedrich, H., Höffgen, S. K., Jöster, M., Köble, T., Lennartz, W., Metzger, S., Pusch, T., Risse, M., Schumann, O., Rosenstock, W., and Weinand, U.: Air-Bound Measurements of Radioactive Material with Swarm-Behaved UAVs – The ANCHORS Project, Proc. 9th Future Security Conference, Berlin, Germany, 16–18 September 2014, 64–70, 2014.

DIN Deutsches Institut für Normung e.V.: DIN EN 61000-6-2: Electromagnetic compatibility (EMC) – Part 6-2: Generic standards – Immunity for industrial environments (IEC 61000-6-2:2005), 2006.

Duan, Y., Ji, X., Li, M., and Li, Y.: Route planning method design for UAV under radar ECM scenario, 12th International Conference on Signal Processing (ICSP), Hangzhou, China, 19–23 October 2014, 108–114, doi:10.1109/ICOSP.2014.7014979, 2014.

Federal Ministry for the Environment, Nature Conservation, Building and Nuclear Safety (BMUB): 26th BImSchV, Amended ordinance on electromagnetic fields (14 August 2013), 3266, 2013.

Gao, C., Gong, H., Zhen, Z., Zhao, Q., and Sun, Y.: Three dimensions formation flight path planning under radar threatening environment, 33rd Chinese Control Conference (CCC), Nanjing, China, 28–30 July 2014, 1121–1125, doi:10.1109/ChiCC.2014.6896785, 2014.

Guo, S., Dong, Z., Hu, Z., and Hu, C.: Simulation of dynamic electromagnetic interference environment for Unmanned Aerial Vehicle data link, China Communications, 10, 19–28, doi:10.1109/CC.2013.6570796, 2013.

Kuntze, H., Frey, C. W., Tchouchenkov, I., Staehle, B., Rome, E., Pfeiffer, K., Wenzel, A., and Wollenstein, J.: SENEKA – sensor network with mobile robots for disaster management, IEEE Conference on Technologies for Homeland Security (HST), Waltham, MA, USA, 13–15 November 2012, 406–410, doi:10.1109/THS.2012.6459883, 2012.

Pölläna, R., Peräjärvia, K., Karhunena, T., Ilandera, T., Lehtinenb, J., Rintalac, K., Katajainenc, T., Niemeläc, J., and Juuselac, M.: Radiation surveillance using an unmanned aerial vehicle, Appl. Rad. Iso., 67, 340–344, doi:10.1016/j.apradiso.2008.10.008, 2009.

Simi, S., Kurup, R., and Rao, S.: Distributed task allocation and coordination scheme for a multi-UAV sensor network, Tenth International Conference on Wireless and Optical Communications Networks (WOCN), Bhopal, India, 26–28 July 2013, 1–5, doi:10.1109/WOCN.2013.6616189, 2013.

Torrero, L., Mollo, P., Molino, A., and Perotti, A.: RF immunity testing of an Unmanned Aerial Vehicle platform under strong EM field conditions, 7th European Conference on Antennas and Propagation (EuCAP), Gothenburg, Sweden, 8–12 April 2013, 263–267, 2013.

White, B. A., Tsourdos, A., Ashokaraj, I., Subchan, S., and Zbikowski, R.: Contaminant Cloud Boundary Monitoring Using Network of UAV Sensors, IEEE Sensors Journal, 8, 1681–1692, doi:10.1109/JSEN.2008.2004298, 2008.

Zhang, T., Chen Y., and Cheng E.: Continuous wave radiation effects on UAV data link system, Cross Strait Quad-Regional Radio Science and Wireless Technology Conference (CSQRWC), Chengdu, China, 21–25 July 2013, 321–324, doi:10.1109/CSQRWC.2013.6657419, 2013.

19

Understanding and optimizing microstrip patch antenna cross polarization radiation on element level for demanding phased array antennas in weather radar applications

D. Vollbracht

Faculty of Electrical Engineering and Information Technology, Technische Universität Chemnitz, 09126 Chemnitz, Germany

Correspondence to: D. Vollbracht (dennis-vollbracht@web.de)

Abstract. The antenna cross polarization suppression (CPS) is of significant importance for the accurate calculation of polarimetric weather radar moments. State-of-the-art reflector antennas fulfill these requirements, but phased array antennas are changing their CPS during the main beam shift, off-broadside direction. Since the cross polarization (x-pol) of the array pattern is affected by the x-pol element factor, the single antenna element should be designed for maximum CPS, not only at broadside, but also for the complete angular electronic scan (e-scan) range of the phased array antenna main beam positions.

Different methods for reducing the x-pol radiation from microstrip patch antenna elements, available from literature sources, are discussed and summarized. The potential x-pol sources from probe fed microstrip patch antennas are investigated. Due to the lack of literature references, circular and square shaped X-Band radiators are compared in their x-pol performance and the microstrip patch antenna size variation was analyzed for improved x-pol pattern.

Furthermore, the most promising technique for the reduction of x-pol radiation, namely "differential feeding with two RF signals 180° out of phase", is compared to single fed patch antennas and thoroughly investigated for phased array applications with simulation results from CST MICROWAVE STUDIO (CST MWS). A new explanation for the excellent port isolation of dual linear polarized and differential fed patch antennas is given graphically. The antenna radiation pattern from single fed and differential fed microstrip patch antennas are analyzed and the shapes of the x-pol patterns are discussed with the well-known cavity model. Moreover, two new visual based electromagnetic approaches for the explanation of the x-pol generation will be given: the field line approach and the surface current distribution approach provide new insight in understanding the generation of x-pol component in microstrip patch antenna radiation patterns.

1 Introduction

Dual polarized phased array antennas for weather radar application in dense radar networks are currently under discussion (Vollbracht, 2014). Such antennas promise even faster three dimensional volume scanning compared to state-of-the-art high power radar systems with reflector antennas and their twin axis mechanical drives. The change from parabolic reflector antennas with fixed beamwidth and gain values to phased array antennas with inherent scan angle dependent co/x-pol patterns is highly challenging.

Weather radar systems with dual polarization capability observe echoes backscattered by hydrometeors from illuminated volumes to classify the precipitation, primarily with horizontally and vertically polarized electromagnetic waves. For accurate polarimetric echo measurements and their interpretation, the antenna CPS performances for both polarization planes are of significant importance. Depending on the CPS, the estimation of the Differential Reflectivity (ZDR) and the Linear Depolarization Ratio (LDR) can be biased. High-quality reflector antennas provide CPS values of up to -30 dB in S- C- X- Band, within the half power beamwidth. A new phased array weather radar antenna should provide the same CPS on broadside, and even more challenging, the same CPS over the complete e-scan range. The e-scan ranges stated from (Vollbracht, 2014) are 120° azimuth and 30° elevation for one planar phased array antenna panel. One 3-D volume radar scanner presented from (Vollbracht, 2014) would be equipped with three of such planar phased array antennas to achieve 360° azimuth volume scanning.

From phased array antenna theory the co- and x-pol antenna radiation pattern can be calculated with the multiplication law that gives the overall antenna pattern as product of the element pattern (EP) and the array factor (AF). Consequently special attention should be paid to the design of the single antenna element and its EP in order to avoid unwanted x-pol radiation right from the very beginning.

Antenna manufacturers typically provide only E and H Cut radiation pattern. For phased array weather radar applications this amount of information is inadequate. If the beam were to be pointed off- broadside- axis, the CPS would be much lower. As a consequence, all antennas for phased array weather applications should be analyzed also in the 45°/135° Cut planes, since the highest x-pol radiation may be expected in this region. A detailed x-pol investigation can only be performed by the 3-D analysis of the complete upper hemisphere of the microstrip patch element pattern.

Dual polarized microstrip patch antennas are suitable candidates for phased array weather radar applications. These antennas are fabricated easily at low cost and have known radiation characteristics. On single antenna element level, the different x-pol reduction methods and their performances are now investigated and discussed. Finally new visual based electromagnetic approaches for the explanation of the x-pol generation are introduced.

2 X-pol suppression techniques on element level known from literature

Only a few suppression techniques on antenna element level have been published. The authors from Liang et al. (2004), Fulton and Chappell (2011) and Zhou and Chio (2011) use differential feed excitation with two signals, 180° out of phase. This method will be explained in detail by the antenna prototype in 2.2 and promises to be a viable solution for phased array weather radar antennas. The defective ground plane solution published in Guha et al. (2005) is useful for single element applications without complex feeding networks and dual polarization capabilities. The radiation from the defective ground area induces amplitude taper disturbances to the array feeding network and would degrade the overall antenna performance. Mohanty and Das (1993) presented a printed thin dielectric substrate with strip grating in front of the radiating antennas for suppressing the x-pol radiation. But strip grating solutions are only useful for single polarization applications. Recently a new investigation to explain x-pol radiation from microstrip antennas was published in Bhardwaj and Rahmat-Samii (2014). Here the x-pol for differential feeding (180° out of phase) and rotational feeding (mirrored antenna elements inside antenna array) are discussed by analyzing the near-field radiation behavior. The major intention of the authors (Bhardwaj and Rahmat-Samii,

2014) is to provide a new understanding of the x-pol radiation generation by observing the near field disturbances. The explanations of the x-pol generation process in Sects. 4.2 and 4.3 can be understood as an additional contribution to the visual based near field approach from Bhardwaj and Rahmat-Samii (2014), but here with field lines and surface currents. A comparison of x-pol performances of circular and rectangular/square microstrip radiators for dual linear polarization applications was surprisingly not found in available literature. Also no capable literature reference could be found where the patch antenna size variation was analyzed for improved x-pol pattern. Consequently the circular and rectangular shape types and the patch antenna size variation are investigated and compared in Sect. 3.3 and 3.5.

2.1 Feeding methods for low x-pol radiation in microstrip array design

There are numerous feeding mechanisms available for microstrip patch antenna elements. Well-known feeding techniques are aperture-coupling, probe-feeding, proximity-coupling and insert-feeding. Aperture-coupled patch antennas are often used to enhance the bandwidth with the radiating slots underneath the patch radiator. To further enhance the bandwidth, a second patch separated with foam material from the first one and with slightly different dimensions and corresponding resonance frequency is implemented. The second patch has also the advantage of reducing the x-pol radiation, since the x-pol radiation generated by the lower patch and the slots are shadowed. The patch antenna design from Liang et al. (2004) shows great CPS performances for horizontal and vertical port of −30 to −35 dB, respectively. But due to the aperture-coupled radiating slots and the back radiation to the feeding network, the mutual coupling aspect becomes critical for array applications. Further disadvantages are the sensitivity against height tolerances of the foam material and the generation of surface waves due to the radiating slots and the cavities between the foam layers. Moreover, the design presented in Liang et al. (2004) is working with the reflector ground plane technique, which is not useful for array operation with multiple elements and feeding networks between.

Proximity coupled patches have similar disadvantages for array configurations as aperture-coupled antennas. The back radiation close to the feeding networks for two polarization planes is undesired in array configurations and the port isolation would become insufficient. Insert fed microstrip patches have their feeding networks on the same layer and unwanted x-pol radiation is generated. In keeping with these observations, insert fed and proximity coupled patch antennas are not suitable for phased arrays with high x-pol requirements.

Probe fed microstrip patch antennas are suitable candidates for the phased array weather radar application with high x-pol requirements. The possibility to drive the single patch antenna element with 180° out of phase and the low

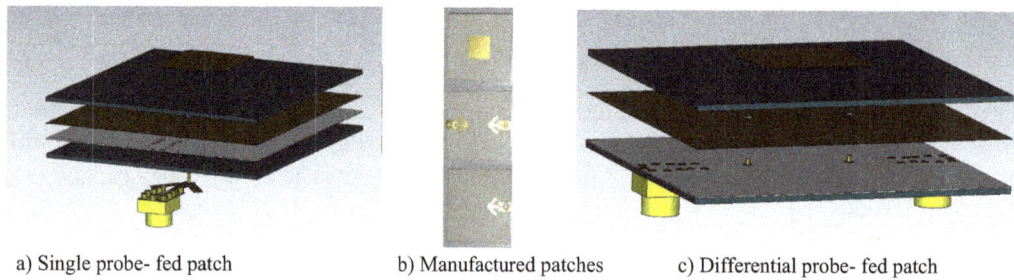

a) Single probe- fed patch   b) Manufactured patches   c) Differential probe- fed patch

Figure 1. Layer set-up of squared microstrip patch antenna design.

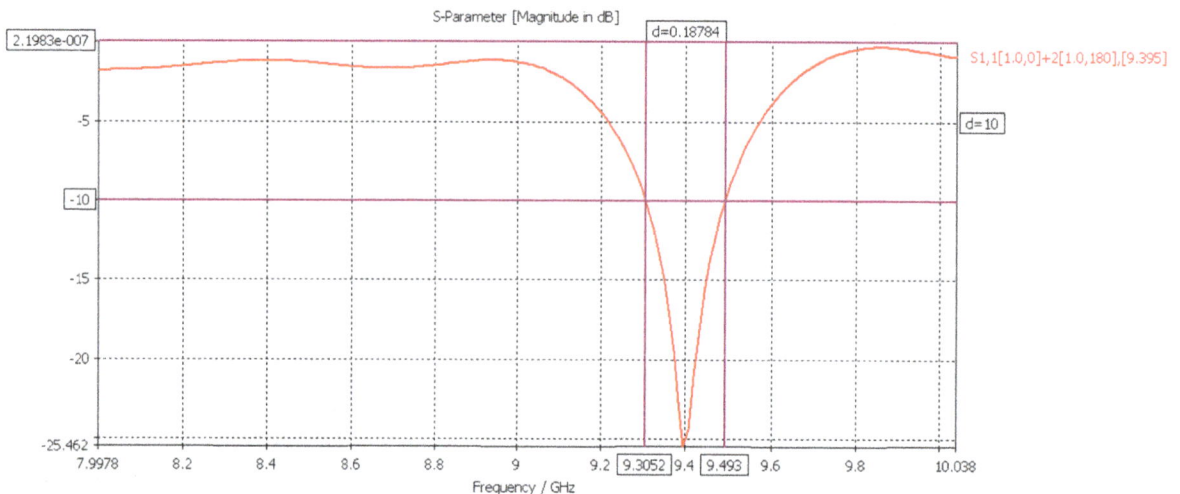

Figure 2. S11 resonating at 9.395 GHz with VSWR 1 : 2 Bandwidth of 188 MHz.

back radiation due to the small feed via holes through the shielding ground plane generates only small x-pol contribution.

2.2 Thorough analysis of the "differential feeding technique" for phased array weather radar applications

One valuable technique to reduce the x-pol radiation of microstrip antennas was presented by Liang et al. (2004) and Fulton and Chappell (2011) and is called "differential feeding". Differential feeding can be established with two RF drive signals exactly 180° out of phase. To demonstrate and to analyze the advantages of differentially fed microstrip patch antennas for phased array weather radar applications two probe fed antenna designs with single and differential phase feeding are compared. It will be further explained with these antenna prototypes how the x-pol radiation is generated from surface currents and field line distributions.

The square probe fed microstrip patch antennas were designed and simulated with CST MWS. The patch antennas are consisting of two RO6002 substrates (dark green in Fig. 1) with $\varepsilon_r = 2.94$, 508 μm height and excellent dissipation factor $\tan \delta = 0.0012$ for adequate antenna efficiency

performances. The grey layer represents in Fig. 1a and b the RO4403 prepreg with $\varepsilon_r = 3.17$ and 100 μm height. The RO4403 prepreg will be used to combine the two RO6002 laminates during the compression process. For suppressing unwanted surface currents or propagation channels within the substrate right from the design beginning, this prepreg shows only small ε_r differences compared to RO6002. The copper has a thickness of 35 μm and the diameter of the isolation circle inside the shielding ground for the probe feed transition is 0.91 mm. The radiating square patches with 10.55 mm edge length, pin-via diameter of 0.4 and 0.416 mm pin- via to edge- distance are designed for a resonance frequency of 9.395 GHz (see Fig. 2).

Both antennas from Fig. 1 are sharing the almost same design parameters. The only differences are the additional port in Fig. 1c with 180° out of phase excitation and the width and length of the matching circuit (arrow) close to the feed. The Rosenberger 3-D connector model was completely integrated in the CST MWS full wave simulation and impedance matched with GND vias and microstrip impedance transformer to the 50Ω line impedance. The microstrip stubs in parallel to the probe feed are used for matching the impedance of the patch antenna. In Fig. 1b the manufactured patches are visualized.

a-1) Single probe- fed patch   b-1)Differential probe- fed patch

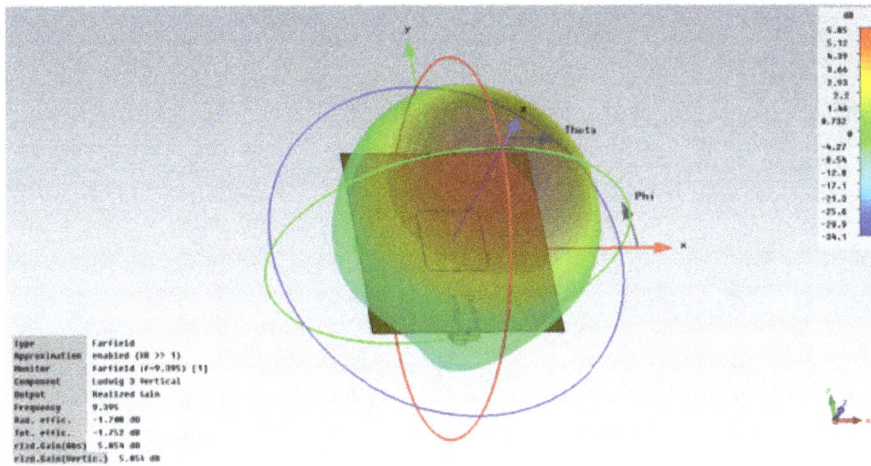

a-2) Single probe- fed co-pol 3D Pattern

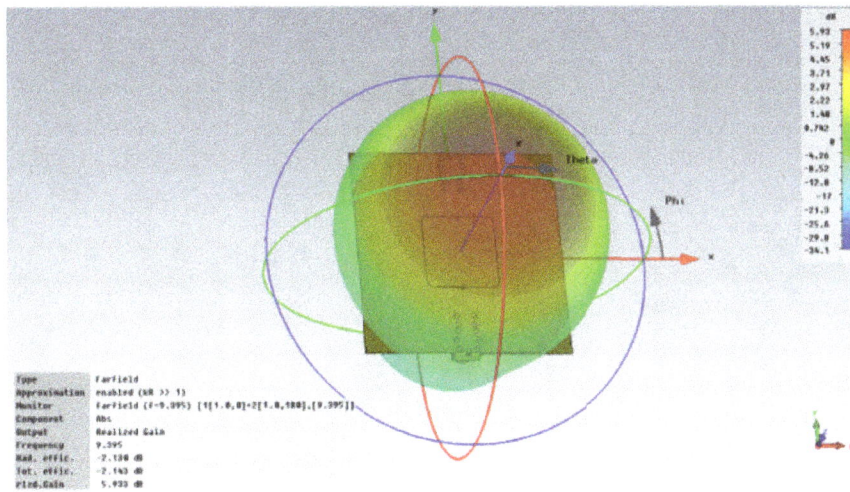

b-2) Differential probe- fed co- pol 3D Pattern

Figure 3.

Figure 2 shows the S11 of the matched connector port with patch antenna. The VSWR 1 : 2 bandwidth of 188 MHz was simulated.

Figure 3 compares the co/x-pol pattern without (Fig. 3a) and with (Fig. 3b) the differential feeding technique. The radiation patterns are plotted with Ludwig's third definition of

x-pol (Ludwig, 1973). Note that the reference parameter "realized gain" in dBi also considers the impedance mismatch losses for the gain calculations. In the left row of Fig. 3 the single probe fed patch antenna radiation patterns are shown together with the co-pol "realized gain" of 5.85 dBi. The right row of Fig. 3 displays the radiation patterns of the dif-

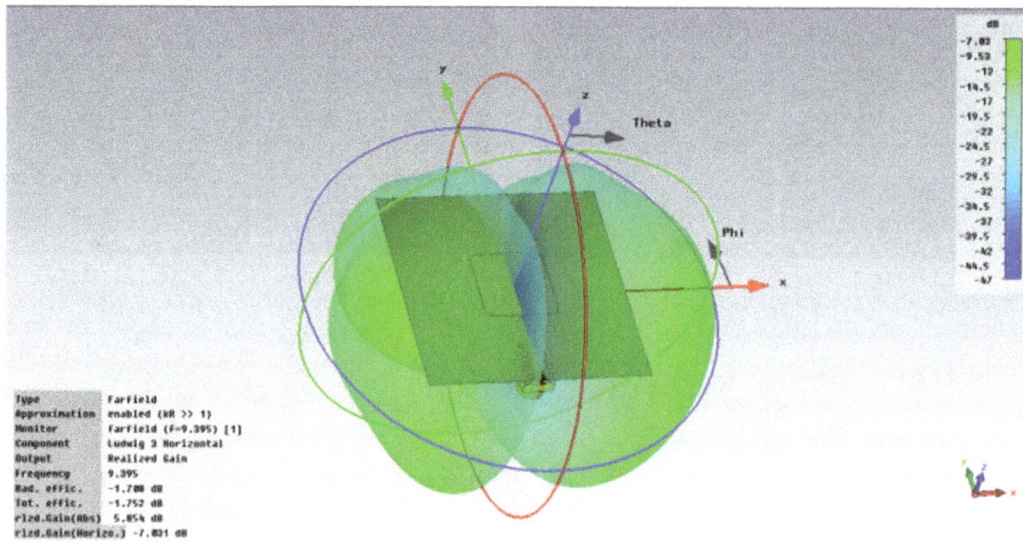

a-3) Single probe- fed x-pol 3D Pattern

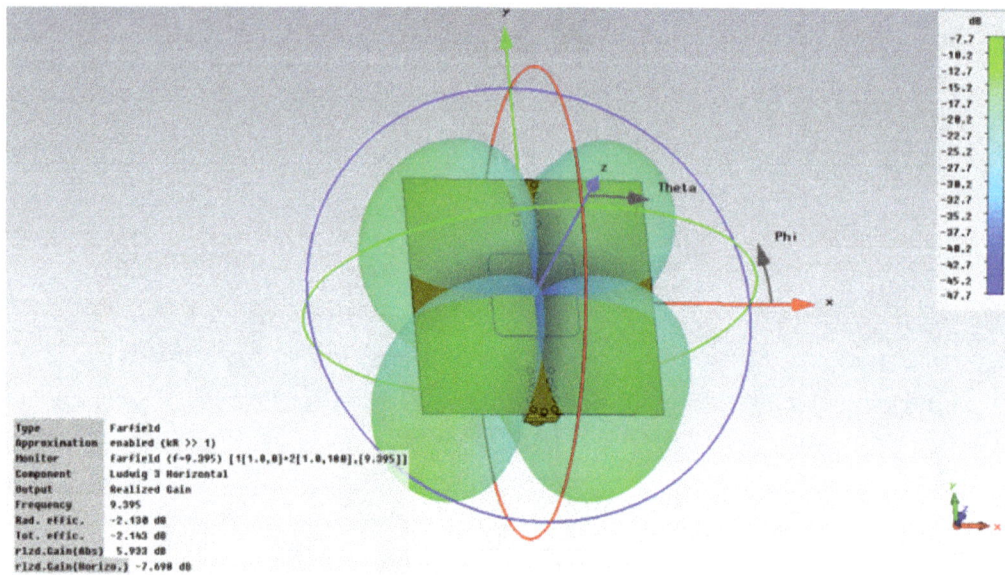

b-3) Differential probe- fed x-pol 3D Pattern

Figure 3.

ferential probe fed patch antenna with 180° out of phase feeding and the co-pol "realized gain" of 5.93 dBi.

The x-pol pattern in Fig. 3a-3 shows two maxima and one null in the E Cut. The radiating edge close to the feed contributes the highest x-pol radiation. The opposite to feed located radiating edge adds much lower x-pol radiation.

By further analyzing the x-pol pattern it is obvious that the E ($\Phi = 90°$) and H Cut ($\Phi = 0°$), typically provided by antenna manufacturer, would not show the maximum x-pol radiation intensities from the upper far field hemisphere. The

E Cut CPS would show almost perfect results and the H Cut would exhibit a high, but unrealistic x-pol contribution from the 3-D antenna radiation pattern. Only the dedicated 45° Cut would show the real antenna performance in CPS. For this reason, it is recommended to use 45° Cuts in any CPS discussions for patch antennas when the integrated CPS is important. Figure 4 shows the comparison of the single and differential probe fed patch antenna CPS performance using the recommended 45° Cuts.

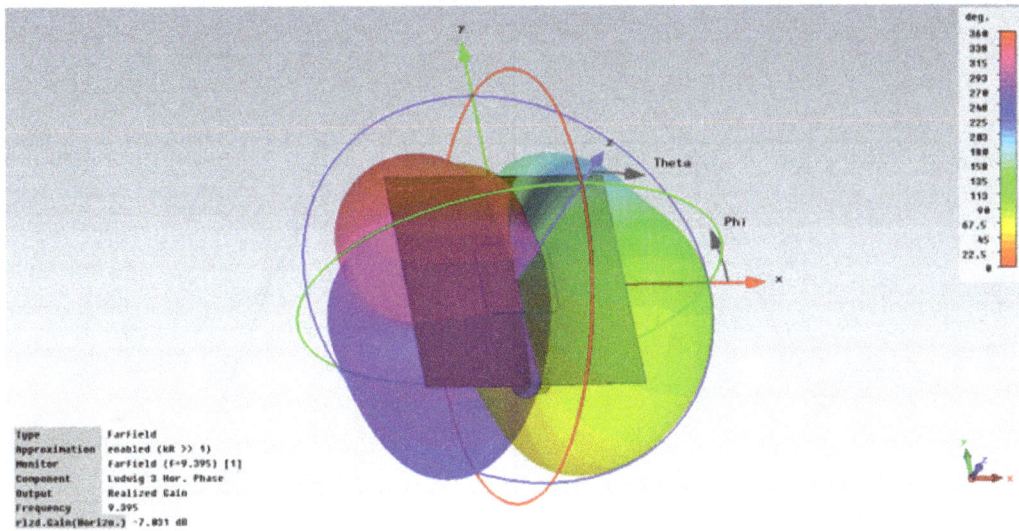

a-4) Single probe- fed phase x-pol 3D Pattern

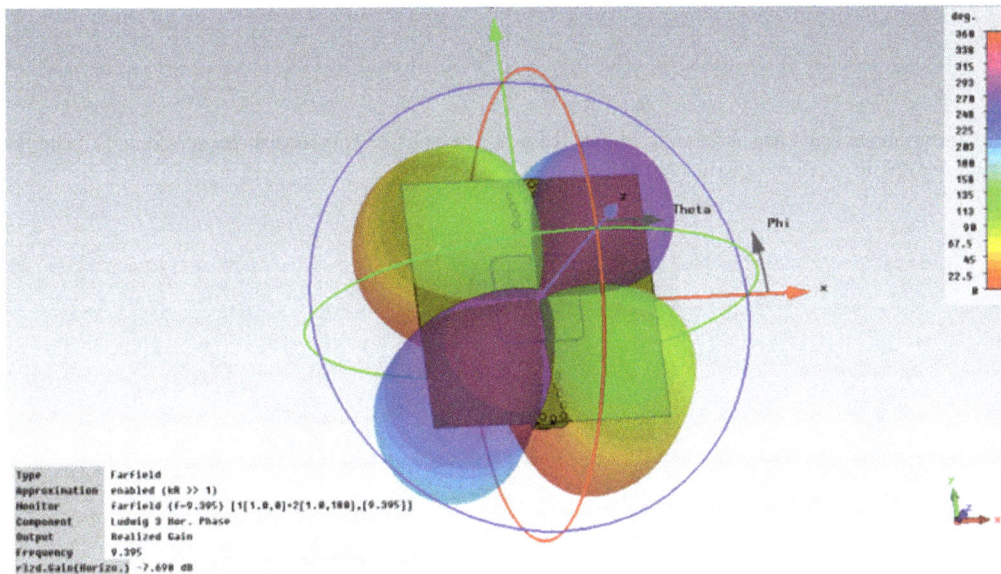

b-4) Differential probe- fed phase x-pol 3D Pattern

Figure 3. Comparison of co/x-pol patterns (**a**) without and (**b**) with differential feeding.

The differences in CPS performances are clearly identifiable by the plots in Fig. 4. The co- and x-pol plots from Fig. 4 are generated by the $\Phi = 45°$ Cuts from Fig. 3a-2, b-2, a-3 and b-3 respectively. Typically phased array antennas are providing azimuth e-scan ranges from +60 to −60°. Within this angular scan range the integrated CPS should be as low as possible. An integrated CPS of 14.2 dB for differential fed patch antennas can be achieved at ±60° AZ position, compared to only 6.6 dB at −60° AZ for the single fed patch antenna. The 3 dB half power beamwidth highlighted with the ellipse on the center top of Fig. 4 shows another advantage of the differential fed antenna for phased array applications: the beamwidth is quiet symmetric and the phased array scan gain loss with respect to the element pattern would also be symmetric for positive and negative scan angles. The single fed antenna shows a slight shift of the beam maxi-

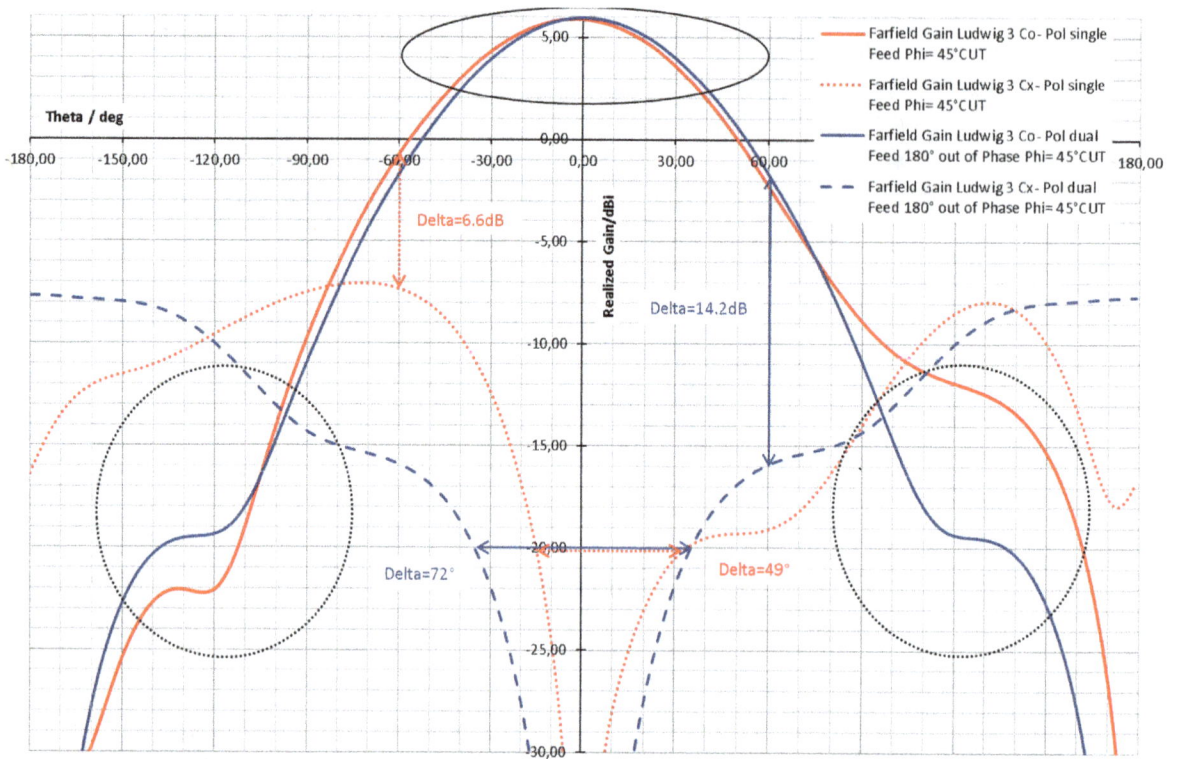

Figure 4. Comparison of single (red) and differential (blue) fed patch antenna CPS performances at 45° CUT. The dashed curves are representing the gain plots of the x-pol radiation patterns.

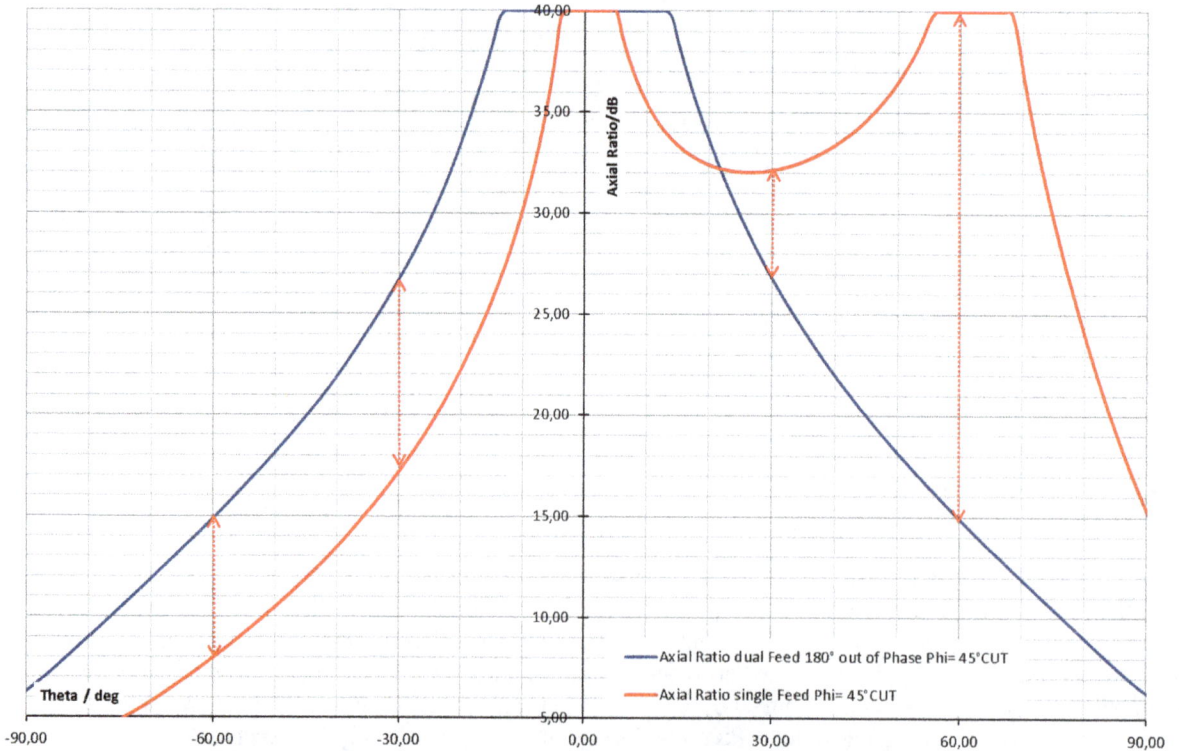

Figure 5. Comparison of single (red) and differential (blue) fed patch antenna axial rations at 45° CUT.

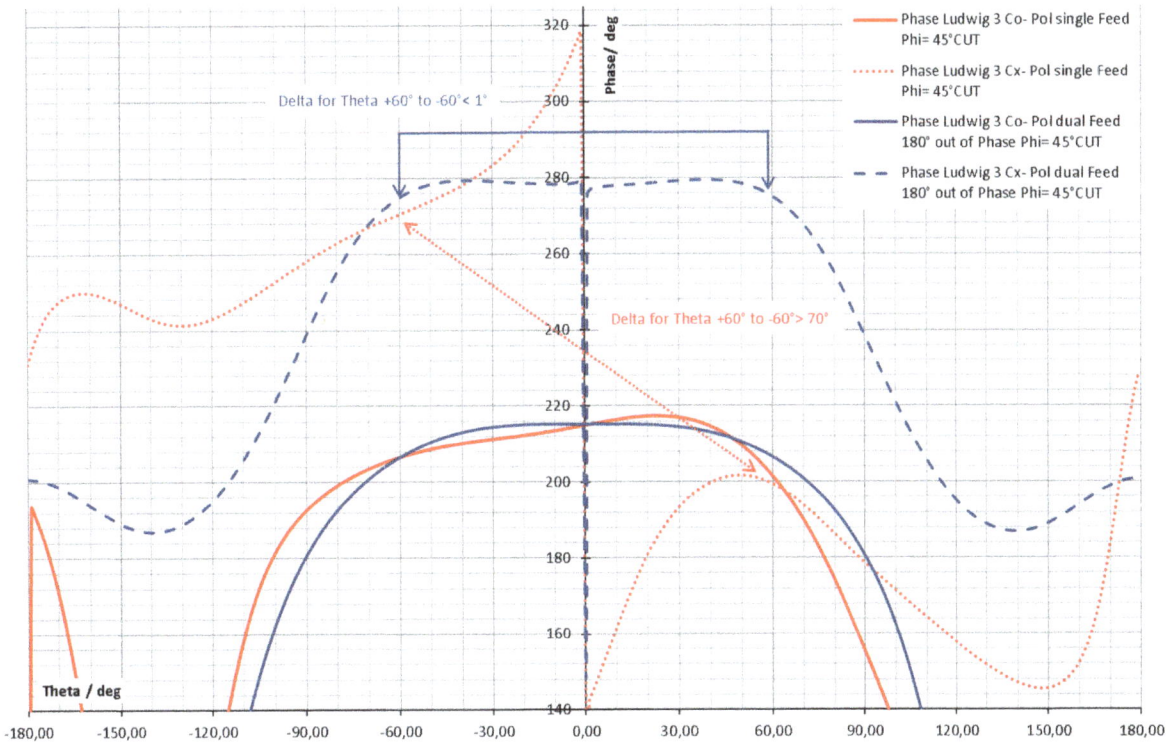

Figure 6. Comparison of Co and X-pol Phase plots from single (red) and differential (blue) fed patch antennas. The dashed curves are representing the phase plots of the x-pol radiation patterns.

mum in negative Theta direction and an asymmetric beam shape. This can be associated to the single feed configuration and the fact that stronger electromagnetic fields are radiated from the radiating edge located close to the feed. The black circles at $\Theta = \pm 120°$ indicate back radiated energy and here again the differential fed antenna back radiation is symmetric and would induce the same amount of energy in both backward directions. The mutual coupling would be equal to the neighbor elements compared to the asymmetric back radiation from single fed antennas. Furthermore, the x-pol free area below the main beam at $-20\,\text{dBi}$ is much broader for differential feeds with 72° azimuth range in contrast to only 49° azimuth range for the single feed configuration. Figure 5 compares the antenna axial ratios of the antennas. CST MWS is following the IEEE-Standard for the calculation of the axial ratio shown in Fig. 5. The axial ratio is the ratio of the major axis to the minor axis of the polarization ellipse. E_1 represents the horizontal polarized field vector and E_2 the vertical polarized field vector. It is calculated as follows:

$$AR = \sqrt{\frac{|E_1^2| + |E_2^2| + |E_1^2 + E_2^2|}{|E_1^2| + |E_2^2| - |E_1^2 + E_2^2|}}. \qquad (1)$$

The axial ratio for the differential feed patch is again symmetric and shows better results for negative and positive scan angles of up to $\Theta = 25°$. For negative angles the differential

feed shows a better axial ratio with smaller decay per degree. The single feed antenna shows for $\Theta > +25°$ better results than the differential feed antenna, but when considering all other disadvantages mentioned before this fact is negligible. It should also be noted here that an asymmetric axial ratio induces non- consistent scan angle biases for polarimetric phased array weather radar measurements.

The x-pol phases from Fig. 6 are generated from the $\Phi = 45°$ Cut of Fig. 3a-4 and b-4 respectively. The co-pol phase plots are not visualized in 3-D. It can be recognized that the co- and x-pol phase of the differential fed antenna is very symmetrical distributed, especially in the desired scan range of $\pm 60°$. The phase is identical over diagonal plane (see Fig. 3b-4). The co- and x- phases of the single fed antenna are asymmetric and the maxima of x-pol phases on diagonal planes are more than 70° out of phase. By analyzing the $\Phi = 135°$ and $\Phi = 45°$ Cuts from Fig. 7, a phase difference of exactly 180° can be observed for the complete Θ range. Each point is exactly 180° out of phase by comparing both diagonal planes. This predictable diagonal phase distribution can be used for array configurations where the neighboring elements x-pol contributions cancel each other in far field.

Furthermore, the two designs from Fig. 3 are redesigned for dual polarization mode to be comparable in their polarimetric port isolation characteristic. State-of-the-art weather

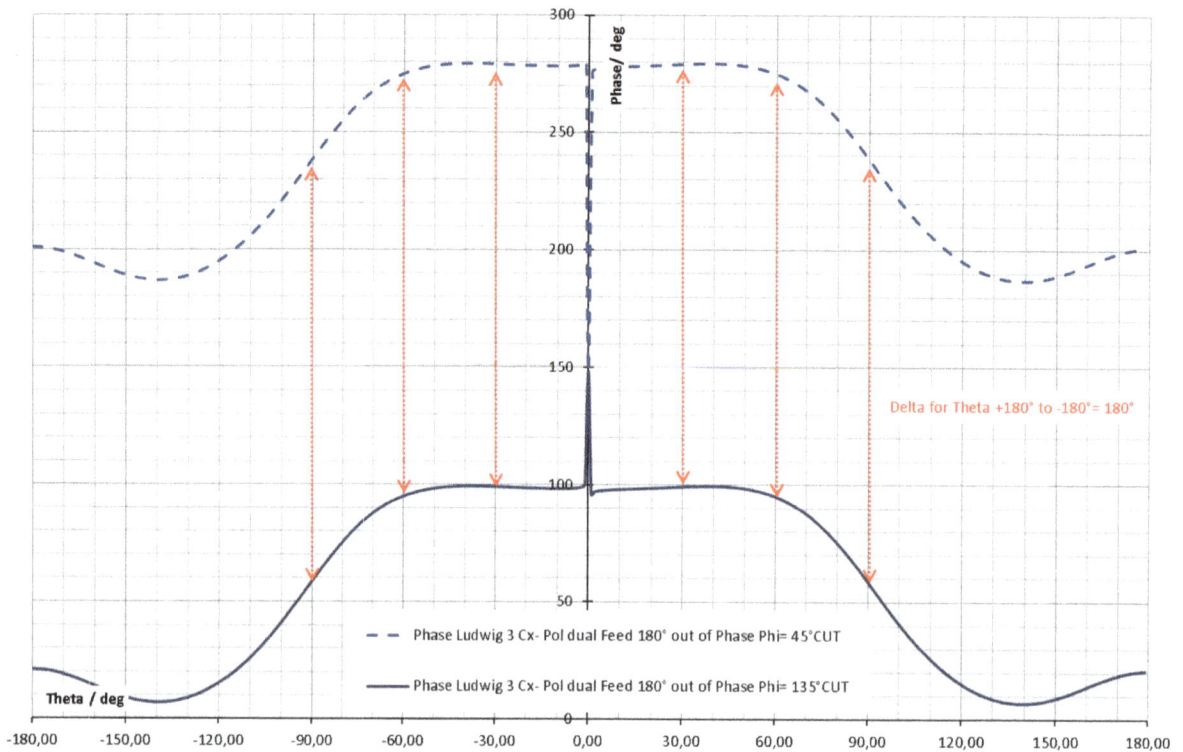

Figure 7. Phase difference of 180° for $\Phi = 45°$ Cut (dashed) compared to $\Phi = 135°$ Cut, established by differential feeding technique. The 180° phase difference is very useful in antenna array configurations for canceling out the x-pol components in far field from neighboring elements.

radars are working in pulse alternating or simultaneously transmitting dual polarization modes. For these dual polarization modes the single fed antenna will be equipped with two feeds; one for horizontal and one for vertical polarization. The differential fed patch antenna accommodates four feeds for dual polarization mode; two probe feeds 180° out of phase for horizontal polarization and two probe feeds 180° out of phase for vertical polarization. The differential fed patch antenna in dual polarization design is graphed in Fig. 8. The isolation and S11 results from CST MWS 3-D full wave simulations are visualized in Fig. 9. It can be recognized that the differential fed dual polarized patch antenna maximum port isolation provides on desired frequency at 9.395 GHz, meanwhile the single fed dual polarized antenna very poor isolation characteristics established. The excellent isolation characteristic of the differential fed dual polarized patch antenna can be explained with Fig. 8. Let us consider an input signal at the vertical port. The signal travels through the T-Junction splitter (red point in Fig. 8) and excites at first the Feed_V_1 and subsequently the Feed_V_2 with 180° phase shift. Both signals exciting the patch antenna to radiate, but small signal parts (dashed lines in Fig. 8) are received by the horizontal ports Feed_H_1 and Feed_H_2. Current probes have shown from simulations that the two signals received by Feed_H_1 and Feed_H_2 are in-phase at the probe feeds

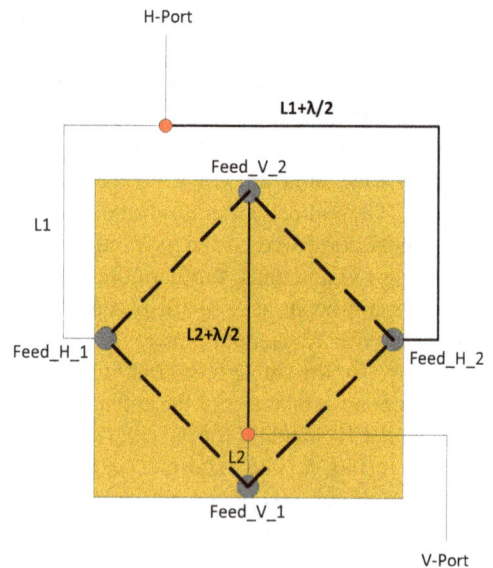

Figure 8. Graphical description for excellent isolation functionality of dual polarized patch antenna with differential feeding.

and cancel each other out at the T-Junction splitter in front of the horizontal splitter due to the length difference of $\lambda/2$.

It turns out, that the differential fed patch antennas with 180° out of phase excitation are suitable for phased array weather radar applications and many valuable advantages compared to single fed patch antennas are evidenced and first time documented within this publication.

3 Analysis of potential x-pol sources in microstrip patch antenna design

Open via holes, rectangular patch edges and the position of connector and matching circuits below the shielding ground are investigated and verified as potential x-pol sources. Additionally, circular and square patch geometries are compared and variations of the patch sizes are analyzed in their x-pol performances.

3.1 Probe fed patches with plugged via holes

In Fig. 10a the cross section of the hole plugged probe feed is graphed. After drilling the via holes into the multilayer substrate the walls of the holes are galvanized with copper to connect the upper and lower RF layer. The holes are plugged using the following two-step process: first the via hole is filled with cured resin to generate a plane surface on both sides and secondly an additional galvanization step takes place to close the via hole and to establish a plane surface for the radiation patch antenna element and the microstrip line. From the hole plugging process one would assume that the probe feed produces lower x-pol radiation because of plane patch antenna surface and lower resistance in the area of the feed, but CST MWS simulations have shown that the CPS is exactly (within ±0.01 dB) the same compared to usual open probe feeds. It was verified that hole plugging does not improve the CPS of microstrip patch antennas and the expenses for the additional plugging process can be saved.

3.2 Bended square patch antenna edges

By analyzing the x-pol radiation pattern in Fig. 3a-3, b-3, two (for single fed), respectively four (for differential fed) maxima above the patch edges are observable. It is assumed that x-pol radiation contributions are generated by the four 90° angles from the patch edges. Performed CST MWS simulations have evidenced that the patch antenna edge bending does not reduce the x-pol radiation. Only the resonance frequency will be shifted due to the change in absolute length of the radiating edge. After impedance re-matching of the bended patch antenna element almost the same x-pol values with the difference of only 0.04 dB was observed. Edge bending does not improve the CPS and the four 90° angles are not contributing significantly to the x-pol radiation.

3.3 CPS comparison of circular and square shaped microstrip patch antenna elements

An intensive literature search did not reveal any adequate reference comparing the CPS of circular and square shaped microstrip patch radiators. Therefore, a comparison of the two patch antenna shapes with respect to their x-pol performances is provided below. The comparison is based on the same resonance frequency, same layer configuration, slightly different parallel stub configuration due to the different patch impedances and the same differential feeding technique.

The circular patch radiator has a diameter of 12.67 mm, the probe via diameter is 0.4 mm and the probe via to edge distance is 0.5 mm. The diameter of the isolation circle inside the shielding ground for the probe feed transition is again 0.91 mm. The resonance frequency is located at 9.395 GHz with a VSWR 1 : 2 bandwidth of 152 MHz. In Fig. 11a the "realized gain" of 6.66 dBi can be recognized and the x-pol pattern from Fig. 11b shows the same results featuring four maxima comparable to Fig. 3b-3 due to the differential feeding technique. The gain is higher compared to the square patch but only due to the larger antenna patch area. The integrated CPS of 14 dB for differential fed circular patch antennas was established at ±60° azimuth position, which is almost equal to the 14.2 dB for differential fed square patch antennas from Fig. 4. By applying longer optimization iterations for impedance matching, the same integrated CPS for square and circular patch antenna geometries can be expected. It should be noted here that the value of 14 dB anyhow reflects the theoretical maximum of CPS at ±60° for microstrip patch antennas.

3.4 Analysis of probe feed and its matching network position below the shielding GND of the patch antenna

Typically, for impedance matching the probe feed positions are changed along the y axis. The maximum impedance is determined by the position of the probe feed at the edge of the patch. The minimum available impedance is zero and is located at the center of the patch antenna element. For phased array application with high x-pol requirements the differential feeding design verified in Sect. 2.2. should be applied. Consequently, only edge feeding can be realized to inject two separated signals with 180° out of phase, especially for dual polarization applications. Straight connections of the edge probe feeds are seldom possible due to the limited space for feeding networks, especially for dual polarization capabilities. In this analysis the direction of the probe feeds and its matching network below the shielding ground is changed with a rotation of 90° (Fig. 12a) and analyzed for x-pol (Fig. 12b) degradation.

For the differential fed square-patch with 90° rotated connectors an integrated CPS of only 6.2 dB was observed at ±60° azimuth position for the 45° Cut of Fig. 12, which

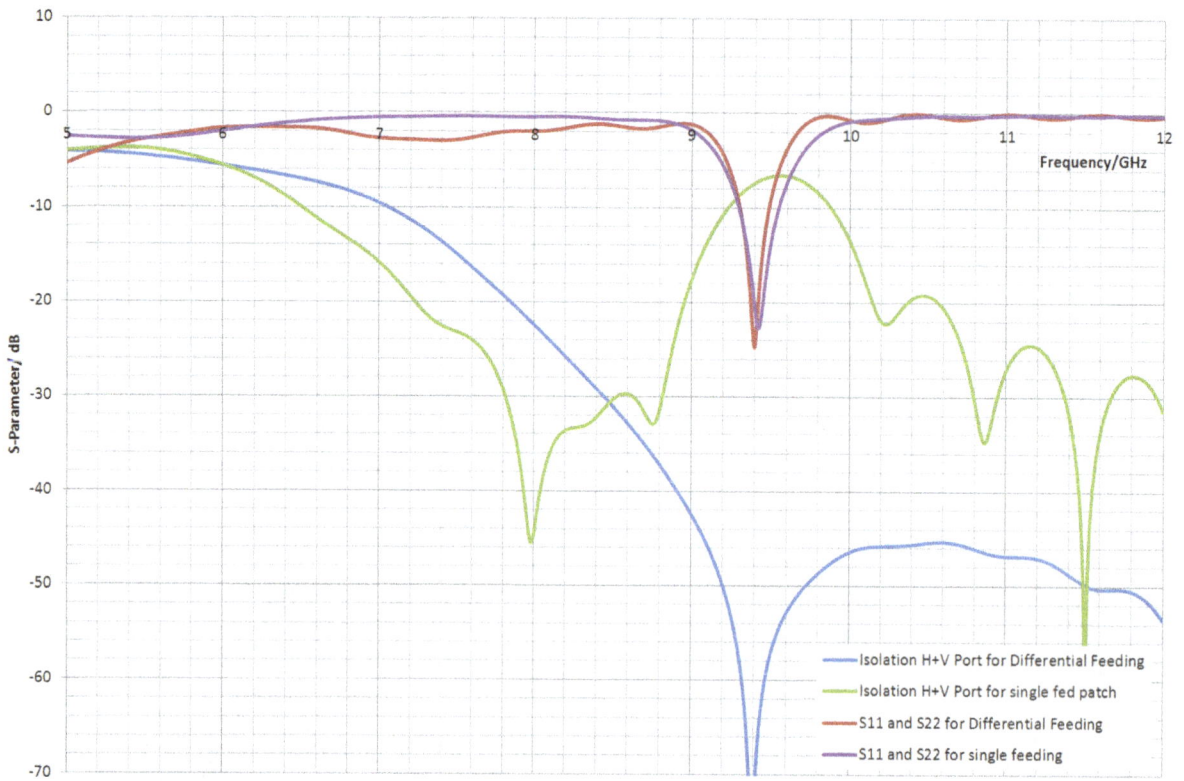

Figure 9. Comparison of port isolation characteristics from differential and single fed patch elements.

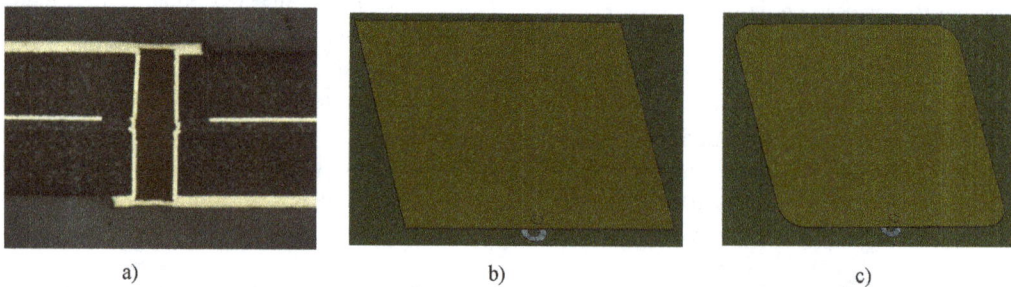

Figure 10. (a) Via filled with fossil resin and closed with cooper; square patches **(b)** without and **(c)** with edge bending.

is no longer comparable to the 14.2 dB for differential fed square patch antennas of Fig. 4, where the probe feeds are straightly oriented (see Fig. 3b-1) and all mirror symmetries are maintained. From the not shown co-pol plot, an undesired main beam drift can also be recognized. The reason for the x-pol degradation is probably related to the asymmetric current distribution on the shielding ground below the patch. The asymmetric current distribution on the shielding ground is generated by the matching stub positions and the current disturbances during excitation process of the antenna. This source of x-pol radiation should be considered by every antenna designer for array designs with high x-pol requirements.

3.5 Analysis of antenna patch size

Surprisingly no capable reference could be found in literature where the patch antenna size was analyzed for improved x-pol pattern. For this reason the differential fed patch antenna design from Fig. 1c was investigated in this regard. Figure 13 shows graphically the results of the co- and x-pol pattern for variable patch sizes from 10.32 to 10.80 mm. Table 1 summarizes all important design parameters, the results of realized gain at broadside direction and the CPS at $\Theta = \pm 60°$.

The design example shows CPS of up to 15.50 dB at $\Theta = \pm 60°$. On the other hand the realized gain stays almost constant between 5.9 and 6.0 dBi. In Fig. 13 the redistribution of the x-pol pattern for increased patch sizes becomes visible. Inside the dashed circles the CPS will be improved,

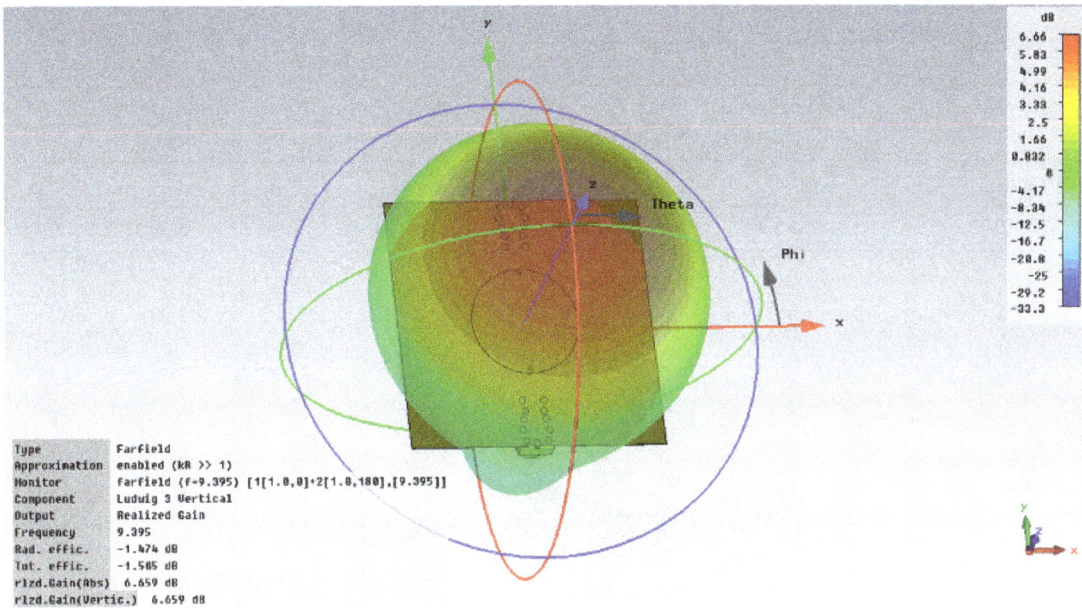

a)

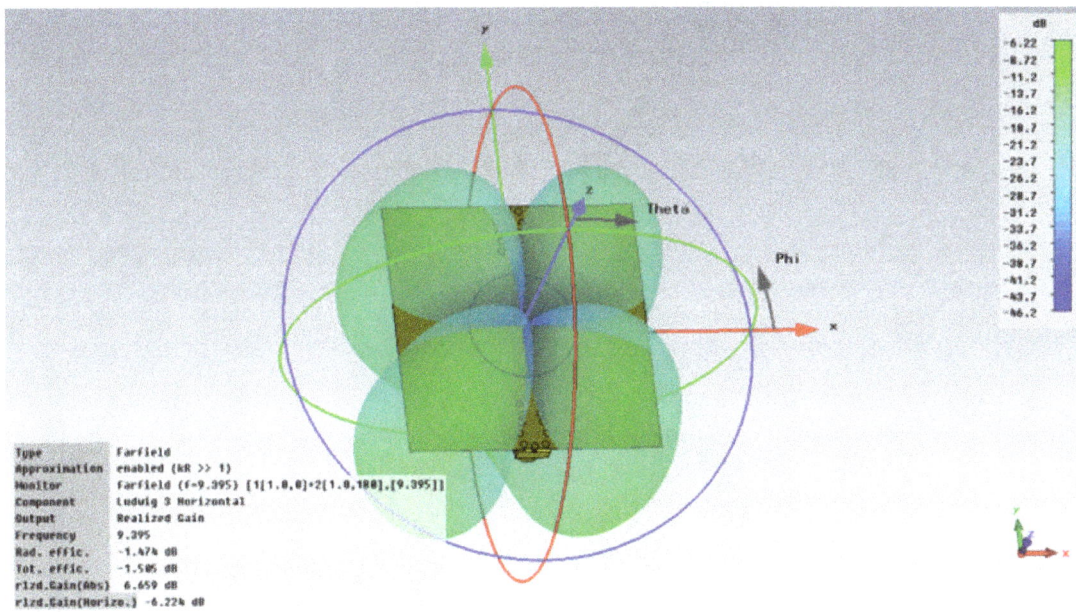

b)

a) Co-pol pattern, b) x-pol pattern of circular shaped X-Band patch antenna

Figure 11. (**a**) Co-pol pattern, (**b**) x-pol pattern of circular shaped X-Band patch antenna.

meanwhile the dashed ellipses in upper right and left corner are showing CPS degradation. This redistribution is acceptable desirable, since the CPS at invisible space behind the antenna is not that important.

The improvement of the CPS by increasing the microstrip antenna patch size is significant and can be identified as valu-able tool for antenna designers to reduce the cross polarization of microstrip patch antennas. The x-pol radiation is moved from the center to the backside of the antenna. Especially for phased array applications the x-pol improvement for intended scan ranges between −90 and +90° is remark-able.

Table 1. Design parameter and CPS results for variable patch sizes.

Patch size	Via diameter	GND Isolation diameter	Stub width	Stub length	Via distance to patch edge	CPS at $\Theta = \pm 60°$	Realized length	GND
mm	mm	mm	mm	mm	mm	dB	dBi	mm
10.32	0.2	0.91	1.47	4.78	0.42	13.20	6.00	30×30
10.55	0.4	0.91	1.31	5.13	0.42	14.20	5.95	30×30
10.66	0.5	1.2	1.42	5.19	0.42	14.75	5.94	30×30
10.80	0.7	1.5	1.49	5.39	0.50	15.50	5.90	30×30
10.66	0.7	1.2	1.42	5.19	0.42	18.34	5.31	40×40

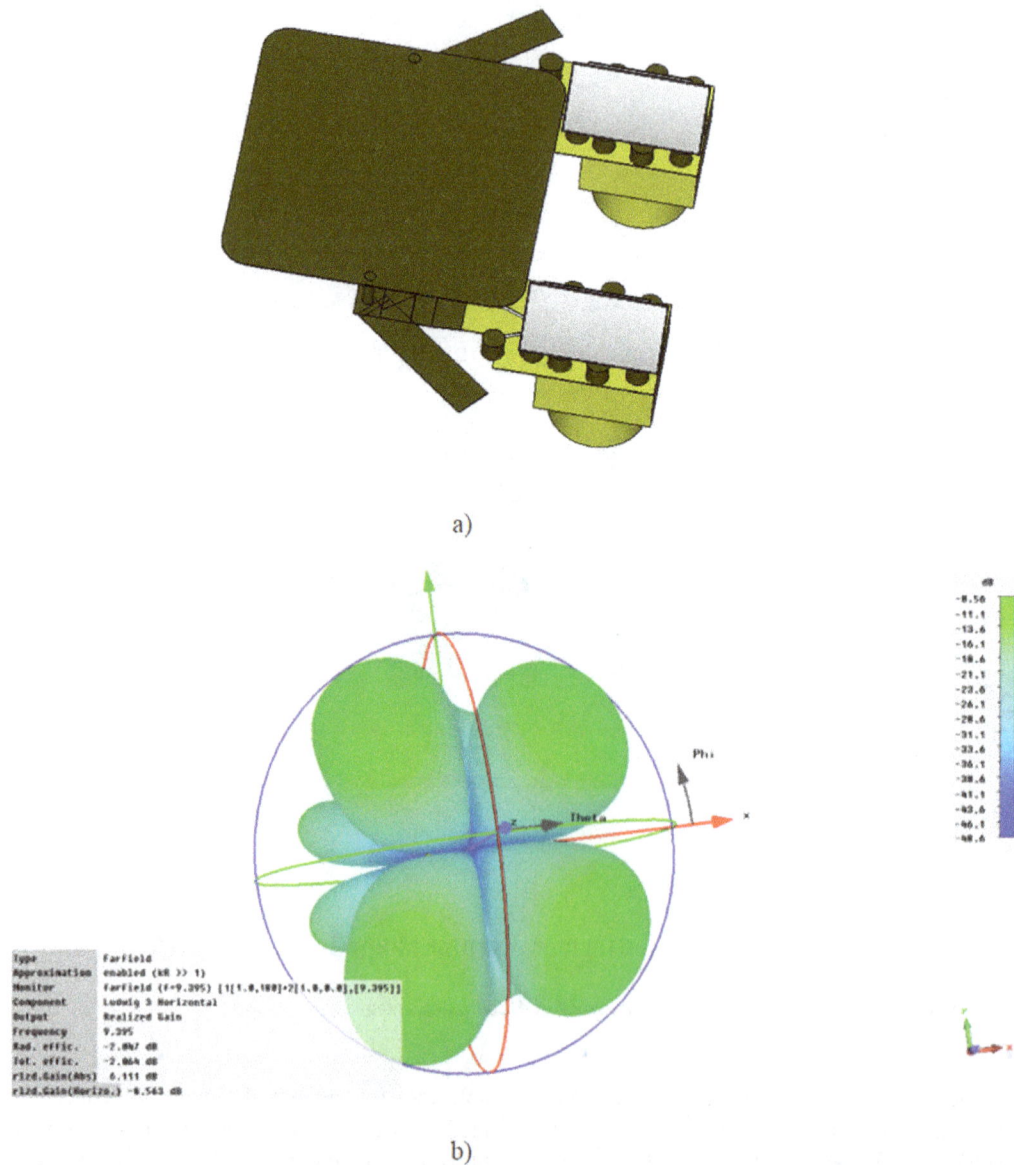

a)

b)

Figure 12. (a) Connectors and matching stubs below the patch and shielding ground with 90° rotated, **(b)** degraded x-pol pattern of the respective patch antenna.

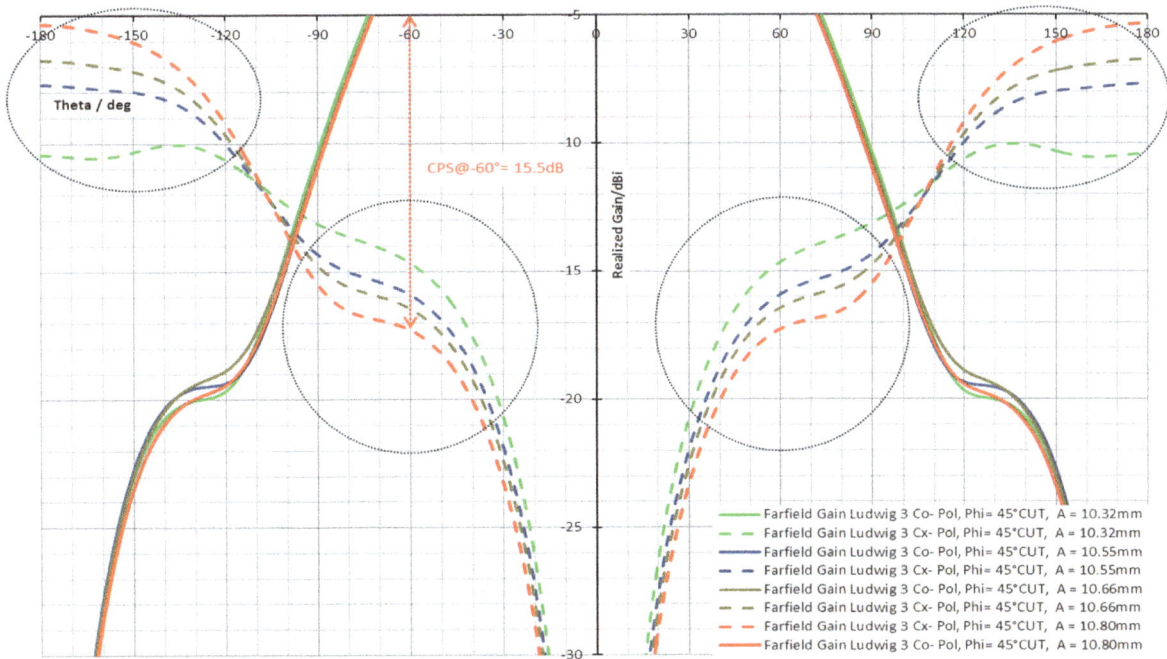

Figure 13. Analysis of co- and x-pol radiation pattern for differential fed patch antennas with variable patch length. The patch length varies from 10.32 to 10.80 mm. A valuable CPS improvement can be recognized in the circles between $\Theta = 90$ and $30°$. The ellipses between 180 and $120°$ show a degradation in CPS, which is acceptable since the CPS at invisible space behind the antenna is not that important.

In a separate experiment the GND layer of the 10.66 mm patch design was extended from 30×30 to 40×40 mm. The realized gain in broadside direction dropped to 5.3 dBi but the CPS at $\Theta = \pm 60°$ became 18.34 dB. This astonishing result is interesting for single antenna applications and their designers but is not less important for phased array applications where the GND layer and the patch to patch distance can mostly not be extended due to the occurrence of entering grating lobes.

From the results in Table 1 an empirical equation can be constructed for the length $L_{\mathrm{CPS}}^{\mathrm{opt}}$ of a quadratic patch antenna element with optimized x-pol performance:

$$L_{\mathrm{CPS}}^{\mathrm{opt}} = K \frac{\lambda}{2\sqrt{\varepsilon_r^{\mathrm{eff}}}} = 1.12 \frac{\lambda}{2\sqrt{\frac{\varepsilon_r+1}{2} + \frac{\varepsilon_r-1}{2}\left[\frac{1}{1+\sqrt{1+\frac{12\,h}{W}}}\right]}}. \quad (2)$$

For patch antennas working at the dominant TM010, the edge to edge distance is typically $\frac{\lambda}{2}$. The x-pol pattern will be optimized if the patch size L is slightly enlarged with $K = 1.12$. The effective permittivity $\varepsilon_r^{\mathrm{eff}}$ for microstrip lines where the conductor width is larger than the substrate height ($w > h$) was found in (Hartley, 2014) and represents the part below the square root in Eq. (2).

4 Understanding the generation of x-pol for microstrip patch antennas by using the new visual based field line and surface current distribution approach

4.1 Cavity model from literature

In several antenna literature references e.g. Balanis (2005) or Grag et al. (2001), the microstrip patch antenna radiation process is explained by the well-known cavity model. The field distribution below the patch as illustrated in Fig. 14a generates the co- and x-pol pattern by assuming two radiating slots with distance L. The uniform field distribution over the width W and the cosine distribution over the length L for the dominant TM010 mode can be recognized. The current and charge distribution of microstrip patch antennas is presented in Fig. 14b. Fringing fields from the cosine distribution along the length L are not considered by the cavity model, due to the assumption of perfect magnetic walls surrounding the volume. Accordingly, this is one source of x-pol radiation only visible in 3-D full wave analysis.

Bhardwaj and Rahmat-Samii (2014) pointed out that the x-pol pattern from the cavity model does not agree with the x-pol pattern from measurements or 3-D full wave simulations. Their simulation and measurement results from single fed antennas have only shown a single null in E Cut and the cavity model predict nulls in E and H Cut.

Here another interpretation is proposed. If the field vectors from Fig. 14a are exactly the same in intensity and $180°$ out

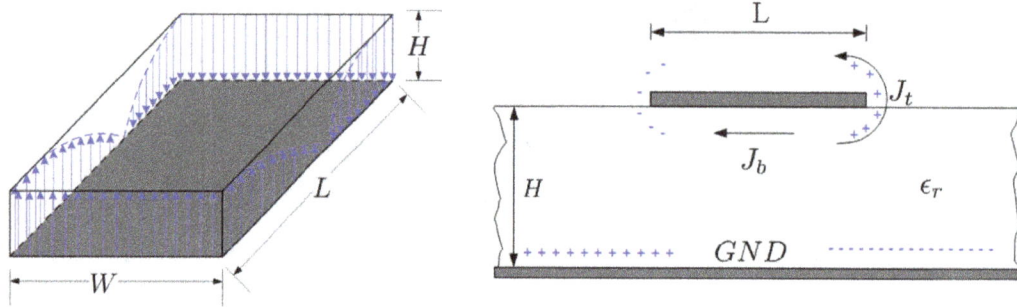

Figure 14. (a) Electrical field distribution below microstrip patch antenna with surrounding, perfect magnetic walls, **(b)** current and charge distribution of microstrip patch antenna (Grag, 2001).

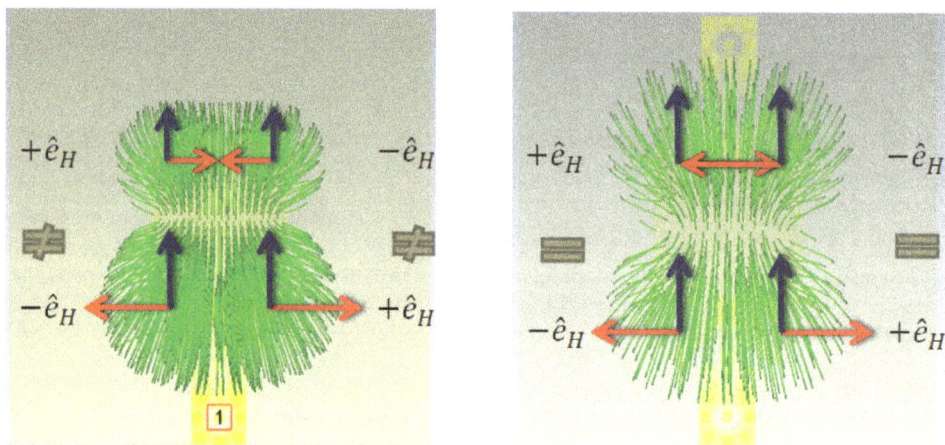

Figure 15. Field-line distributions: **(a)** single probe fed, **(b)** differential probe fed patch.

of phase at the two radiating edges, the E and H Cut can be established with two nulls like predicted from cavity model. Here the electrical field vectors have the same amplitude and show 180° phase difference, indicated by the up and down direction of the electrical field vectors from Fig. 14a. On the contrary, if a patch antenna will be fed by a single probe feed like in Fig. 3a-1 the field intensities across the length L are unequal (see also Fig. 15), so that a x-pol pattern with only one null in E Cut as illustrated in Fig. 3a-3 will be generated. From both figures it is visible that the x-pol contribution is much higher on the radiating edge side, where the probe feed is located. This effect will now be analyzed and discussed in detail by the field line and surface current distribution approach in the following Sects. 4.2 and 4.3.

4.2 Visual based field line distribution approach for x-pol explanation

The field line approach uses the field line generation process from 3-D full wave analysis visualized with CST MWS. Figure 15 illustrates the field line distribution for single probe feed excitation (Fig. 15a) and differential excitation with two signals 180° out of phase (Fig. 15b). As mentioned before,

the field intensities below the patch, across the length L is unequal for single excitation (Fig. 16a) and equal for the differential excitation (Fig. 16b). The insertion loss of the path from the single probe feed to radiating edge on the opposite side is very low and the difference in field intensity is not explained herewith. A time delay causes the field intensity differences for a given point of time. As a consequence, the field line development process on the probe feed side is at advanced stage, with respect to the opposite sided radiating edge without the feed probe. The difference in field line strength between the upper and lower half of the patch at a given point of time generates the higher x-pol pattern and no null in H Cut will be produced in far field. If the x-pol contribution (generated by the field lines which are not exactly vertical (co-pol) oriented for the example in Fig. 15 of the electrical field vectors on upper and lower half of the patch) are equal in strength and 180° out of phase the cancellation in far field takes place and the null in H Cut will be generated. From Fig. 15b one can observe a very symmetrical and time synchronized field line development process. 3-D animations of the field line development process over a complete RF phase cycle are showing that the opposite 180° out of phase field lines are always synchronized in time and of

a)

b)

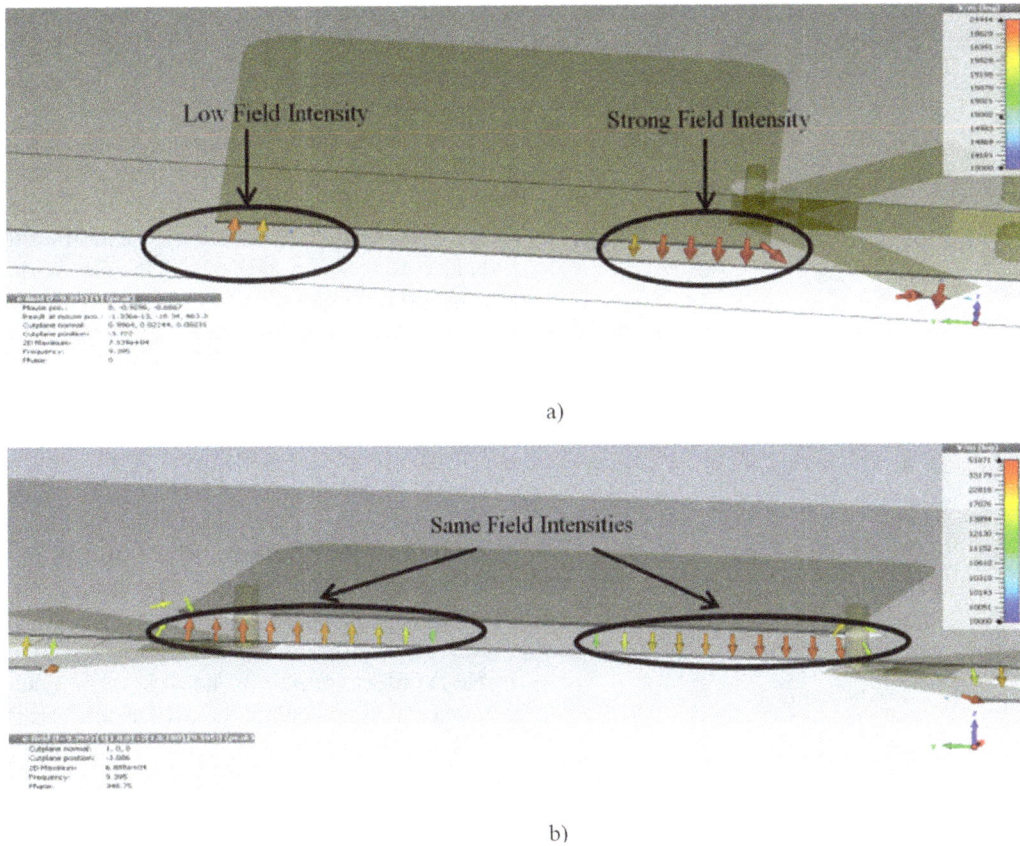

Figure 16. (**a**) Field distribution below the patch element (**a**) is different in strength and phase for single feed patch antennas (**b**) is equal in strength and exactly 180° out of phase for differential fed patch antennas.

the same intensity. By further observing Fig. 15a and b, the two and four typical x-pol maxima visible in 45 and 135° Cuts can be derived. Figure 15b shows that the field lines with the strongest horizontal electrical vector contribution (x-pol contribution) are located exactly above the 4 patch corners. Correspondingly, the very symmetric x-pol pattern with the typical four x-pol maxima over 45 and 135° Cut will be generated as highlighted in Fig. 3b-4. The visualization of the field vectors helps to understand that the opposite x-pol vectors $\hat{e}_H$ (in red) must be the same of strength and exactly 180° out of phase to cancel each other out in the antenna phase center of the H Cut. This is certainly only the case for the differential fed antenna in Fig. 15b.

4.3 Visual based surface current distribution approach for x-pol explanation

Another approach for understanding the generation process of x-pol is the observation of surface currents from microstrip patch antennas. For this approach every single current vector from Fig. 17 can be interpreted as an elementary electric dipole (Hertzian Dipole) located on the patch antenna surface. By applying the right hand law, small dipole patterns are generated which add in the far field. As already discov-

ered by the analysis of the field line approach, the mirror symmetry of field vectors is of significant importance for the reduction of x-pol radiation. The same is true for the distribution of the currents on the patch surface. Figure 17a shows the surface current distribution of the single fed microstrip patch antenna. Figure 17b shows the current distribution on the differential fed microstrip patch antenna, excited with two signals 180° out of phase. By thoroughly investigating the current vector directions from Fig. 17a, the different vector orientations close to the two radiating edges are unambiguous. The patch surface at the feed side shows much stronger horizontally oriented E field vectors (x-pol). This explains the stronger x-pol radiation pattern values from Fig. 3a-3 and the stronger field lines in Fig. 15a. Furthermore, the current distribution on the surface of the patch is not symmetrical, so that no perfect cancellation in far field can be reached for the single feed case. On the contrary, the current vectors from Fig. 17b and c are antipodal in direction in the region close to the two opposite probe feeds and occur with exactly the same current strength. Because of the mirror symmetry of the current distributions at the feed areas, the radiated x-polarized electrical fields (here with horizontal content) with 180° phase difference cancel each other in the far field. This explains the x-pol radiation pattern from Fig. 3b-3

a)

b)

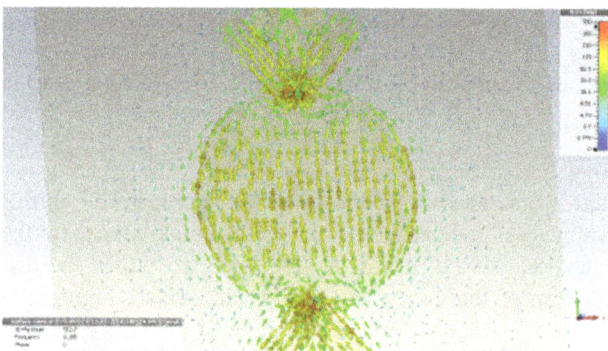

c)

Figure 17. Surface currents on (**a**) single fed square patch antenna; (**b**) differential fed square shaped patch with two signals 180° out of phase (**c**) differential fed circular shaped patch with two signals 180° out of phase.

and the symmetric field line distribution in Fig. 15b. It should be stated here, that the current distribution for circular and square patches are different. Interestingly, the different surface current distributions from circular and square patch generate almost the same CPS in far-field. Both current distributions are mirror symmetrical on the patch surfaces and the x-pol contributions cancel in the far field.

5   Conclusions

The differential feeding method was verified as a valuable x-pol reduction solution from literature research for phased array weather radar applications. Numerous feeding methods were discussed for phased array applications. The most beneficial feeding method is the probe fed antenna with its relatively small ground plane via hole transition to the radiating patches. The analysis of the geometric shape of the x-pol pattern revealed that only 45°/135° Cuts are able to provide sufficient information for accurate cross polarization suppression measurements, since the x-pol maxima are located on the diagonal planes. The differential feeding method was verified and compared to single probe feeding by two CST MWS modeled X-Band patch antennas. Several advantages from differential probe fed patch antennas for phased array applications were documented; the co-pol main beam becomes symmetric and the integrated CPS of 14.2 dB compared to only 6.2 dB at $\Theta = \pm 60°$ was discovered. Furthermore, the x-pol pattern with its four maxima and exactly 180° phase difference of the two diagonal planes is advisable in order to reach reasonable x-pol far field cancellations in antenna array configurations. The analysis of potential x-pol sources in patch antenna element design has shown that resin filled, copper plated probe feed vias and the bending of patch edges do not improve the x-pol radiation pattern. For the first time, circular and square shaped patch geometries were investigated and qualified as almost equal in CPS performance. The thorough investigation of the position of the probe feed and its matching network below the shielding ground of the patch has shown significant x-pol degradation due to shielding ground current disturbances during the patch antenna radiation process. The improvement of the CPS by increasing the microstrip antenna patch size can be identified as valuable tool for antenna designers to reduce the cross polarization of microstrip patch antennas. The x-pol radiation is moved from the center to the backside of the antenna. Especially for phased array applications the discovered x-pol improvements for intended scan ranges stated in Vollbracht (2014) is remarkable.

Starting from the discussion of the cavity model, two new visual based approaches with field line and surface current distributions were developed as a contribution for better understanding of the generation process of x-pol radiation from microstrip patch antennas. Finally, the reason for two and four maxima in the x-pol radiation pattern is explained with the new visual based approaches for single and differential probe fed microstrip patch antennas.

Acknowledgements. I express my gratitude to my supervisors Madhukar Chandra and Frank Gekat for the guidance and the fruitful discussions on weather radar topics and antenna theory.

References

Balanis, C. A.: Antenna Theory, Analysis and Design, 3rd Edn., 1136 pp., 2005.

Bhardwaj, S. and Rahmat-Samii, Y.: Revisiting the Generation of X-pol in Rectangular Patch Antennas: A Near-Field Approach, IEEE Antenn. Propag. M., 56, 14–38, 2014.

Fulton, C. and Chappell, W.: A dual-polarized patch antenna for weather radar applications, IEEE International Conference on Microwaves, Communications, Antennas and Electronics Systems (COMCAS), Tel Aviv, Israel, 7–9 November 2011, 1–5, 2011.

Grag, R., Bhartia, P., Bahl, I., and Ittipiboon, A.: Microstrip Antenna Design Handbook, ARTECH HOUSE, London, UK, Boston, USA, 2001.

Guha, D., Biswas, M., and Antar, Y. M. M.: Microstrip Patch Antenna With Defected Ground Structure for X-pol Suppression, IEEE Antenn. Wirel. Pr, 4, 455–458, 2005.

Liang, X.-L., Zhang, Y.-M., Zhong, S.-S., and Wang, W.: Design of Dual-Polarized Microstrip Patch Antenna With Excellent Polarization Purity, Proc. 3rd International Conference on Computational Electromagnetics and Its Applications, Beijing, China, 1–4 November 2004, 197–199, 2004.

Ludwig, A. C.: The Definition of X-pol, IEEE Transactions on Antennas and Propagation, AP-21(1), 116–119, 1973.

Mohanty, A. and Das, N. K.: Characteristics Of Printed Antennas And Arrays Covered With A Layer Of Printed Strip-Grating For Suppression Of X-Pol, International Symposium of Antennas and Propagation Society, Ann Arbor, MI, USA, 28 June–2 July, 1993.

Vollbracht, D.: System specification for dual polarized low power X-Band weather radars using phased array technology, TPH22, International Radar Conference, Lille, France, 13–17 October, 2014.

Zhou, S.-G. and Chio, T.-H.: Dual Linear Polarization Patch Antenna Array with High Isolation and Low X-pol, International Symposium on Antennas and Propagation (APSURSI), Spokane, WA, USA, 3–8 July 2011, 588–590 2011.

20

Comparison of electromagnetic solvers for antennas mounted on vehicles

M. S. L. Mocker[1], S. Hipp[2], F. Spinnler[1], H. Tazi[3], and T. F. Eibert[1]

[1]Technische Universität München, Lehrstuhl für Hochfrequenztechnik, Arcisstrasse 21, 80333 Munich, Germany
[2]CST AG, Bad Nauheimer Str. 19, 64289 Darmstadt, Germany
[3]Audi AG, August-Horch Str., 85055 Ingolstadt, Germany

Correspondence to: M. S. L. Mocker (marina.mocker@tum.de)

Abstract. An electromagnetic solver comparison for various use cases of antennas mounted on vehicles is presented. For this purpose, several modeling approaches, called transient, frequency and integral solver, including the features fast resonant method and autoregressive filter, offered by CST MWS, are investigated. The solvers and methods are compared for a roof antenna itself, a simplified vehicle, a roof including a panorama window and a combination of antenna and vehicle. With these examples, the influence of different materials, data formats and parameters such as size and complexity are investigated. Also, the necessary configurations for the mesh and the solvers are described.

1 Introduction

For solving electromagnetic problems in complex environments, the choice of the most appropriate method does not only determine the time efficiency, but has an influence on the accuracy of the gained results, as well. There is not a single combination of a numerical method and algebraic solver, in the following called solver, which can fulfill all requirements. Moreover, many parameters, such as size in relation to wavelength, complexity and resonating behavior must be considered. In the following, the transient (T), the frequency (F) and the integral (I) solver offered in CST MWS (Weiland, 1996), (CST, 2015) are investigated.

The T solver is based on the finite integration technique. The geometrical model is here divided into hexahedra (Yee, 1966) and a time signal is propagated through the structure (Weiland, 2008). In general the hexahedral mesh is a very robust way of meshing for complicated structures, but

has some disadvantages, for example in case of curved geometries. In these cases, the mesh must either be extremely dense or is meshed by utilizing the perfect boundary approximation technique, where sub-cellular information is taken into account for curved elements (Krietenstein, 2001). An improved mesh can be achieved by subgridding (Podebrad, 2003), where critical areas are meshed with more lines than the rest. For highly resonant structures such as antennas, the simulation duration may be very high or resonances may even not be simulated correctly at all due to an insufficient decay of energy within the system. This can be solved by an autoregressive (AR) filter (Percival, 1993) which drastically reduces the simulation time as the spectral properties can be retrieved from rather few time steps.

The F solver uses the finite element method. A limit for this method is the availability of random access memory (RAM) which is used mainly dependent on the number of mesh cells. The resulting matrix is sparsely populated as elements are only non-zero if nodes in the discretized geometry are neighboring. The numerical system size can be reduced by a model order reduction technique (MOR) (Ilic, 2004). In a first step, the structure is meshed with surface triangles and only in case there is a thickness, the volume is meshed with tetrahedra. At critical points, the mesh is corrected for the highest simulation frequency usually mainly by further refinements (Cendes, 1985; Pinchuk, 1985).

The I solver uses the Method of Moments. As only the surface must be meshed, the method is well suitable for large solution domains. Dielectrics are not meshed for this solver method in CST MWS.

In the following, the mentioned solvers and methods are investigated in order to simulate a complex roof antenna

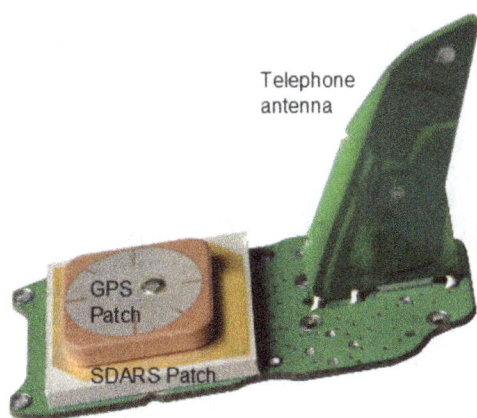

Figure 1. Photograph of the antenna structures.

Figure 2. Photograph of the antenna plastic cap.

Table 1. Services joined within one antenna assembly.

Antenna	S Parameter	Frequencies
SDARS patch	S_{11}	2.33 GHz
GPS patch	S_{22}	1.58 GHz
Telephone fin	S_{33}	824–894 MHz
		1.85–1.99 GHz
		1.71–1.755 GHz

mounted on a vehicle accurately and efficiently. Therefore, several simulations with the roof antenna itself are conducted. In a second step, the solvers are compared for the purpose of simulating extended simulation domains as vehicles and roofs. In these models, monopoles are used as simplified antennas in order to isolate the problems from each other. Finally, a vehicle including the roof antenna is simulated. Also, the influence of data formats and materials is taken into consideration. The values of interest for the feasibility and efficiency of a simulation are majorly the RAM and time consumption. Especially the time consumption is only a rough value. The benchmark computer has 2 processors of the type Intel(R) Xeon(R) CPU with E5640@2.67 GHz and 24 GB RAM. Each of the processors consists of 4 cores and moreover the Intel(R) Hyper-Threading Technology is enabled. Some simulations could not be performed on this computer so the necessary time was estimated. Simulations for the purpose of time comparability were started in order to estimate the differences in computation speed. Finally, the given times can be seen as benchmarks.

2   Roof antenna

The first part of the investigation is a roof antenna itself. The antenna designed for the north American market consists of a SDARS patch, a GPS patch and a telephone antenna, which are contained in one antenna assembly as shown in Fig.1. It is built by thin metal sheets and several dielectrics. All covered services are listed in Table 1. The antenna assembly is adapted to a mounting consisting of metal and plastic for the purpose of sealing and is covered by a housing of plastic as shown in Fig. 2.

All solvers explained above, except the I solver, are evaluated for the antenna model and finally compared to measurement results. All connections are modeled as coaxial structures. The antenna needs to be slightly modified for each solver. For the transient solver all ports were implemented as perfect conducting wires between two points realizing a source, called edge ports, whereas with the frequency solver a face is used instead of a thin wire, called discrete face ports, were used. This port modification does not relevantly change the simulation behavior as the results for the Global Positioning System (GPS) and Satellite Digital Audio Radio Services (SDARS) antennas do correspond well to each other. For the simulations with the T solver, the antenna is meshed using hexahedra as shown in Fig. 3. To ensure that all metalizations are correctly identified in the hexahedral mesh they are thickened in order to ensure at least 2 mesh lines for each material, even though this was only mandatory for dielectrics. For the F solver, the antenna is meshed with tetrahedra as shown in Fig. 4. The automatic discretization process is more stable for the hexahedral mesh in comparison to the tetrahedral mesh, especially if the model contains material jumps in combination with complicated structures.

The simulated and measured reflection parameters of the SDARS antenna are shown in Fig. 5 and of the telephone antenna in Fig. 6. The resonance frequencies are well met for the SDARS and GPS simulations and roughly for the simulation of the telephone antenna. In Table the RAM and time consumption are listed for all simulations. Efficient and accurate simulations are possible with the F solver. After simplifications of the model by neglecting details and a manual optimization of the mesh by assigning discretization densities to different materials in the model, the number of cells can be reduced to 200 000 cells (Mocker, 2014). With the adaptive mesh, the results are not completely correct for resonance frequencies as shown in Fig. 6. The adaptive mesh of the F solver does not change the results of the SDARS and the

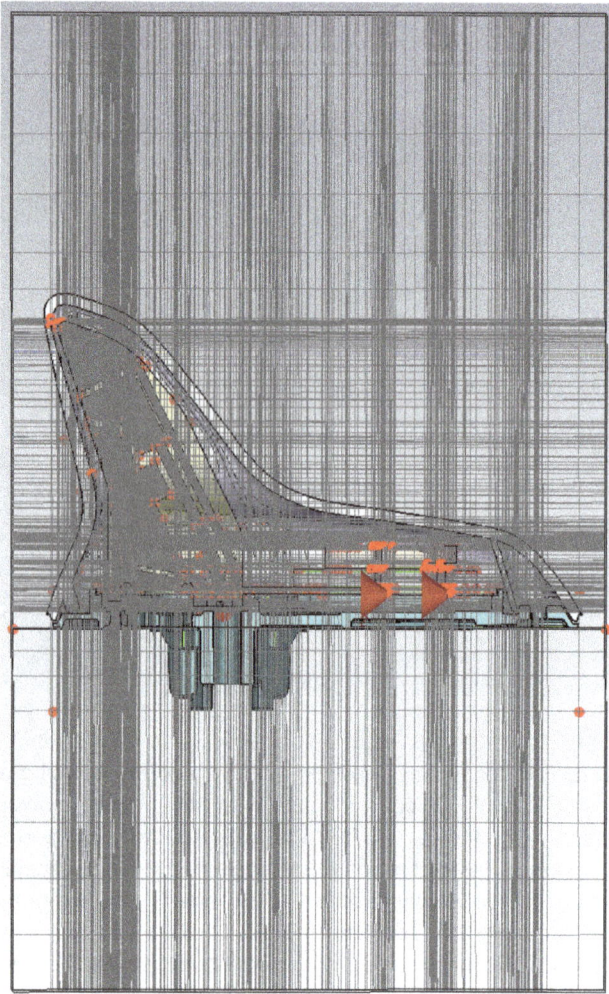

Figure 3. Hexahedral mesh.

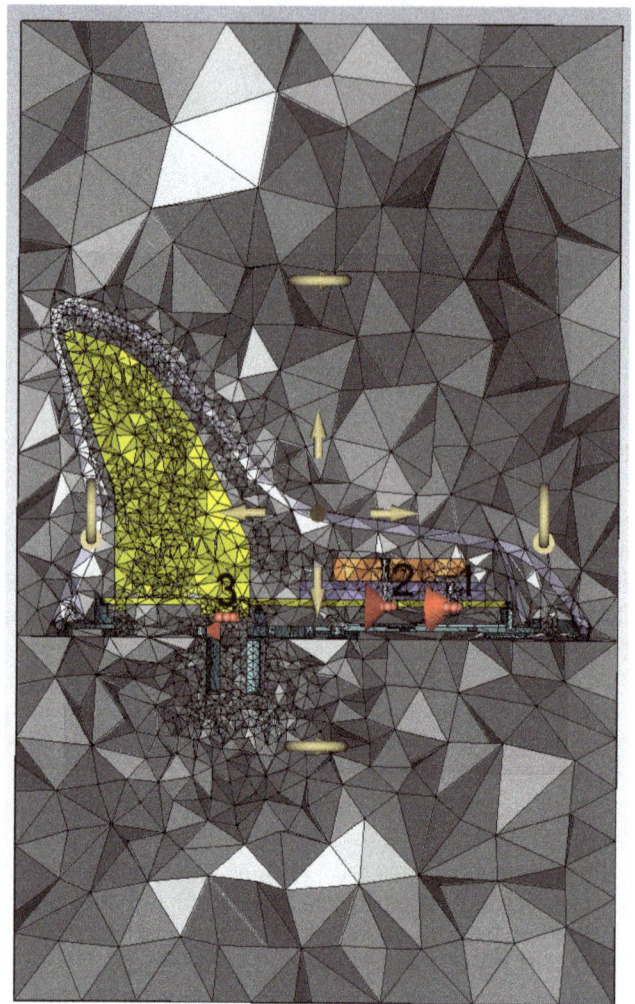

Figure 4. Tetrahedral mesh.

GPS antenna, but for the telephone antenna. For the frequencies over 1.5 GHz the reflection parameters change by using the adaptive mesh. If the simulation bandwidth is reduced, the newly arising resonances do not exist. A reason for this is that adaptive meshing algorithms in time and frequency domain analyze the simulated structure to increase the mesh density at high field values in order to decrease the simulation error. While in time domain the time pulse transports the energy, yielding a broadband adaption of the mesh, the frequency solver solves the equations at distinct frequency points and thus adapts at single frequency values. By default the highest possible frequency is used in order to assure the best resolution. However, maximum field values might occur at other points in the structure, namely where resonances take place. In order to account for this effect, it is possible to set the adaption to the resonances explicitly. If this option is not chosen, simulation results may differ for varying frequency bands due to different maximum frequencies. Using the resonant fast S Parameter method based on MOR does

not relevantly improve the scattering parameters or the time consumption, but drastically increases the maximum RAM consumption. The MOR in this case is very costly so that the advantages do not carry weight. In all cases the resonances are more distinct in the simulations with the F solver, than with the T solver, because the energy only decays very slowly when propagating through the structure at the appearance of resonances. The standard configuration is to abort the simulation after a time according to 20 times the length of the input impulse. At this point of time the energy only decays to $\approx -40\,\mathrm{dB}$ and ripples still exist in the scattering parameters. As well the adaptive mesh refinement cannot be used under these conditions as first results are necessary for the refinement of the mesh. A way to circumvent this problem is the AR Filter. With the AR Filter, the resonances can be estimated before the energy is decayed completely, thus, the simulation duration is reduced to less than one hour. The number of required adaptive mesh refinements cannot be given in general as it is dependent on the initial mesh.

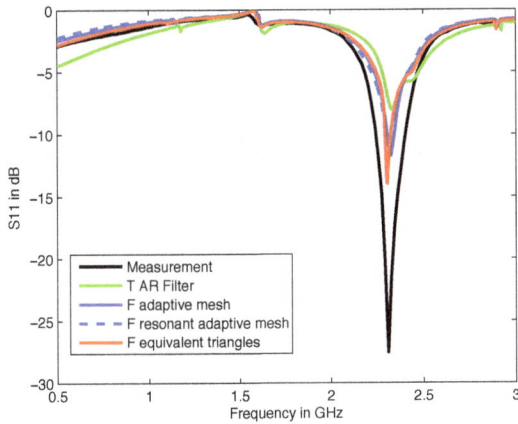

Figure 5. S_{11} of the SDARS patch simulated with different solvers.

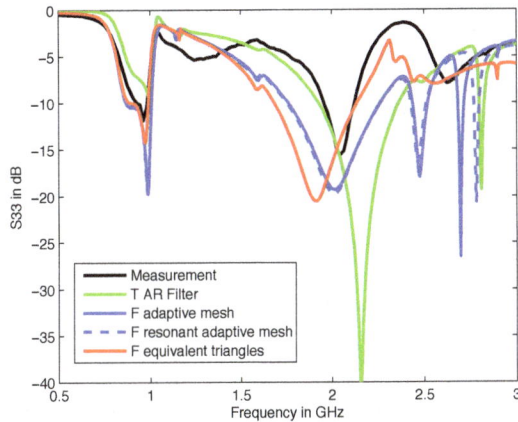

Figure 6. S_{33} of the telephone antenna simulated with different solvers.

For useful investigations with the T solver, the AR filter is necessary. Once some experience with the meshing of the structure could be achieved, the most efficient simulations still can be undertaken with the F solver. A further advantage of the F solver is the fact that single frequencies can be simulated at frequency points of interest after the solver run has finished without performing adaptive meshing.

3 Extended simulation domains

Vehicles feature an extended and at the same time complex environment which strongly influences the far field patterns of roof antennas. A common data format for vehicles is the Computer Aided Three-Dimensional Interactive Application (CATIA) format. In this format, every detail is included and the total amount of data is by far too extensive for the import into electromagnetic field solver programs. To reduce the amount of data and for reasons of compatibility the data is simplified to Nasa Structural Analysis System (NAS-TRAN) data, in which the surface is represented by triangles

Table 2. Comparison of different solver for the roof antenna.

Solver	F	F	F	T
Configuration	–	–	MOR (res)	AR
Mesh	Optimized	Adaptive	Adaptive	one cycle
Elements	200 000	370 000	267 982	6 373 600
RAM	2.7 GB	4.6 GB	27 GB	4.6 GB
Time ≈	2 h	4 h, 15 min	1 h, 21 min	49 min

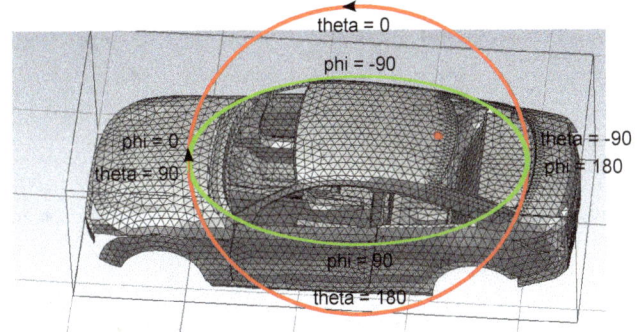

Figure 7. Nastran mesh of a vehicle.

as shown in Fig. 7. For the investigation of the efficiency and the accuracy of the solvers, the antenna is simplified to a monopole which is located in the rear part of the roof.

The simulation with the T solver is carried out in a frequency range from 1 to 2.5 GHz and the impulse is propagated through the structure until the energy level decreased to −30 dB. The mesh configuration is 10 lines per wavelength with a mesh line ratio limit of 999, which indicates the ratio of the largest cell size to the smallest cell size. The adaptive mesh refinement is an automatic refinement process in order to improve the mesh quality, especially in areas with high levels of electromagnetic energy the discretization is refined. It is deactivated in the simulations described in the following, as each refinement step approximately takes as long as the simulation duration given in Table 3. The simulation with the I solver is carried out for one single frequency point at 2 GHz. The I solver is configured with first solver order, an accuracy of 0.001 and 10 and 5 mesh cells per wavelength λ are used. The F solver meshing is configured with the default values allowing curved elements. The far field patterns simulated with all solvers are approximately similar as shown in Fig. 8. The comparability of the RAM and time consumption in Table 3 is only possible taking into consideration the varying frequency bandwidth and maximum frequency. The T solver is simulated in a broad frequency bandwidth with a maximum frequency of 2.5 GHz whereas the I and F solver are started at one single frequency point at 2 GHz. The decreased maximum frequency means that there are less mesh cells necessary in total.

The time efficiency of the solvers is dependent on the number of frequencies of interest. In case only one frequency

theta / Degree vs. dBi (Phi=0°),

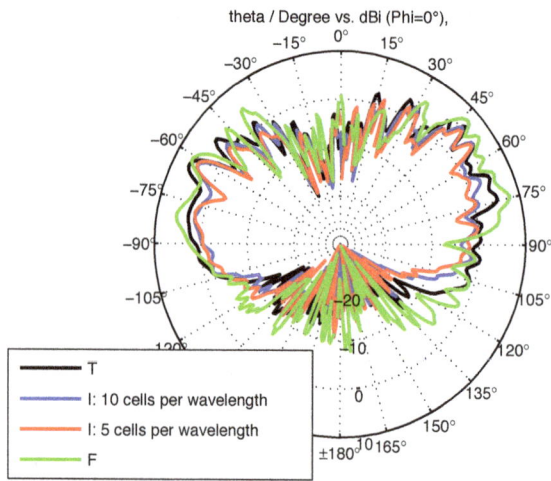

Figure 8. Far field simulation results at 2 GHz for T, F and I solver with different accuracies.

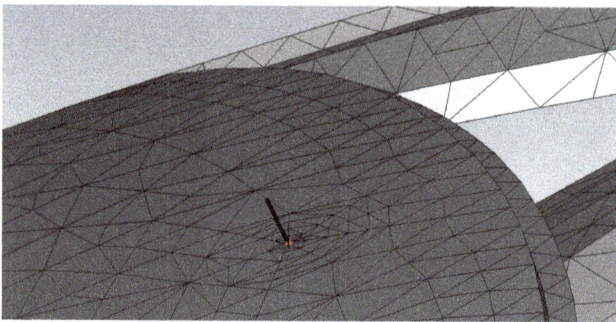

Figure 9. Meshing of NASTRAN structure with triangles.

Table 3. Comparison of different solvers for the monopole on a metallic vehicle modeled in NASTRAN.

Solver	T	I (10 cells/λ)	I (5 cells/λ)	F
Frequency	1–2.5 GHz	2 GHz	2 GHz	2 GHz
Elements	48 341 870	278 649	75 000	496 000
RAM	6.6 GB	5.1 GB	1.7 GB	19.3 GB
Time ≈	10 h, 40 min	2 h, 40 min	2 h, 20 min	20 min

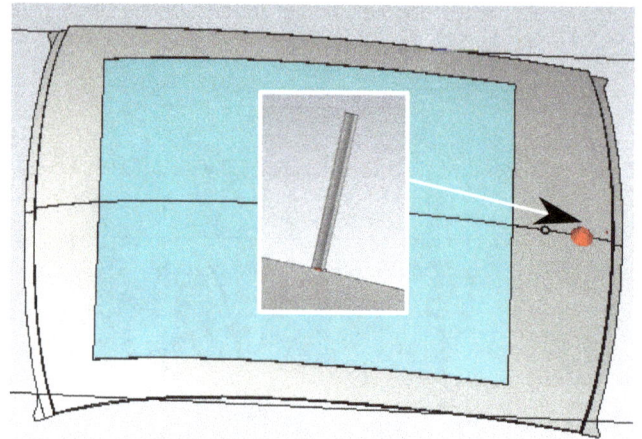

Figure 10. Roof with panorama glass window and monopole as simplification of an antenna.

point is investigated, the I solver is faster than the T solver. As soon as scattering parameters should be simulated at the same time, a larger bandwidth is necessary for reasonable investigations and the time consumption with the I solver will increase. Additionally it must be considered that windows are important for the far field behavior which were not considered in the I solver as they are dielectrics.

For the I and F solver a reduction of the overall model by deleting parts which do not influence the far field patterns, brings advantages as there are less triangles. With the T solver this effect is less distinctive because the whole box including air is meshed. For this reason, in the following only the roof is taken into consideration. Another vehicle model had to be used for the roof comparisons. Usually vehicle models are prepared in NASTRAN format at AUDI AG. The disadvantage of the NASTRAN format in CST MWS is that the mesh gets unnecessary fine as the triangles cannot be loaded as the mesh itself but are meshed a second time as shown in Fig. 9. Even if the triangles could be loaded as the final mesh, the limitation to a specific frequency by the size of the triangles makes this process inflexible. Originally,

the vehicles are saved in CATIA format which represents the geometry as non-uniform rational basis splines (NURBS).

In Table 4, the comparison of a metallic roof imported in NASTRAN and CATIA format is shown. With the CATIA format, reasonable results could be gained in the T solver with a mesh configuration of 10 lines per λ, whereas the NASTRAN format needs 3 times more mesh cells to meet the resonance frequency and the expected far field. At points, where the structure is discontinuous or at the end of straight lines describing the surface, the automatic meshing detects fixpoints, where discretization lines are applied. It is important to switch off the fixpoints as the hexahedral mesh would be by far too dense to even start the solver. Still some fixpoints around the antenna and the port are necessary. The mesh configuration was set equal to the configuration giving good results for the CATIA model. Within 3 cycles of adaptive mesh refinement the configuration was changed to 34 lines per λ. The first cycle takes 5 h, the second cycle 8 h and the third cycle, finally giving the expected resonance frequency of the monopole takes 10 h. In case the mesh configurations are known, the adaptive mesh can be skipped and the values as given in Table 4 can be expected. Another disadvantage of the NASTRAN format is the thickening of the triangular surface in order to prepare the model for the meshing with hexahedra. The far field pattern in vertical cut is shown in Fig. 12.

theta / Degree vs. dBi (phi=90)

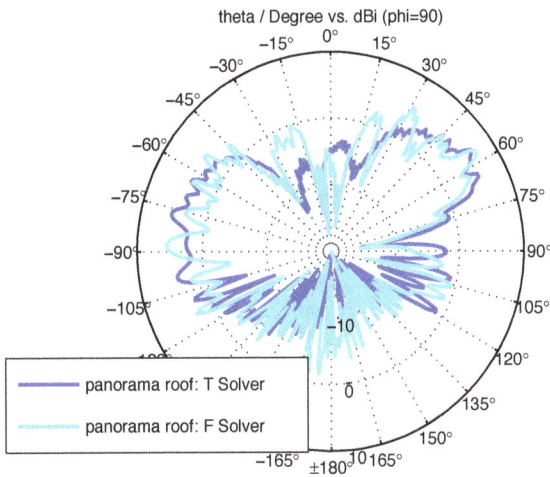

Figure 11. Far field patterns at 2 GHz simulated with T and with F solver.

theta / Degree vs. dBi (phi=90)

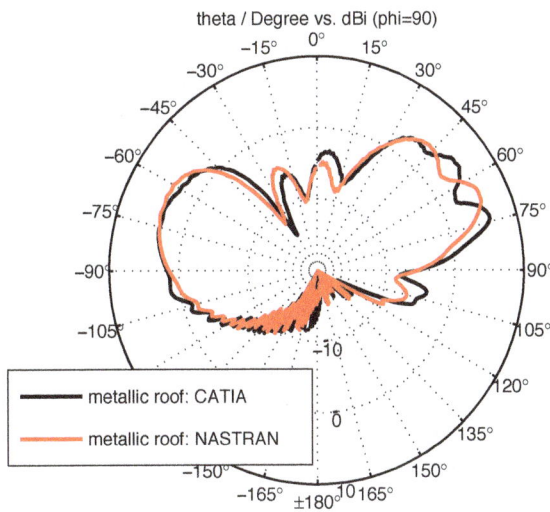

Figure 12. Simulated far field results at 2 GHz comparing the simulation using a roof model in CATIA and NASTRAN format.

The differences between the two plots can be explained by the deviations of the models which result from the conversion to NASTRAN. The problem changes when dielectrics as glass are introduced. For that a rectangular glass window is introduced into the roof as shown in Fig. 10. The vehicle roof with and without glass is investigated with the T solver. The simulations were carried out for a frequency range from 1 to 6.5 GHz. The metallic roof has a size of approximately $1.35\,m \times 1.15\,m$. The introduced window has a size of $1\,m \times 0.9\,m$ which means that approximately 60 % are then consisting of glass. The wavelength λ in glass, with an ϵ_r equal to 7, for the highest simulated frequency of 6.5 GHz is 17.4 mm. In free space the wavelength is 46 mm. This explains the increased number of mesh cells as shown in Table 4 which also leads to an increase of RAM and time con-

Table 4. Comparison of different solvers for the monopole on a roof and influence of a panorama glass window.

Solver	T	T	T	F
Data Format	NASTRAN	CATIA	CATIA	CATIA
Material	Metal	Metal	Panorama	Panorama
Surface	Sheet	Volume	Volume	Volume
Elements	99 059 100	33 885 108	51 251 112	1 591 851
RAM	11.7 GB	7.7 GB	11.6 GB	73 GB
Time $\approx$	10 h	15 h	50 h	20 h

sumption. The dramatic increase of time consumption in the simulation with the panorama glass window can be explained by the fact that data was swapped from the RAM to the hard disk, as only 12 GB of RAM were available.

The results of the simulated far field patterns correspond to typically observed results. A typical effect with panorama windows is the damping of the far field in the horizontal direction (Kwoczek, 2011). This effect is only observable in the simulation in case the glass has a sufficient thickness which was in this case set to approximately 4 mm in order to guide the wave through the glass. The CATIA model with the panorama window was additionally simulated with the F solver. The frequency in this simulation ranges from 1 to 6.5 GHz which is the same range as with the simulations with the T solver. There are some deviations between the simulations with the T and the F solver. The reflection parameters with the F solver shows a more broadband resonance and the far field at 2 GHz shown in Fig. 11 also shows some deviations. Overall the results of the T solver are more credible. The F solver is faster, but still needs more RAM.

4 Roof antenna on vehicle

The far field pattern is, in contrast to the scattering parameters, not only dependent on a small area around the antenna. This is why a simulation of the model including the vehicle from Fig. 7 and the roof antenna from Fig. 4 and Fig. 3 is necessary. The previous investigations and comparisons of the different solvers show that only the T solver can perform this simulation and high-performance computers are required. So the simulation is conducted on a workstation with 4 Tesla K40 graphic processing units (GPU) (NVIDIA, 2014). For the meshing, 10 lines per λ, a lower mesh limit of 10 and a mesh line ratio limit of 600 is used. To ensure that the antenna is meshed in the same way as before fixpoints were used. Still they must be ignored for the vehicle in NASTRAN format. These configurations lead to 513 218 568 mesh cells in total. The accuracy is set to $-30\,dB$.

With these settings the same reflection parameters as with the T solver in Figs. 5 and 6 are achieved. For the simulation 61 GB of RAM and an adaptive meshing is necessary with 3 cycles each taking 27 to 43 h. Altogether the simulation duration aggregates to 118 h.

5　Conclusions

In this paper, the theoretical background of the F, T and I solver and their accuracy and efficiency for a roof antenna mounted on a vehicle were discussed. The results show that the choice of the solver is not only dependent on the structure of the simulation domain, but also on the demanded results. The scattering parameters are more dependent on the structure itself, whilst the far field is strongly dependent on the environment. For the simulation of the roof antenna itself the T solver under usage of the AR filter and the F solver give good results whereas the vehicle is most efficiently simulated using the T solver, especially in case it contains dielectrics as glass. For this reason, the roof antenna including the vehicle was simulated with the T solver using the AR filter. The meshing of both the vehicle and the antenna works out the best when importing the data in CATIA format. The scattering parameters were validated with measurements and the far field patterns agreed with experiences from similar measurements. By comparing the different ways of simulations, an efficient way for investigating further antenna systems concerning scattering parameters as well as far field patterns could be described.

Acknowledgements. The authors wish to thank AUDI AG for providing CAD Data which are used in the simulation models and for the measurement data which serve to validate the simulation results. Also, a special thanks goes to the company CST AG for parts of the investigations and the simulation support.

References

Cendes, Z. J. and Shenton, D. N.: Adaptive Mesh Refinement in the Finite Element Computation of Magnetic Fields, IEEE Transactions on Magnetics, vol. MAG-21, 1811–1816, September 1985.

CST Computer Simulation Technology AG: CST MWS Description, available at: https://www.cst.com/Products/CSTMWS (last access: November 2014), 2015.

Ilic, M. M., Ilic, A. Z., and Notaros, B. M.: Higher Order Large-Domain FEM Modeling of 3-D Multiport Waveguide Structures With Arbitrary Discontinuities, IEEE Trans. Microw. Theory Tech., 52, 1608–1614, 2004.

Krietenstein, B., Schuhmann, R., Thoma, P., and Weiland, T.: The perfect boundary approximation technique facing the big challenge of high precision field computation, 19th International Linear Accelerator Conference, 2001.

Kwoczek, A., Raida, Z., Lacik, J., Pokorny, M., Puskely, J., and Vagner, P.: Influence of car panorama glass roof antenna on car2car communication, Vehicular Networking Conference, 2011.

Mocker, M. S. L., Engelmann, S., Tazi, H., and Eibert, T. F.: Finite Element Model Generation for Efficient and Accurate Electromagnetic Vehicle Roof Antenna Simulations, Loughborough Antennas and Propagation Conference, November 2014.

NVIDIA GmbH: Tesla K40 GPU, available at: http://www.nvidia.com/object/tesla-servers.html, last access: November 2014.

Percival, D. B. and Walden, A. T.: Spectral Analysis for Physical Applications: Multitaper and Conventional Univariate Techniques, Cambridge University Press, 1st Edn., 1993.

Pinchuk, A. R. and Silvester, P. P.: Error Estimation for Automatic Adaptive Finite Element Mesh Generation, IEEE Trans. Magnetics, MAG-21, 2551–2554, 1985.

Podebrad, O., Clemens, M., and Weiland, T.: New Flexible Subgridding Scheme for the Finite Integration Technique, IEEE Trans. Magnetics, 39, 1662–1665, 2003.

Weiland, T.: Time Domain Electromagnetic Field Computation with Finite Difference Methods, Int. J. Numer. Model., 9, 295–319, 1996.

Weiland, T., Timm, M., and Munteanu, I.: A Practical Guide to 3-D Simulation, IEEE Microwave Maga., 62–75, 2008

Yee, K. S.: Numerical solution of initial boundary value problems involving maxwell's equations in isotropic media, IEEE Trans. Ant. Propagation, 15, 802–907, 1966,

21

A programmable energy efficient readout chip for a multiparameter highly integrated implantable biosensor system

M. Nawito[1], **H. Richter**[1], **A. Stett**[2], **and J. N. Burghartz**[1]

[1]Institut für Mikroelektronik Stuttgart, Stuttgart, Germany
[2]NMI Naturwissenschaftliches und Medizinisches Institut an der Universität Tübingen, Reutlingen, Germany

Correspondence to: M. Nawito (nawito@ims-chips.de)

Abstract. In this work an Application Specific Integrated Circuit (ASIC) for an implantable electrochemical biosensor system (SMART implant, Stett et al., 2014) is presented. The ASIC drives the measurement electrodes and performs amperometric measurements for determining the oxygen concentration, potentiometric measurements for evaluating the pH-level as well as temperature measurements. A 10-bit pipeline analog to digital (ADC) is used to digitize the acquired analog samples and is implemented as a single stage to reduce power consumption and chip area. For pH measurements, an offset subtraction technique is employed to raise the resolution to 12-bits. Charge integration is utilized for oxygen and temperature measurements with the capability to cover current ranges between $30\,nA$ and $1\,\mu A$. In order to achieve good performance over a wide range of supply and process variations, internal reference voltages are generated from a programmable band-gap regulated circuit and biasing currents are supplied from a wide-range bootstrap current reference. To accommodate the limited available electrical power, all components are designed for low power operation. Also a sequential operation approach is applied, in which essential circuit building blocks are time multiplexed between different measurement types. All measurement sequences and parameters are programmable and can be adjusted for different tissues and media. The chip communicates with external unites through a full duplex two-wire Serial Peripheral Interface (SPI), which receives operational instructions and at the same time outputs the internally stored measurement data. The circuit has been fabricated in a standard 0.5-μm CMOS process and operates on a supply as low as 2.7 V. Measurement results show good performance and agree with circuit simulation. It consumes a maximum of $500\,\mu A$ DC current and is clocked between 500 kHz and 4 MHz according to the measurement parameters. Measurement results of the on-chip ADC show a Differential Non Linearity (DNL) lower than 0.5 LSB, an Integral Non Linearity (INL) lower than 1 LSB and a Figure of Merit (FOM) of 6 pJ/conversion.

---

1 Introduction

An integral part of all types of active implantable medical devices, such as cochlear and brain implants, is the electronic module. For monitoring of neuronal and metabolic activity a readout chip has to be implemented, which controls the data acquisition and management. In case of biosensor applications like subcutaneous metabolic monitoring, the electrochemical detection of ions, oxygen and pH requires a precise setting and measurement of voltage and currents at the metallic microelectrodes (Kubon et al., 2010; Jafari et al., 2014). For applications where large batteries and cabling is not suitable, stringent requirements on the readout chip in terms of size and energy efficiency are placed.

The SMART Implant consortium develops highly integrated implantable biosensor systems (Stett et al., 2014). As shown in Fig. 1a, the system contains a Read-Out Application Specific Integrated Circuit (RO-ASIC, or ROIC) which controls the measurement electrodes connected directly to the tissue or material to be characterized. A microcontroller sends instructions to the ROIC, receives the results and relays them to the power and data management ASIC. This frontend chip is responsible for power supply regulation of the implant and for the transmission of data via an inductive interface to an external reader unit placed outside the body. In this paper, the read out ASIC developed for this system is

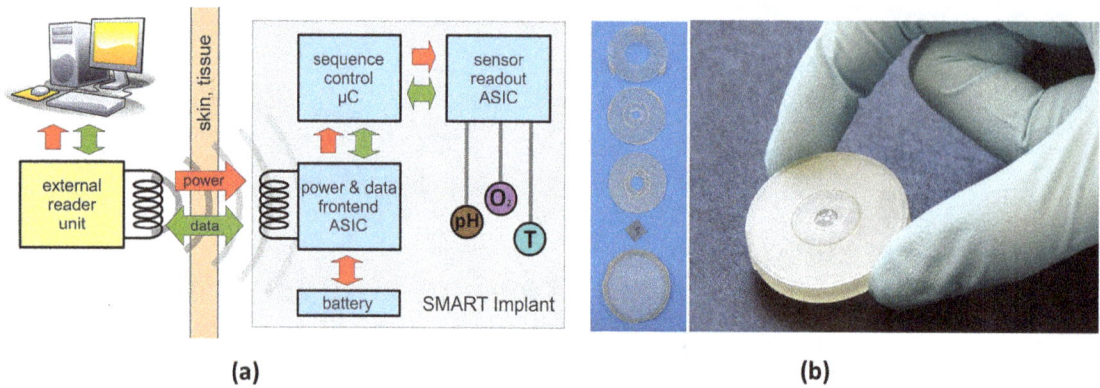

Figure 1. (a) Block diagram of the SMART Implant system highlighting readout ASIC, (b) photograph of actual implant and housing structure.

presented. The aim here is not to go through all the numerous analog and digital circuit blocks individually, but rather to give a description of the chip's functionality and structure, highlighting issues of energy efficiency, programmability and reliability of operation, in addition to the design techniques employed to approach these aspects.

2 Measurement sequence

The ROIC drives the measurement electrodes and performs amperometric measurements for determining the oxygen concentration, potentiometric measurements for evaluating the pH-level as well as temperature measurements. In order to increase the data integrity of the measurement process, a sequential approach has been adopted, hence avoiding any disturbance that might occur due to simultaneous sampling and processing of different signals. To further improve the quality of the acquired data, a number of up to 128 "single measurements" are preformed and then averaged in order to obtain a measurement sample. This way random spikes or erratic data points are eliminated, which are generated due the fluctuating nature of the chemical reaction taking place between the electrode surface and the connected tissue (Lindner et al., 1986). The measurement samples form a "measurement sequence", which would eventually settle to a final value. The number of samples and the final value are determined by the external microcontroller according to the criteria set by the International Union of Pure and Applied Chemistry (IUPAC; Lindner et al., 1986). The aforementioned concept is illustrated in Fig. 2. As shown, it is a requirement for single measurements, whether they are of temperature, pH or O_2 type, to be performed at least every $128\,\mu s$, but the time scale for a final value to be reached is in the order of seconds. The chip performs a complete measurement sequence each 15 min. Figure 2 also emphasizes the necessity for an energy efficient design since the ROIC is supposed to remain operational inside the battery operated implant for duration up to 4 weeks.

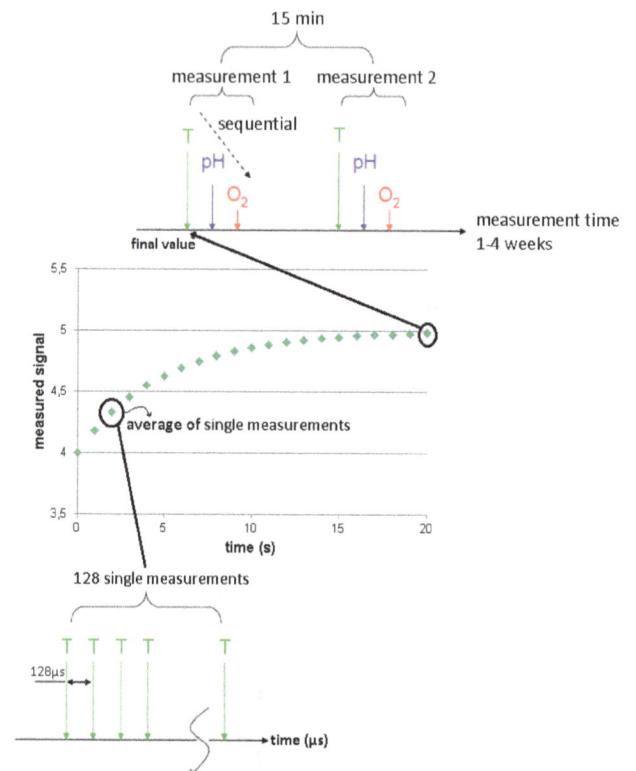

Figure 2. Measurement cycle as performed by the ROIC.

Since it is an essential requirement for the implant to be able to measure and characterize different types of tissues and materials, the ROIC must be designed to allow for a programmable and reliable operation. For this reason a parameterized approach has been adopted, where all measurement sequences and types can be carried out with different run times, boundary conditions and settings. Also an obstacle facing in vivo biosensors is the lack of access to the measured media, or to put it simply, the sensors are measuring in the dark, with no possibility to detect, observe or see the mea-

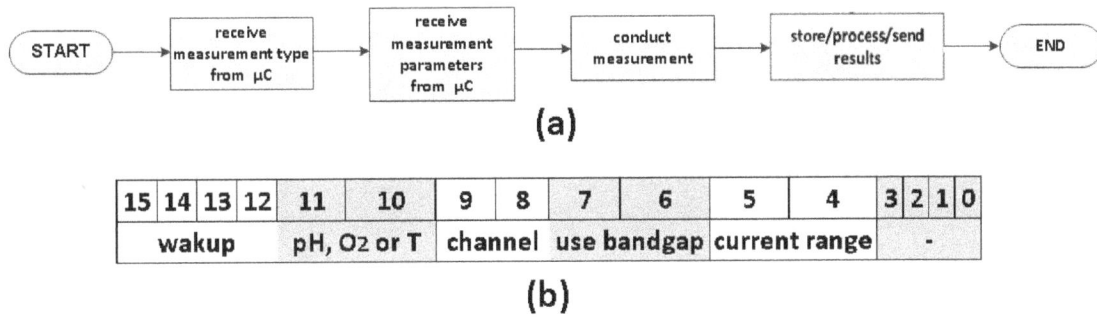

(a)

15	14	13	12	11	10	9	8	7	6	5	4	3	2	1	0
wakup				pH, O₂ or T		channel		use bandgap		current range		-			

(b)

Figure 3. (a) Flow chart of the measurement sequence, **(b)** wakeup opcode and relevant parameters. Upon wakeup the measurement type is set, the input channels and current range are set and the bandgap circuit is either turned on or off.

surement conditions. For this reason the calibration and presetting of the ROIC is imperative to allow for meaningful interpretation of the results. For example when performing O_2 measurements, the settling time required for the electrodes to start producing measurable currents differs from one material to another, hence the settling time is calibrated according to the tissue to be characterized.

Figure 3a shows the implemented flow chart of the ROIC. The chip receives a 16 bit operational code or opcode, where the first 4 bits constitute the instruction to be carried out (measure, calibrate, sleep etc.) and the relevant parameters are packed in the remaining 12 bits. As an example, Fig. 3b illustrates the structure of the wakeup command and the parameters passed to the chip during wakeup process. As seen during wakeup the measurement type and measurement channels are chosen, in addition to the setting of the current measurement range for O_2 and temperature measurements. Also for testing and calibration purposes, reference voltages of the ADC could be either controlled by a bandgap circuit or generated directly from a simple resistive divider, hence the option "use bandgap" is available during wakeup.

3 ASIC structure

To implement the functions described in the previous section, the ASIC has been divided into digital and analog sections, as shown in Fig. 4. For both O_2 and pH measurements, two measurement channels have been implemented respectively.

3.1 Digital circuits

For communication, a full duplex Serial Peripheral Interface (SPI) is implemented allowing the chip to simultaneously receive the 16 bit opcode from the external microcontroller and to send measurement results as a 16 bit output data word. The digital controller interprets the received opcodes and activates the analog circuitry which drives the external electrodes connected to the tissue and carry out the actual measurement. In case of a data transmission error or a faulty opcode, the chip sends an interrupt signal to request a resending

of the instructions. To realize an energy efficient operation, all measurements are conducted with only the needed analog blocks turned on. For example since the charge integrator, as will be discussed shortly, is only required for amperometric sensing, it is turned off during potentiometric measurements. In the case of idle times where the chip is not measuring, all components are turned off except for the SPI interface which continues to listen for incoming instructions.

3.2 ADC

A central component of the readout path is a 10 bit cyclic ADC based on the pipelined principle, which converts the measured signal and sends it to the digital core for processing and storage. The ADC is designed to convert input voltages between 0.5 and 2.5 V with a LSB of 1.96 mV. The pipelined architecture has been chosen due to its relatively simple circuitry, reliable operation and capability of achieving the required 10 bit resolution needed for temperature and O_2 measurements. A typical structure for such an ADC would consist of 9 stages in series, each producing 1.5 bits, and some form of digital correction. However, the large chip area occupied by such an arrangement in addition to its high power consumption would be unacceptable for this application, which necessitated a modified design. Figure 5 shows the implemented ADC which consists of a single 1.5 bit stage followed by a sample and hold circuit. The input analog signal is converted by the sub ADC, the residue is sampled by the sample and hold stage and then same circuit is reused again until the 10 bit word is produced. In other words instead of converting the analog input through 9 stages, a single stages is reused 9 times, reducing the area and power consumption by almost an order of magnitude.

3.3 pH measurement

For the pH measurements, a 12 bit resolution was required. To achieve this without redesigning the entire ADC and increasing the complexity of the design, an offset subtraction technique is introduced along with a couple of extra components to the 1.5 bit stage circuit. Specifically the voltage gen-

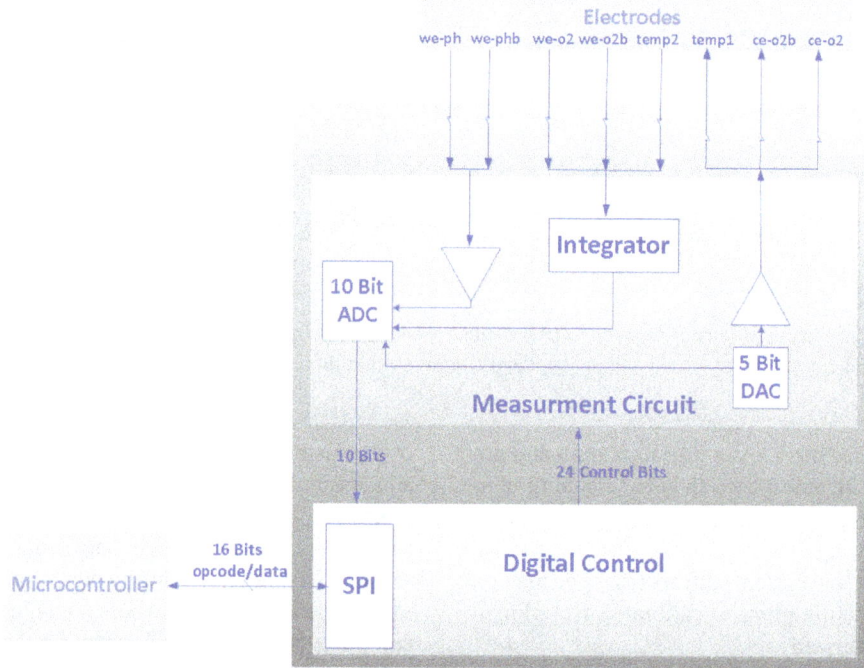

Figure 4. ROIC block diagram.

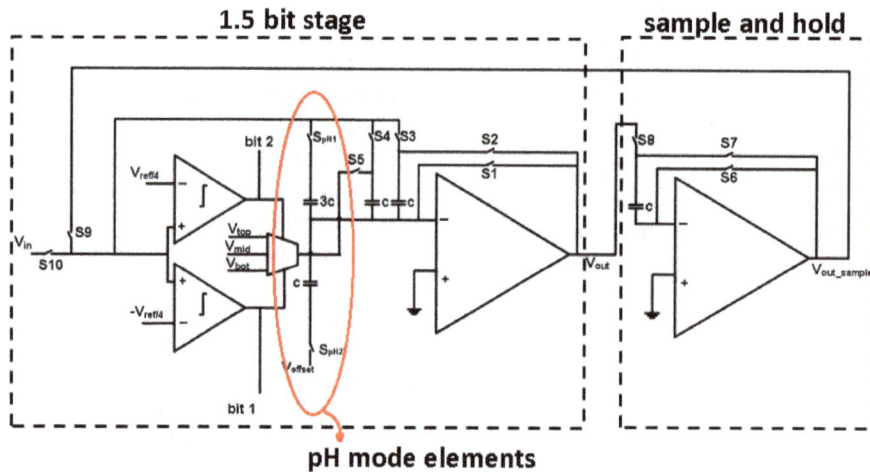

Figure 5. ADC circuit with special pH mode switches and capacitors.

erated by the measurement electrodes, which corresponds to the measured pH value and lays between 0.7 and 2.3 V, is first converted to a 10 bit "coarse" word, then according to the conversion result, a known offset generated by the on-chip 5-bit Digital to Analog Converter (DAC) (shown in Fig. 4) is subtracted from the original input. The difference is multiplied by 4 and converted to a 10 bit "fine" word. Summing the value of the offset and the fine results gives the final value with the required 12 bit resolution. To realize the aforementioned process, switches Sph1 and Sph2 , in addition to capacitors with the values C and $3C$ are added as a new modi-

fication to the circuit, as highlighted in Fig. 5, since they are specifically used to carry out the offset subtraction and difference multiplication. Specifically, in the first phase of the pH mode operation, switches are clocked so that the multiplying DAC connecting these elements with is connected as shown in Fig. 6a, where in that case the total charge of the system is given by

$$Q_{\text{phase1}} = V_{\text{in}} \times 3C. \tag{1}$$

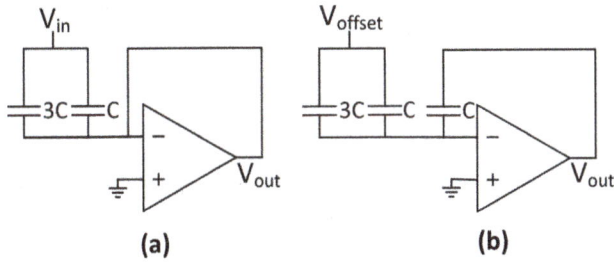

Figure 6. (a) pH mode phase 1, (b) pH mode phase 2.

In phase 2, the circuit is configured as shown in Fig. 6b and the total charge is given by

$$Q_{phase2} = V_{out} \times C + V_{offset} \times 4C. \tag{2}$$

Since Q_{phase1} equals Q_{phase2} the final output voltage is given by

$$V_{out} = 4(V_{in} - V_{offset}). \tag{3}$$

3.4 O₂ and temperature measurement

In the case of O_2 and temperature measurement, the measured analog signals are currents, where for the former a three electrode measurement setup is implemented and current flowing between the working electrode and the counter electrode is of interest (Kubon et al., 2010). For temperature measurements, an external Schottky diode is reversed biased and used as a transducer, where the reverse current is a measure of the temperature and the sensitivity of the sensor is controlled by the reverse bias voltage. The reverse voltage is also produced by the internal DAC and is one of the parameters to be set by the user.

To convert currents into voltages, a necessary step given that the ADC operates on voltage inputs, the charge integrator shown in Fig. 6. is implemented. Following the basic equation of charge integration which states that

$$V_{out} = \frac{I_{in} \times t}{C}, \tag{4}$$

where V_{out} is the output voltage, I_{in} is the input current, t is the integration time and C is the integration capacitance, it is clear that by adjusting t and C, various current ranges can be measured. For this reason, different integration capacitances are added in parallel, as illustrated in Fig. 7. Furthermore, the integration time can be set as an input parameter, allowing the implemented integrator to cover current ranges between 30 nA and 1 μA with a 10 bit resolution.

4 Fabrication

The design has been implemented using IMS GATE FOREST® 0.5 μm, 2 Metal CMOS technology

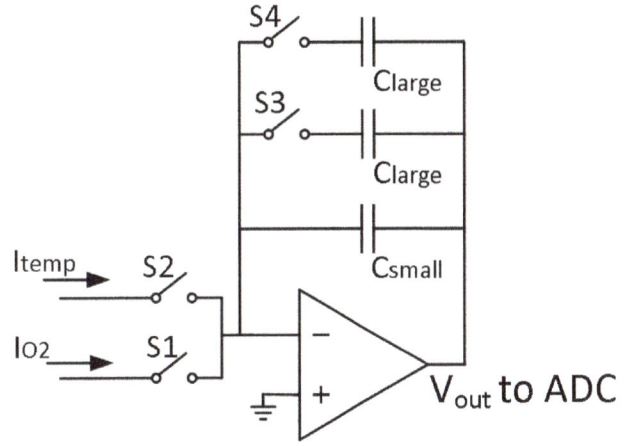

Figure 7. Programmable charge integrator.

Figure 8. (a) Photograph of fabricated chip in housing, (b) size comparison between final chip and a coin.

(http://www.ims-chips.de/content/pdftext/White_paper_MS_Array_09_11.pdf). This sea of transistors technology offers transistors and passive circuit elements implemented in a certain number of fixed dimensions. The digital and analog circuits are then realized by connecting these elements with metal wires creating semi custom designs. A major advantage of this approach is that the devices are very well characterized and modeled, enhancing the process yield and the reliability of the ASIC.

In Fig. 8a, a photograph of the fabricated ASIC bonded inside a QFN48 housing is shown, where Fig. 8b shows a size comparison between the final sealed housing to be used in the actual implant and a coin.

5 Measurement results

In Fig. 9a and b, the measured Differential Nonlinearity (DNL) and measured Integral Nonlinearity (INL) of the on chip ADC are shown, respectively. The test setup for producing the measured results involved connecting a voltage sources to the working electrodes of the pH circuits, sweeping the voltage and evaluating the digital output data sent via the SPI interface. The ADC also demonstrated a Figure of Merit (FOM) of 6 pJ/conversion.

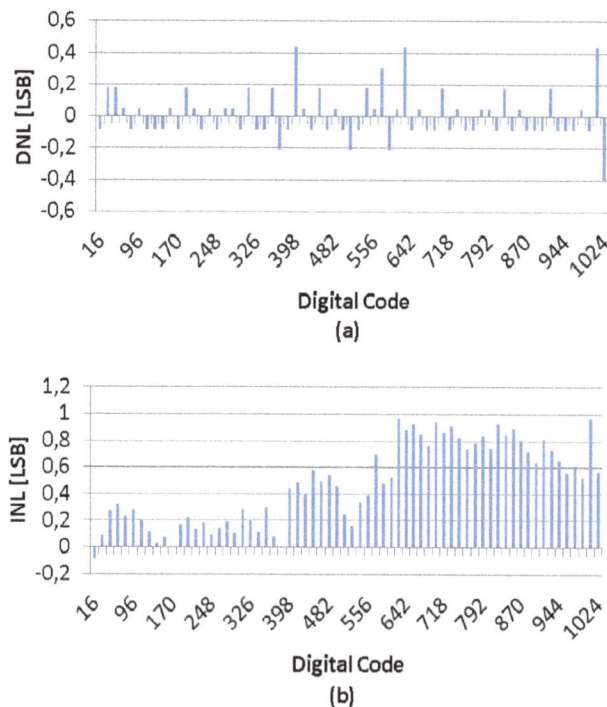

Figure 9. (a) Measured ADC DNL, (b) measured ADC INL.

Table 1. Performance parameters of the ROIC.

Technology	0.5 µm, 2M CMOS, see of gates
Supply	2.7–3.3 V
Temperature	20–80 °C
Max. total DC current	$\sim 500\,\mu A$ at 3.3 V, 80 °C
Min. total DC current	$\sim 250\,\mu A$ at 2.7 V, 20 °C
Nominal operating frequency	500 kHz
Max. operating frequency	4 MHz
Number of analog devices	432
Number of digital gates (NAND equiv.)	1185

The chip operates at supply voltages between 2.7 and 3.3 V and at temperatures between 20 and 80 °C, where the temperature measurement circuits have been optimized for the range between 20 and 43 °C with an accuracy of 0.1 °C. The maximum total DC current that all circuits could consume is about 500 µA nevertheless, this value is never reached since that would require all digital and analog building blocks to be turned on at the same time. As mentioned before, this case is avoided by activating the relevant components for a certain measurement type only. In order to reduce dynamic power consumption and at the same time fulfilling the requirement of obtaining a single measurement point every 128 µs given the cyclic nature of the ADC, a nominal operating frequency of 500 kHz is chosen. However, the chip can function reliably at a maximum frequency of up to 4 MHz. Table 1 summarizes the performance parameters of the ROIC.

6 Conclusion

The successful design and implementation of a programmable read out ASIC for miniature active implants comprising high functionality is an interesting and multifaceted undertaking. The core philosophy driving the development process should be the consideration of energy requirements and the fulfilment of flexible operation during all phases of the design. Efficient scheduling of the measurement sequence, implementing an elaborate set of operations and the time multiplexing of internal components, in addition to the optimization of power consumption on the circuit level are all important methods in their own right, but yield the most interesting results when employed in combination.

Acknowledgements. This work was funded by the German Federal Ministry of Education and Research, BMBF as part of the MicroTEC Südwest Cluster project SMART Implant, grant no. 16SV5979K, 16SV5980 and 16SV5982-86. The authors would also like to sincerely thank all members and partners of the SMARTImplant consortium. Without their valuable efforts, this work would not have been possible.

References

Jafari, H. M., Abdelhalim, K., Soleymani, L., Sargent, E. H., Kelley, S. O., and Genov, R.: Nanostructured CMOS Wireless Ultra-Wideband Label-Free PCR-Free DNA Analysis SoC, IEEE Journal of Solid-State Circuits, 49, 1223–1241, 2014.

Kubon, M., Moschallski, M., Link, G., Ensslen, T., Werner, S., Burkhardt, C., Nisch, W., Scholz, B., Schlosshauer, B., Urban, G., and Stelzle, M.: A Microsensor System to Probe Physiological Environments and Tissue Response, IEEE Conference on Sensors, Kona, Hawaii, 1–4 November 2010, 2607–2611, 2010.

Lindner, E., Toth, K., and Pungor, E.: Definition and determination of response time of ion selective electrodes, Pure Appl. Chem, 58, 469–479, 1986.

Stett, A., Bucher, V., Cihova, M., Gutoehrlein, K., Kubon, M., Link, G., von Metzen, R., Stamm, B., Stelzle, M., Pojtinger, A., Schneider, K., Mintenbeck, D., Rossbach, D., Richter, H., Nawito, M., Boven, K.-H.,; Moeller, A., Jeschke, C., Paetzold, J., Goettsche, T., Bludau, O., Haas, N., Tompkins, D., Lebold, T., and Kokelmann, M.: SMART Implant: Electronic Implants for Diagnosis and Monitoring, Energieautarke Sensorik (GMM-FB 79), Contributions of the 7th GMM Workshop, Magdeburg, Germany, 24–25 February, 2014.

22

Investigation of the strong turbulence in the geospace environment

O. Kharshiladze[1] and K. Chargazia[1,2]

[1]Ilia Vekua Institute of Applied Mathematics, Ivave Javakhishvili Tbilisi State University, Tbilisi, Georgia
[2]M. Nodia Institute of Geophysics, Tbilisi, Georgia

Correspondence to: K. Chargazia (khatuna.chargazia@gmail.com)

Abstract. Plasma vortices are often detected by spacecraft in the geospace (atmosphere, ionosphere, magnetosphere) environment, for instance in the magnetosheath and in the magnetotail region. Large scale vortices may correspond to the injection scale of turbulence, so that understanding their origin is important for understanding the energy transfer processes in the geospace environment. In a recent work, turbulent state of plasma medium (especially, ionosphere) is overviewed. Experimental observation data from THEMIS mission (Keiling et al., 2009) is investigated and numerical simulations are carried out. By analyzing the THEMIS data for that event, we find that several vortices in the magnetotail are detected together with the main one and these vortices constitute a vortex chain. Such vortices can cause the strong turbulent state in the different media. The strong magnetic turbulence is investigated in the ionosphere as an ensemble of such strongly localized (weakly interacting) vortices. Characteristics of power spectral densities are estimated for the observed and analytical stationary dipole structures. These characteristics give good description of the vortex structures.

1 Introduction

In a dispersive medium, especially in space, astrophysical and laboratory plasmas, the various nonlinear localized wave structures are generated and developed easily enough (Horton, 1990; Aburjania, 2006). Investigation of nonlinear interaction of wave structures with each other and with medium is important. Nonlinear interaction of wave structures can be described by interaction of the localized structures or separate wave harmonics. At certain conditions this interaction leads to chaotization of phases of the structures or waves. As a result of chaotic dynamics of phases of the wave structures the macroscopic motions occur usually named as turbulent motion.

Following its own logic of development, in the sixty years of the last century the plasma turbulence theory was based on the weak turbulence model when the weak interaction between the modes due to nonlinearity was considered. Within the framework of this model solution of the wide range of questions and explanation of a number of the important nonlinear phenomena (Horton, 1990; Galeev and Sagdeev, 1976) was possible. The weak turbulence theory is constructed via decomposition of the initial equations for plasma with respect to a small parameter – relation between the fluctuations' energy and full energy of plasma.

According to the existing representations, in each certain situation the strong turbulence at some extent represents a set of interacting waves and the ordered nonlinear structures (vortices). Depending on interaction between free (weakly turbulent) waves and structures, the strong turbulence can be either mainly wave, or structural (vortical, granular) turbulence (Diamond and Carreras, 1987). Herewith, these structures absorb free energy of plasma more effectively, than the linear waves (Galeev and Sagdeev, 1976). So the strongly localized vortex structures containing the trapped particles, kneading in plasma, can raise the strong turbulence and increase the transport of heat and particles.

However, the strongly localized vortex structures are common in the Earth's magnetosphere and ionosphere. They prevail in the nightside plasma sheet (Keiling et al., 2009). Plasma vortex-like flows have also been observed on the middle to high-latitude boundary of the outer radiation belt by the Cluster spacecraft fleet (Snekvik et al., 2007). The plasma flow vortex found in the magnetotail is characterized by pronounced vortical motion in the plane that is approximately parallel to the ecliptic plane. Vortex structures in the plasma sheet are thought to be important in transportation of the kinetic energy from fast flow or bursty bulk flows (BBFs), which are interpreted as a consequence of reconnection, in the magnetosphere to the ionosphere (Snekvik et al., 2007).

The aim of this paper is development of the theoretical and experimental study of the stationary strong vortex turbulence in space plasma, in particular, the self-consistent model and interpretation of experimentally observed frequency and spatial spectra of turbulent pulsations presented in works (Sahraoui et al., 2004, 2006; Alexandrova et al., 2008; Narita, 2007; Keiling et al., 2009).

2 Model of strong structural (vortical) turbulence

The turbulent state described by model, given in (Aburjania et al., 2009) consists from the small amplitude modes of wide continuous spectrum according to wave numbers (weakly turbulent spectrum), and also from the ensembles of the vortex structures considered. Herewith, each vortex, moving with velocity u gives the certain contribution to a frequency spectrum $\omega = ku$ of fluctuations of density and electromagnetic fields of plasma. As the vortex velocity depends on amplitude, the frequency spectrum of a vortex set with various amplitudes can be wider, than a corresponding spectrum of small amplitude waves poorly correlating with each other. Experimental magnetospheric observation (Sahraoui et al., 2004; Zimbardo, 2006; Narita et al., 2007) and laboratory plasma observations (Gekelman, 1999) have shown, that a frequency spectrum width of fluctuations of density, electric and magnetic fields greatly exceeds that predicted by the theory of weak turbulence (Horton, 1985). Therefore it is possible to assume, that the basic state to a fluctuations spectrum in magnetized plasma is given by solitary waves, vortex solitons. As regards of the weak turbulent parts of a spectrum, its role will be assumed negligible small and can be added up, if necessary, with soliton part of turbulence (we shall partially consider its influence expressed in stochastization of a vortex structures' spectrum).

Because of strong localization of vortex structures in space they do not have long-range action and consequently they are distributed randomly, similarly to molecules of gas. Herewith, random position and a phase of the vortex structures are caused by collisions among themselves. All this allows constructing the model of strong turbulence of plasma in the form of ensemble of the vortex structures, with vortices of various amplitudes randomly distributed in space, and due to this to apply the statistical approach for their description.

Thus, we shall consider, that strong turbulence of plasma represents ensemble of weakly interacting vortex structures (the basic condition), each of which is characterized by equal distribution of energy of system between N identical vortices (N is a parameter of state). Herewith, each vortex represents a separate degree of freedom of system. Then quasi stationary turbulent state can be expressed during each given moment of time on the basic states (on ensembles).

In the model of the strong vortex turbulence (Aburjania et al., 2009) the basic state of plasma turbulence represents an ensemble of two-dimensional drift- Alfvén vortices of the electron skin size: each active area of the plasma medium with a size $L \times L$ is covered by the N randomly distributed vortices of identical amplitude. We shall notice, that in a real turbulent state the different kind vortices are mixed up, but numerical simulations (Birn, 2004), laboratory experiments (Nezlin and Snezhkin, 1993) and space observations (Chmyrev et al., 1991; Alexandrova et al., 2006; Keiling et al., 2009) have shown that the vortices with essentially different amplitudes pass through each other without an interaction. So, we can assume that only the number of the vortices with the same scale is essential in plasma, and the system of the basic conditions is closed. At different amplitude the vortices also differ by width, in this case one represents for another simply quasi classic hole, therefore their merging is hardly possible, though energy pumping is admitted.

It is known, that the strong turbulence theory traditionally is built according to the theories of Richardson (1922) and Kolmogorov (1941), which is based firstly on isotropy and homogeneity of the turbulent state and secondly somehow on a forced averaging (Kingsep, 1990). This theory does not include any proper, even very important solutions of the initial nonlinear dynamical equations. Contrary to works (Richardson, 1922; Kolmogorov, 1941; Iroshnikov, 1963; Kraichnan, 1965), in the model (Aburjania et al., 2009) the turbulence is supposed to be anisotropic and from the very beginning a solution of the initial dynamical equations is built in the form of a stationary strongly localized two dimensional vortices with fully definite scales, amplitudes and the velocities. Further, on the basis of these vortices, as on the turbulent perturbations' carriers, the model of the strong turbulence is developed. But, as well as in the mentioned works, we suppose too the turbulent motion energy density W to be constant for the different spectral range – energy of the large scale pulsations will be transferred to the small scale ones so that energy dissipation does not happen in this region.

A very important theory of anisotropic magnetohydrodynamic (MHD) turbulence was proposed by Goldreich and Sridhar (1995). They supposed that MHD turbulence is strongly anisotropic due to the external magnetic field so that the turbulent transport structures (in our case: the vortices) are elongated in its direction. Correspondingly, they supposed that the energy transfer time to smallscales within the system τ_{tr} is of the order of the nonlinear time or eddy turnover time τ_{NL}. The Goldreich-Sridhar picture, however, does not fully agree with numerical simulations (Birn, 2004). From these works it is clear that it is not necessary that these two times to be equal, as was supposed by Goldreich-Sridhar. The equality of the ratio $\chi = \tau_{tr}/\tau_{NL}$ to unity seems to be very restrictive and does not correspond to some of the results stemming from direct numerical simulations where χ can be smaller than unity, as observed by Müller et al. (2003). We as well as other authors (for example, Galtier et al., 2005), suppose χ to be a constant for all scales but not necessarily equal to unity (the critical balance condition). Solar wind (Matthaeus et al., 1994) and magnetosheath data (Alexandrova, 2008), where χ seems to be smaller than unity, support the validity of our assumptions.

When the energy flux unit volume W_B is considered to be constant and its transfer rate – scale independent, from the initial equations (Aburjania et al., 2009), the following relation obtains:

$$k_{\parallel} \sim \frac{k_{\perp}^{1/3}}{B_0^{2/3}}, \tag{1}$$

where $k_{\perp} = 2\pi/l_{\perp}, k_{\parallel} = 2\pi/l_{\parallel}$ are the so called "wave vector analogues" of the vortex perpendicular and parallel linear scales, B_0- mean magnetic field.

This relation shows that the considered turbulence is intrinsically anisotropic. These scaling have been confirmed by numerical simulation of electron MHD (EMHD) turbulence (Cho and Lazarian, 2004). It is obvious that turbulence develops more freely in background magnetic field direction. The size of anisotropy and the strength of the expressed orthogonal spreading of turbulence change in accordance with that of the magnetic field induction. So, the magnetic field induces the anisotropies of compressible MHD turbulence. Anisotropy increase with the scale decrease was predicted for Alfvenic motion by Goldreich and Sidhar (1995) and confirmed numerically for compressible MHD in Cho and Lazarian (2004).

Analogously, when the energy flux unit volume W_B is considered to be constant and it is transfer rate scale independent, using the density scaling $-\rho_{l_{\perp}} \sim (l_{\perp})^{-3\mu}$, where $\rho_{l_{\perp}}$ is a perturbation of the medium density, $l_{\perp} \sim \lambda_s \sim \rho_i$-ion Larmur Radius, and the following relation is obtained:

$$B_{l_{\perp}} \sim l_{\perp}^{2/3-\mu}. \tag{2}$$

Thus, a medium compressibility significantly influences the spatial spectra of the turbulence.

The relation obtained above determines the energy spectra $E(k_{\perp})$ of the strong vortex turbulence as a function of the transversal "wave vector" $k_{\perp}$ similar to the work (Alexandrova, 2008).

$$E(k_{\perp}) \sim \frac{B_{l_{\perp}}^2}{k_{\perp}} \sim k_{\perp}^{-7/3+2\mu}. \tag{3}$$

In incompressible plasma limit ($\mu = 0$), this phenomenology predicts a $k^{-7/3}$ spectrum. Such a spectrum has been observed both in direct numerical simulations of an incompressible EMHD turbulent system (Biskamp et al., 1999) and in the EMHD limit of the incompressible Hall MHD shell model (Galtier and Buchlin, 2007). In the case of isotropic compressions toward smaller scales ($\mu = 1$), which can take place in interstellar medium, the spectrum is $E(k) \sim k^{-1/3}$. If isotropic compression is going on toward large scales

($\mu = -1$), the spectrum will be $E(k) \sim k^{-13/3}$, which was confirmed by the numerical calculations for the conditions of the solar wind (Alexandrova, 2008). Recently, new energy spectra of turbulence $E(k) \sim k^{-8/3}$ were found by the Cluster mission in the magnetosheath (Sahraoui et al., 2004), in the foreshock-region (Narita et al., 2007), and in the solar wind (Howes et al., 2008) correspond to a value of plasma compressibility degree $\mu = -1/6$. Generally, the value of μ for a certain medium has to be determined by appropriate observations and measurements or on the basis of corresponding numerical modeling.

3 Experimental detection of the space vortices as elements of strong vortex turbulence

The isolated magnetospheric substorm starts with a growth phase when a southward interplanetary magnetic field (IMF) merges with the Earth's dayside magnetic field and transfers energy from the solar wind to the magnetosphere. This energy is transported to the tail lobe magnetic field where it is stored and eventually released by reconnection (expansion phase) in the near-Earth magnetotail causing the strong shear of the plasma flow velocity. Velocity shear instability leads to formation of the strongly localized vortex structures in the plasma medium. The THEMIS (The Time History of Events and Macroscale Interactions during Substorms) mission has detected vortices in the magnetotail in association with the strong velocity shear of a substorm plasma flow (Keiling et al., 2009), which have conjugate vortices in the ionosphere (see Fig. 1). THEMIS mission is the fifth NASA Medium-class Explorer (MIDEX), launched on 17 February 2007 to determine the trigger and large-scale evolution of substorms. The mission employs five identical micro-satellites (hereafter termed "probes") which line up along the Earth's magnetotail to track the motion of particles, plasma and waves from one point to another and for the first time resolve space–time ambiguities in key regions of the magnetosphere on a global scale (Angelopoulos, 2008; McFadden et al., 2008). The probes are equipped with comprehensive in-situ particles and fields instruments that measure the thermal and superthermal ions and electrons, and electromagnetic fields from DC to beyond the electron cyclotron frequency in the regions of interest.

On 19 February 2008 the substorm occurred at approximately 05:25 UT observed from THEMIS satellite mission (Keiling et al., 2009). The four THEMIS spacecraft (TH-A, TH-B, TH-C, TH-D) were located in the nightside magnetosphere inside the plasma sheet and close to the neutral sheet monitoring in situ the conjugate space vortices. TH-A, D, and E were closely clustered which were separated in a triangular-like constellation (1–2 RE (radius of the Earth)), allowing an unambiguous identification of a counterclockwise flow. The associated clockwise vortex was tentatively

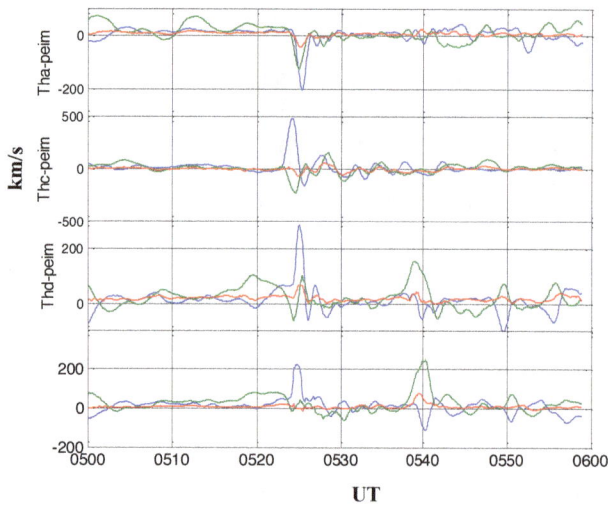

Figure 1. Plasma flow velocity components (blue – V_x, green – V_y, red – V_z) of TH-A, C, D, and E in GSM coordinates after smoothing with moving average method.

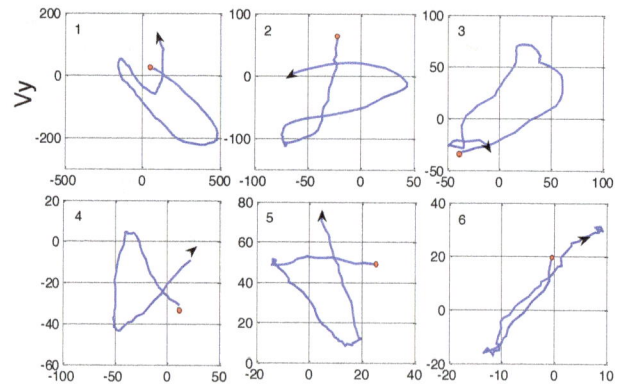

Figure 2. Hodograms of the plasma flow velocities corresponding to the TH-C data.

inferred from the single spacecraft located further west (see Fig. 1; Keiling et al., 2009).

The THEMIS mission measured both the magnetic field and plasma flow velocity fluctuations. As far as the changes in the magnetic field are sharp and the vortex structures are difficult to analyze, in this paper only the data of the plasma flow velocity components will be referred. The event was better catch by the TH-C spacecraft, so in further analysis data only from this one should be used.

The substorm onset caused strong fluctuations of the plasma flow velocity. This sharp change of the field parameters was detected by all of the spacecrafts (Keiling et al., 2009). The flows of the clustered spacecraft (TH-A, D, and E) show characteristics of a counterclockwise vortex while the spacecraft TH-C detected clockwise rotational field (see Figs. 1, 4).

Figure 1 shows plasma flow velocity components. It is obvious that the main vortex is accompanied with the smaller but essential peaks. They are suggested to be the vortex chain or the secondary vortices generated by the first main ones. The idea was that the substorm associated reconnection, which is a strong source of plasma velocity shear – BBF, generates more than one vortex at interaction with the solar wind plasma – a main vortex with the small amplitude satellite vortices, which generate the vortex chain in flow.

Based on dataset, analyzed in Keiling et al. (2009), after providing the hodograms of the TH-C data of the plasma flow velocity components, (Fig. 2) several structures (at least, six structures, having rotational sense of motion) are indeed revealed. But this method is not enough to distinguish the nature of these structures. For further analysis vorticity of the plasma flow was estimated also. Vorticity in the magnetosphere is of importance because it has been associated

with field-aligned current (FAC) systems flowing during substorms. For this purpose, we seek for the minimal conditions, which give us possibility of determination of the flow vorticity by means of spacecraft measurements. In a steady state, vorticity is conserved along the field lines, in which case one might infer much about the ionospheric vortex from magnetospheric vortex observations. For investigation of the ionospheric vortices, information about their vorticity is given by the magnetospheric ones. For this purpose, vorticity is estimated using the flow velocity components for different satellites, linear approximation of which is possible to describe by taking into account the satellite position coordinates and the flow velocity components. Such calculations give a first approximation of the vorticity characteristics.

In a linear approximation for the derivatives of the velocity the following coupled set of equations is given:

$$V_{x_i} = V_{x0} + \frac{\partial V_x}{\partial x} \mathrm{d}x_{i0} - \frac{\partial V_x}{\partial y} \mathrm{d}y_{i0}, \qquad (4)$$

$$V_{y_i} = V_{y0} + \frac{\partial V_y}{\partial x} \mathrm{d}x_{i0} - \frac{\partial V_y}{\partial y} \mathrm{d}y_{i0}, \qquad (5)$$

where $i = A, B, D$ is used to represent the closely clustered satellites; index 0 represent the satellite, according to which the calculations are made. Solving the set of equations, we get z component of the flow vorticity

$$\Omega_z = \frac{\partial V_y}{\partial x} - \frac{\partial V_x}{\partial y}. \qquad (6)$$

The vorticity estimated this way is given in Fig. 3. Also, by means of this method plasma compression divV may be calculated. As it is obvious from the Fig. 3, the vorticity is different from zero, but it also has the strong peaks near the event. This figure shows the existence of the vortex chain in flow. The vorticity was also calculated for the regions with strong peaks separately, corresponding to each structure, revealed by the hodogram analysis. As far as the vorticity changes the

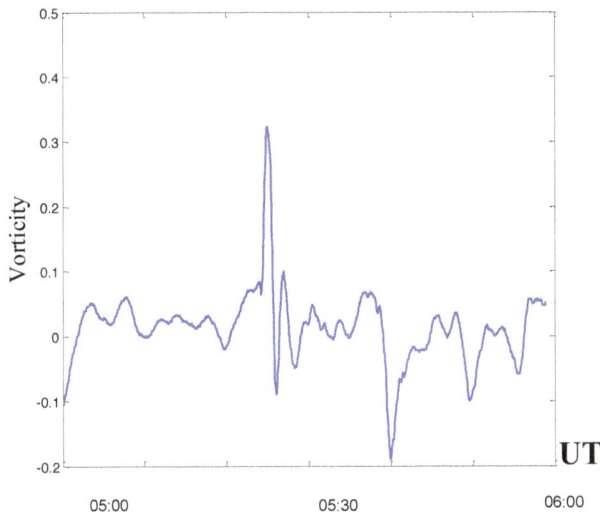

Figure 3. The vorticity of the plasma flow calculated using TH-A, D, E satellites.

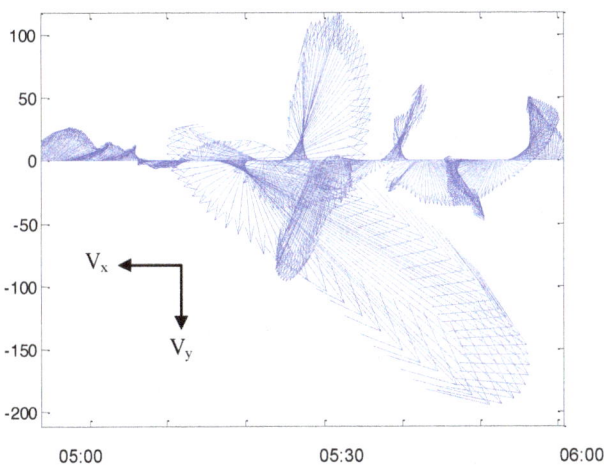

Figure 4. Vector plots of the flow velocity components (projected onto the GSM X–Y plane). Data from TH-C.

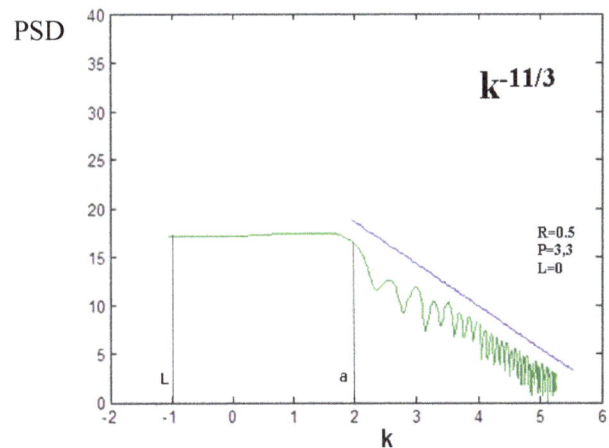

Figure 5. Spectral properties of magnetic fluctuations of a substorm event between 05:00 and 06:00 UT on 19 February 2008.

sign fast, one can assume that the structures, revealed in the experimental data, represent the vortices of different nature (monopole, dipole).

Figure 5 shows the total power spectrum density (PSD) of the magnetic fluctuations (Alexandrova, 2008). The power spectral density of the magnetic field components is calculated using the Morlet wavelet transform. One can see that the high frequency part of the spectrum follows a well defined power law $k^{-11/3}$.

4 Numerical simulation

The strong turbulence model represents a system of nonlinear partial differential equations with inhomogeneous coefficients and vector Jacobean type nonlinearity (Aburjania et

al., 2002, 2009). The complexity of the numerical analysis of this system is caused by nonlinear terms. The perturbations, obtained by the finite difference schemes for the reduced dimensionless quantities ψ, h in some time interval (Aburjania et al., 2006) will be self organized into vortex structures, which coincide with analytical ones for such flows. This indicates that the solutions obtained at numerical simulation in finite time interval contain also the stationary structures, which together with energetic estimations simplifies an estimation of simulation accuracy.

The initial and boundary value problem for time dependent dimensionless equation was solved numerically by using an implicit finite difference scheme described in (Aburjania et al., 2011). The computations were carried out on a 200×200 mesh in the x and y coordinates. The correctness of the computations and the stability of the scheme were controlled by solving model problems and also by checking the conservation of the mass of the structure and perturbation energy. The mass and energy were conserved with an accuracy of no worse than 10^{-2}.

Figure 6 represents evolution of initial monopole and random perturbations of the plasma flow, localized in a circle of radius 0.2 and magnetic field in a Gaussian inhomogeneous flows, where $V_0(y) = V_m \exp(-y^2/L_y^2)$, where L_y is transversal scale of the flow $L_y = (2-3)R$. Interaction of the localized disturbances with the background flow forms the vortex chain, the size of which depends on amplitude and distribution of the background flow. As it is obvious, the background inhomogeneous flow can form the vortex chain; such structures are observed in the experiments. It is clear, that the isolated chain will not occur. It depends on the initial distribution of perturbations and on the structure of BBF. Hence, their spatial distribution depends on BBF width in the magnetotail. As the numerical simulations show, the large scale vortex structures interact with each other and BBF, which defines their dynamics and life length. The size of the structures

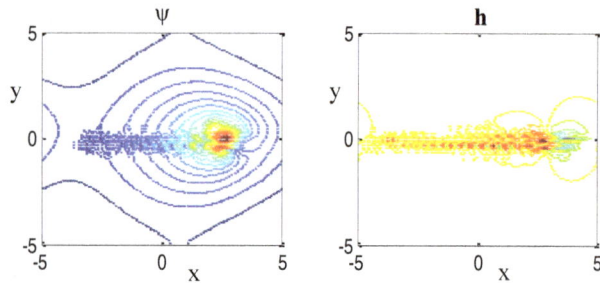

Figure 6. Evolution of localized randome disturbances of the stream function ψ and the magnetic induction h in time $t - 3$.

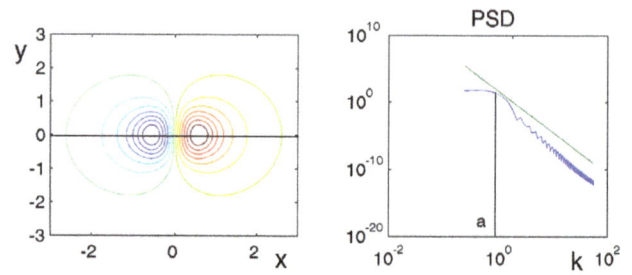

Figure 7. Dipolar vortex and corresponding PSD, describing the characteristics of this dipole: vortex radius (a), distance between the vortices (l) and power low (P): $a = 0.83$, $l = 0$, $P = -6$. Black line on the left figure is the vortex separatrix.

are dependent on the characteristics of BBF, its profile, amplitude and inhomogeneity and as far as the form of the BBF is dependent on solar wind dynamics, generated structures are defined by the solar wind and magnetic field interaction. Numerical analysis shows, that chain generation changes as the characteristics of the flow also magnetic field parameters. So, generated magnetic field perturbations depend on the structures generated in the flow and it is clear from Fig. 6, that, the structures will also be generated when the initial magnetic perturbation is zero. The structure is more complicated, which is in good agreement with experimental data (Keiling et al., 2009).

It is also very interesting to estimate a PSD) of analytical stationary dipole solution (Aburjania et al., 2009). The vortex structures form energetic spectra and participate in BBF flow energy balance. In this case we simulated a scenario of a magnetic probe moving along the x axis with a constant velocity and a distance of closest approach to the vortex axis x. The Fig. 7 shows the power spectral densities (PSD) of these signals calculated via the Morlet Wavelet Transforms. The power spectra of dipole have a knee around the wave vector $k = 1$, corresponding to a radius $a = 0.83$. The dipolar vortex spectrum on this case follows power law k^{-6}.

Note that these spectra are not completely independent of the trajectory of the virtual probe through the vortices. Along some particular trajectories, the magnetic field components are equal to zero and then the spectrum vanishes. These trajectories are vortex separatrices. Actually, the probability that the satellite crosses the vortex along a separatrix is small and the spectra of Fig. 7 can be considered as quasi-universal. The vortex spectra presented above can partially explain the magnetic spectrum presented in Fig. 5. Dipole vortex model reproduces the spectral knee, it appears to be around $k = a^{-1}$. The rather steep power laws of the dipole structures can explain the important steepening of the spectrum.

5   Conclusions and discussions

In this work, we investigated collective processes in magnetized plasma, caused by nonlinear regular structures. It is

shown that in space plasmas, the electromagnetic vortices are significantly elongated in the direction of the mean magnetic field, which well correlates with satellite observation data. These electromagnetic small-scale vortex structures, carrying trapped particles and spreading in plasma, generate the strong turbulence having granular character. Turbulence is represented by a gaseous ensemble of N strongly localized, weakly interacting identical vortices forming the background state. Turbulence excites appreciable fluctuations of density, velocity, magnetic and electric fields and intensifies the transfer processes. Thus, the width of a strong vortical turbulence spectrum is much larger than the value predicted by the weak turbulence theory. The turbulence develops more effectively in transverse direction to the local magnetic field. This anisotropy is essentially as strong as this magnetic field. This theory is in good agreement with experimental observations.

We have studied the experimental data from THEMIS mission on 19 February 2008, related to substorm event and revealed the localized electromagnetic vortex structures, forming the strong turbulent state in the plasma media. The substorm on 19 February 2008 developed a substorm surge that propagated poleward, westward, and eastward which is typical for substorms. Four THEMIS spacecraft monitored in situ the conjugate space vortices. One space vortex engulfed the three clustered spacecraft, which were separated in a triangular-like constellation (1–2 RE), allowing an unambiguous identification of a counterclockwise flow. The associated clockwise vortex was tentatively inferred from the single spacecraft located further west. By the experimental data treatment we revealed several smaller vortices with the main one, associated with the substorm onset. Further analysis has shown that the rotational senses of these smaller space vortices were consistent with the main ones (Fig. 3). This fact was also verified by the numerical simulation in Sect. 4. 3-D modeling (Birn et al., 2004) and observations (Keika et al., 2009; Keiling et al., 2009; Panov et al., 2010) have revealed that the possibility for a plasma to move in the azimuthal direction allows vortex formation. The earthward and tailward flow bursts form vortices with opposite sense of rotation. Figure 5 shows formation of the vortices

Table 1. Characteristics of the vortex structures.

TH-C	1 (70) Structure	2 (70) Structure	3 (70) Structure	4 (65) Structure	5 (80) Structure	6 (50) Structure
Time Scale	210 s	210 s	210 s	195 s	240 s	150 s
Average velocity	148 km s^{-1}	47.4 km s^{-1}	12 km s^{-1}	33 km s^{-1}	36 km s^{-1}	7 km s^{-1}
Size	31 080 km	9870 km	2520 km	6435 km	8640 km	1050 km

during the earthward flow bursts. The tailward-directed flow burst forces vortex chain formation on the two sides of the BBF funnel similar to "von Karman street" (see Fig. 6) with change their sense of rotation. Indeed, numerical 2-D MHD simulations revealed presence of both configurations during the oscillatory BBF braking (see Figs. 6 and 7). Therefore, the results of this analysis are important for understanding of the magnetosphere – ionosphere coupling phenomena.

The substorm on 19 February 2008 developed a substorm surge that propagated poleward, westward, and eastward which is typical for substorms (Akasofu, 1976). Four THEMIS spacecraft monitored in situ the conjugate space vortices. One space vortex engulfed the three clustered spacecraft, which were separated in a triangular-like constellation (1–2 RE), allowing an unambiguous identification of a counterclockwise flow. The associated clockwise vortex was tentatively inferred from the single spacecraft located further west. The experimental data treatment revealed several smaller vortices with the main one, associated with the substorm onset. Further analysis has shown that the rotational senses of these smaller space vortices were consistent with the main ones. Their characteristic sizes and time scales are also estimated for all the spacecrafts, but only for TH-C is given in Table 1.

Regarding the mapping of a flow vortex from the magnetosphere to the ionosphere, Borovsky and Bonnell (2001) showed theoretically that the ionospheric footprint of a positive (downward current) vortex is larger than the mapped footprint of the corresponding magnetospheric vortex because of a spreading of the associated electric potential from high to low altitude. Hence, the larger EIC vortex (600–800 km) versus the smaller mapped footprint (180 km) could be explained by this spreading and/or because the THEMIS spacecraft did not enclose the entire space vortex (Keiling et al., 2009). Assuming conservation of angular speed, the flow speed of 300–900 km s^{-1} corresponds to 4–12 km s^{-1} in the ionosphere. However, this assumption is most likely not valid because of ionospheric drag, and thus the mapped speeds should only be considered as an upper limit. Additional analysis of the ionospheric experimental satellite and ground-based data is necessary for magnetosphere-ionosphere coupling processes in a separate study.

Acknowledgements. Shota Rustaveli National Science Foundation's Grant no 31/14.

References

Aburjania, G., Khantadze, A., and Kharshiladze, O.: Nonlinear planetary electromagnetic vortex structures in F region of the ionosphere, Plasma Phys. Rep., 28, 633–638, 2002.

Aburjania, G. D.: Self-Organization of the Nonlinear Vortex Structures and the Vortical Turbulence in the Dispersive Media: Kom-Kniga, Editorial URSS, Moscow, Russia, 2006.

Aburjania, G. D., Chargazia, Kh. Z., Zelenyi, L. M., and Zimbardo, G.: Model of strong stationary vortex turbulence in space plasmas, Nonlin. Processes Geophys., 16, 11–22, doi:10.5194/npg-16-11-2009, 2009.

Akasofu, S.-I.: Physics of Magnetospheric Substorms, D. Reidel Publ. Co., Dordrecht, the Netherlands, 1976.

Alexandrova, O.: Solar wind vs magnetosheath turbulence and Alfvén vortices, Nonlin. Processes Geophys., 15, 95–108, doi:10.5194/npg-15-95-2008, 2008.

Angelopoulos, V.: The THEMIS Mission, Space Sci. Rev., 141, 5–34, doi:10.1007/s11214-008-9336-1, 2008.

Birn, J., Raeder, J., Wang, Y. L., Wolf, R. A., and Hesse, M.: On the propagation of bubbles in the geomagnetic tail, Ann. Geophys., 22, 1773–1786, doi:10.5194/angeo-22-1773-2004, 2004.

Biskamp, D., Schwarz, E., Zeiler, A., Celani, A., and Drake, J. F.: Electron magnetohydrodynamic turbulence, Phys. Plasmas., 6, 751–758, 1999.

Borovsky, J. and Bonnell, J.: The DC electrical coupling of flow vortices and flow channels in the magnetosphere to the resistive ionosphere, J. Geophys. Res., 106, 28967–28994, 2001.

Chmyrev, V. M., Marchenko, V. A., Pokhotelov, O. A., Stenflo, L., Streltsov, A. V., and Steen, A.: Vortex structures in the ionosphere and the magnetosphere of the Earth, Planet. Space Sci., 39, 1025–1037, 1991.

Cho, J. and Lazarian, A.: The anisotropy of magnetohydrodynamic turbulence, Astrophys. J., 615, L41–L44, 2004.

Diamond, P. H. and Carreras, B. A.: On mixing length theory and saturated turbulence, Comm. Plasma Phys. Contr. Fus., 10, 271–278, 1987.

Galeev, A. A. and Sagdeev, R. Z.: Nonlinear Theory of plasma, in: Reviews of Plasma Physics, edited by: Leontovich, M. A., Consultant Bureau, New York, USA, Vol. 7, 1976.

Galtier, S., Pouquet, A., and Mangeney, A.: On spectral scaling laws for incompressible anisotropic magnetohydrodynamic turbulence, Phys. Plasmas, 12, 092310, doi:10.1063/1.2052507, 2005.

Gekelman, W.: Review of laboratory experiments on Alfven waves and their relationship to space observations, J. Geophys. Res., 104, 14417–14435, 1999.

Goldreich, P. and Sridhar S.: Toward a theory of interstellar turbulence, II. Strong Alfvenic turbulence, Astrophys. J., 438, 763–775, 1995.

Horton, W.: Nonlinear Drift Waves and Transport in Magnetized Plasma, Review, Institute for Fusion Studies the University of Texas at Austin, IFSR, Austin, Texas, USA, 1990.

Howes, G. G., Cowley, S. C., Dorland, W., Hammettm G. W., Quataertm E., and Schekochihinm A. A.: A model of turbulence in magnetized plasmas: Implications for the dissipation range in the solar wind, J. Geophys. Res., 113, 05103, doi:10.1029/2007JA012665, 2008.

Iroshnikov, R. S.: The turbulence of a conducting fluid in a strong magnetic field, Astr. Zh., 40, 742–750, 1963.

Keiling, A., Angelopoulos, V., Runov, A., Weygand, J., Apatenkov, S. V., Mende, S., McFadden, J., Larson, D., Amm, O., Glassmeier, K.-H., and Auster, H. U.: Substorm current wedge driven by plasma flow vortices: THEMIS Observations, J. Geophys. Res., 114, A00C22, doi:10.1029/2009JA014114, 2009.

Keika, K., Nakamura, R., Volwerk, M., Angelopoulos, V., Baumjohann, W., Retinò, A., Fujimoto, M., Bonnell, J. W., Singer, H. J., Auster, H. U., McFadden, J. P., Larson, D., and Mann, I.: Observations of plasma vortices in the vicinity of flow-braking: a case study, Ann. Geophys., 27, 3009–3017, doi:10.5194/angeo-27-3009-2009, 2009.

Kingsep, A. S.: Introduction to the Nonlinear Plasma Physics, in: Reviews of Plasma Physics, edited by: Kadomtsev, B. B., Consultant Bureau, New York, USA, Vol. 16, 1990.

Kolmogorov, A. N.: Local structure of turbulence in the noncompresible viscous fluid at very high Reynolds number, Dokl. Akad. Nauk. SSSR, 30, 299–303, 1941.

Kraichnan, R. H.: Inertial range spectrum hydromagnetic turbulence, Physics of Fluids, 8, 1385–1387, 1965.

McFadden, J., Carlson, C. W., Larson, D., Ludlam, M., Abiad, R., Elliott, B., Turin, P., Marckwordt, M., and Angelopoulos, V.: Space Sci. Rev., 141, 277–302, 2008.

Matthaeus, W. H., Oughton, S., Pontius, D. H., and Zhou, Y.: Evolution of energy-containing turbulent eddies in the solar wind, J. Geophys. Res., 99, 19267–19287, 1994.

Müller, W.-C., Biskamp, D., and Grappin, R.: Statistical anisotropy of magnetohydrodynamic turbulence, Phys. Rev. E., 67, 066302, doi:10.1103/PhysRevE.67.066302, 2003.

Narita, Y., Glassmeier, K.-H., Fränz, M., Nariyuki, Y., and Hada, T.: Observations of linear and nonlinear processes in the foreshock wave evolution, Nonlin. Processes Geophys., 14, 361–371, doi:10.5194/npg-14-361-2007, 2007.

Nezlin, M. V. and Snezhkin, E. N.: Rossby Vortices, Spiral Structures, Solitons, Springer-Verlag, Heidelberg, Germany, 1993.

Panov, E. V., Nakamura, R., Baumjohann, W., Angelopoulos, V., Petrukovich, A. A., Retinò, A., Volwerk, M., Takada, T., Glassmeier, K.-H., McFadden, J. P., and Larson, D.: Multiple overshoot and rebound of a bursty bulk flow, J. Geophys. Res., 37, L08103, doi:10.1029/2009gl041971, 2010.

Richardson, L. F.: Wheather Prediction of "Numerical Method", Cambridge University Press, Cambridge, UK, 1922.

Sahraoui, F., Belmont, G., Pinçon, J. L., Rezeau, L., Balogh, A., Robert, P., and Cornilleau-Wehrlin, N.: Magnetic turbulent spectra in the magnetosheath: new insights, Ann. Geophys., 22, 2283–2288, doi:10.5194/angeo-22-2283-2004, 2004.

Sahraoui, F., Belmont, G., Rezeau, L., and Cornilleau-Wehrlin, N.: Anisotropic turbulent spectra in the terrestrial magnetosheath as seen by the Cluster spacecraft, Phys. Rev. Lett., 96, 075002, doi:10.1103/PhysRevLett.96.075002, 2006.

Snekvik, K., Haaland, S., Østgaard, N., Hasegawa, H., Nakamura, R., Takada, T., Juusola, L., Amm, O., Pitout, F., Rème, H., Klecker, B., and Lucek, E. A.: Cluster observations of a field aligned current at the dawn flank of a bursty bulk flow, Ann. Geophys., 25, 1405–1415, doi:10.5194/angeo-25-1405-2007, 2007.

Zimbardo, G.: Magnetic turbulence in space plasmas in and around the Earths magnetosphere, Plasma Phys. Contr. Fusion, 48, B295–B302, 2006.

23

Investigations and system design for simultaneous energy and data transmission through inductively coupled resonances

C. Schmidt, E. Lloret Fuentes, and M. Buchholz

Research Group RI-ComET at the University of Applied Sciences Saarbrücken, Hochschul-Technologie-Zentrum, Altenkesselerstr. 17/D2, 66115 Saarbrücken, Germany

Correspondence to: C. Schmidt (christian.schmidt@htwsaar.de)

Abstract. Wireless Power Transfer (WPT) with simultaneous data transmission through coupled magnetic resonators is investigated in this paper. The development of this system is dedicated to serve as a basis for applications in the field of Ambient Assisted Living (AAL), for example tracking vital parameters remotely, charge and control sensors and so on. Due to these different scenarios we consider, it is important to have a system which is reliable under the circumstance of changing positioning of the receiving device. State of the art radio systems would be able to handle this. Nevertheless, energy harvesting from far field sources is not sufficient to power the devices additionally on mid-range distances. For this reason, coupled magnetic resonant circuits are proposed as a promising alternative, although suffering from more complex positioning dependency.

Based on measurements on a simple prototype system, an equivalent circuit description is used to model the transmission system dependent on different transmission distances and impedance matching conditions. Additionally, the simulation model is used to extract system parameters such as coupling coefficients, coil resistance and self-capacitance, which cannot be calculated in a simple and reliable way.

Furthermore, a mathematical channel model based on the schematic model has been built in MATLAB©. It is used to point out the problems occurring in a transmission system with variable transmission distance, especially the change of the passband's centre frequency and its bandwidth. Existing solutions dealing with this distance dependent behaviour, namely the change of the transmission frequency dependent on distance and the addition of losses to the resonators to increase the bandwidth, are considered as not inventive. First, changing the transmission frequency increases the complexity in the data transmission system and would use a dispro-

portional total bandwidth compared to the actually available bandwidth. Additionally, adding losses causes a decrease in the energy transmission efficiency.

Based on these facts, we consider a system that changes the channel itself by tuning the resonant coils in a way that the passband is always at a fixed frequency. This would overcome the previously described issues, and additionally could allow for the possibility to run several independent transmission systems in parallel without disturbing each other.

1 Introduction

The interest in WPT with coupled inductive resonators has grown in the last years, especially after the publications by Kurs et al. (2007) in Science. The nearfield coupling of resonating inductors promises the possibility to transfer an amount of energy over a mid-range distance quite efficiently compared to farfield approaches. Important aspects regarding the design of such a system have been in the focus of recent research, like coil design by Kim et al. (2012) and Lee et al. (2013), impedance match by Sample et al. (2010) and Cheon et al. (2011), coupled resonator's passband dependency on transfer distance by Sample et al. (2010) and Zhang et al. (2014) et cetera. So far, WPT systems have been described and investigated in depth, and solutions to overcome certain issues have been presented.

One of the most challenging properties is the transfer distance dependent passband characteristic of two or more coupled resonators. Different distances lead to changes in the coupling coefficient of the resonators, over-critical, critical and under-critical regions can be defined, similar to coupled, analog filters. The difficult thing here is that the transfer characteristic is not static. In principle, when thinking of a real application with freely movable wirelessly powered devices, it is a highly sophisticated task to predict the changes of the transfer function in a dynamic system.

This paper addresses the adaption of the system behaviour by changing the resonant coils parameters to keep the transfer maximum constant at one fixed frequency at first instance. Additionally, the data transmission as a new attribute, not considered in depth so far, adds some more restrictions to the system design, which are discussed based on the proposed channel stabilization method. Finally, the possibilities and challenges, especially concerning how to tune the coils, are highlighted. The study of the proposed system behaviour serves as a basis for further development regarding channel estimation and adaption. The proposed solution is adoptable for both, power and simultaneous data transfer. Other strategies like shifting the carrier frequency by Sample et al. (2010), which would complicate the data transmission, are considered as not useful in the target application. Also switching between a data and a power transfer mode as suggested by Kim et al. (2013) is assessed as a too complex solution in terms of synchronising both entities, especially in a very dynamic environment. Multi-band systems using two resonances like in Joanh et al. (2013) and Dionigi et al. (2012) are an interesting modification, using one frequency band for energy, another one for data transmission. The fact that resonators with two very well defined resonances must be used and, in our case, also controlled depending on the transfer distance, makes the implementation of a dynamically changeable system even more complex.

2 Prototype measurements, parameter extraction and system modelling

The prototype shown in Fig. 1 is used as a reference system in this paper. It is designed as a simple 4-coil-system similar to those described in Kurs et al. (2007); Kim et al. (2012, 2013); Cheon et al. (2011) and Awai et al. (2010). Additionally to the resonant coil, a single turn coupling coil is used on both source and load side to form an impedance matching network described in Sect. 2.2. This implementation of the impedance match is easy to build and to tune, and it has been shown to be a good compromise between transfer distance and efficiency (Hui et al., 2013).

Mechanically, each resonant coil has a coupling coil attached to it. The distance between both can be varied to change the coupling factor and, due to this, the impedance match. Both resonant coils with their coupling coil can be moved on a fixed rail, making it possible to change the transfer distance while keeping the impedance match untouched. All coils are arranged concentrically.

The resonant coils are each wound on a piece of polypropylene tube with a radius of 55 mm, in which a notch of 0.5 mm was milled to keep the windings in a constant in-

Figure 1. Wireless Power Transfer prototype used in this work.

clination of 3.5 mm per turn. The copper wire diameter is of 1.12 mm, resulting in a total coil radius of 55.63 mm.

Similar to the system presented in Kurs et al. (2007), the resonant coils are initially used in their self-resonance. The design goal was to achieve a self-resonant frequency of approximately 20 MHz. To ensure this, each coil consists of ten windings, promising enough self-capacitance for a low resonant frequency. CST Microwave Studio© was used to validate this choice. The simulation shows a resonance at 21.78 MHz and an inductance of 14.775 μH at 10 MHz. At frequencies closer to the resonance, the inductance values extracted from the simulated Z parameters are not meaningful because the coil's behaviour near its resonance frequency can no longer be considered as inductive. The inductance value was validated using Rayleigh and Niven's formula, Eq. (1), for coils short compared to their diameter; afterwards, Rosa's current sheet correction, ΔL, is applied (Rosa et al., 1916):

$$L_s = 4\pi a n^2 \left[\log \frac{8a}{b} - \frac{1}{2} + \frac{b^2}{32a^2} \left(\log \frac{8a}{b} + \frac{1}{4} \right) \right] \qquad (1)$$

$$\Delta L = 4\pi a n (A + B). \qquad (2)$$

Finally, this results in the corrected inductance value L:

$$L = L_s + \Delta L. \qquad (3)$$

In these formulas, a is the winding diameter, b is the total height of the coil, n is the number of turns, A and B are correction values tabulated by Rosa et al. (1916). For our resonant coils, this formula gives a corrected inductance value of 14.7 μH, which is in absolute agreement with the simulated value. Using the well-known Thomson formula, we calculate the coil's self capacitance by setting $f = 21.7$ MHz, neglecting any loss, and $L = 14.7 \mu$H resulting in a capacitance of

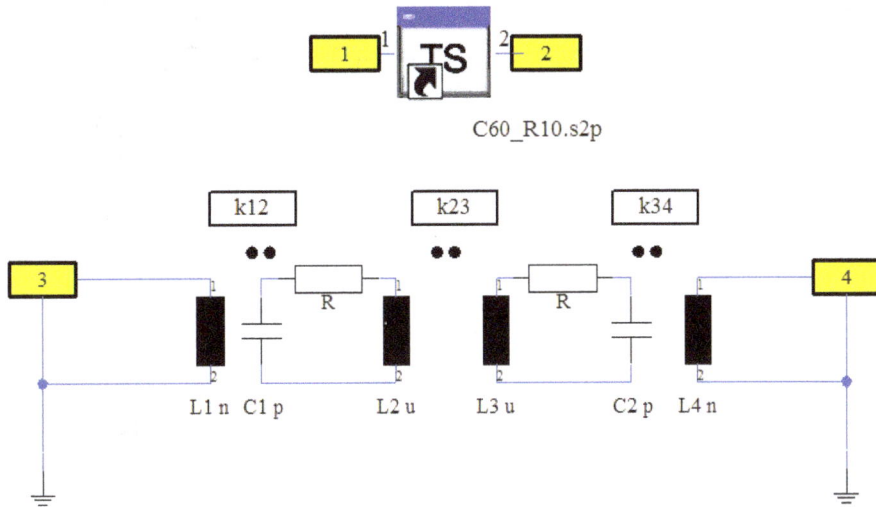

Figure 2. CST Design Studio model used for parameter extraction.

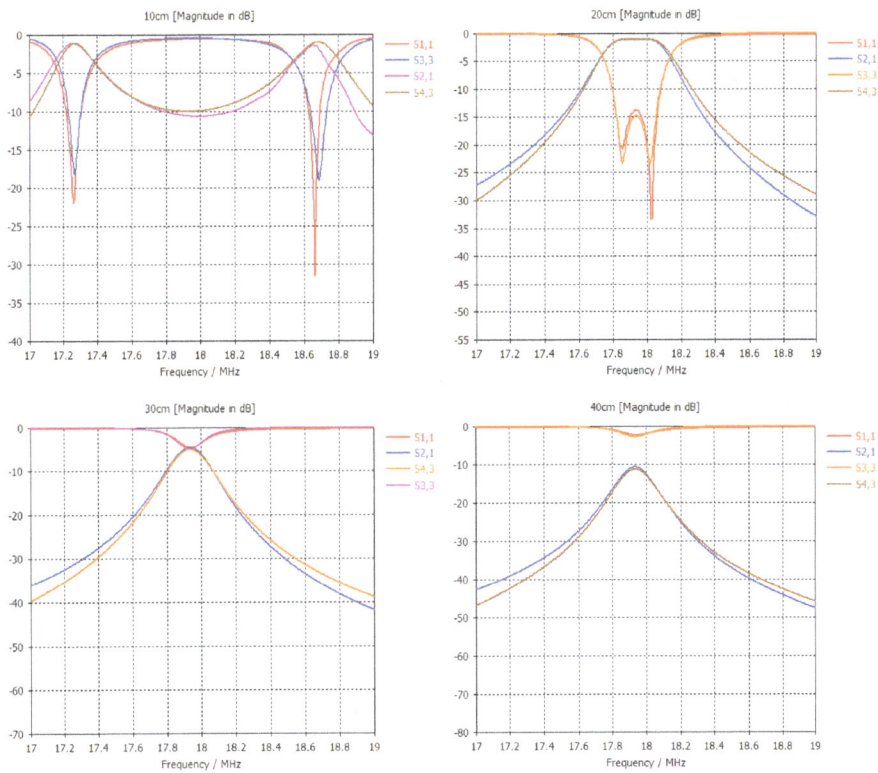

Figure 3. Measured ($S_{1,1}$, $S_{2,1}$) and simulated ($S_{3,3}$, $S_{4,3}$) transmission and reflection amplitudes.

3.63 pF.

$$f = \frac{1}{2\pi \sqrt{LC}} \qquad (4)$$

In order to be able to access the resonant coils directly, SMA connectors are attached to each of them. Thus, the coils can directly be measured and additional variable capacitors can be attached to tune them.

2.1 Measurements and parameter extraction

First measurements on the prototype showed that the expected resonant frequency of 21.78 MHz was shifted to 17.92 MHz for both coils. The measured inductance of the resonant coils is 14.56 μH at 1 MHz, which is a tolerable result compared to the calculated and simulated value of 14.7 μH. As a consequence, the shifted resonant frequency

Table 1. Lumped elements values.

L_2, L_3	C_1, C_2	R	$k_{1,2}, k_{3,4}$	L_1, L_4
$14.56\,\mu\mathrm{H}$	$5.458\,\mathrm{pF}$	$2.8\,\Omega$	0.151	$312\,\mathrm{nH}$

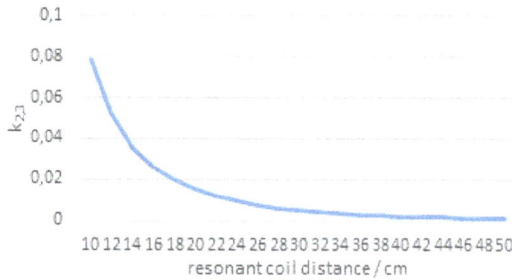

Figure 4. Coupling factor $k_{2,3}$ over transfer distance in cm.

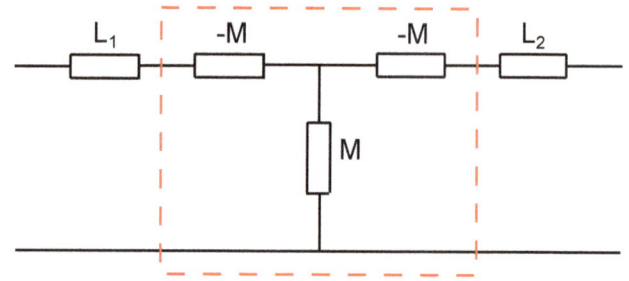

Figure 5. Impedance inverter.

must be the result of an increased capacitance value of approximately 5.4 pF and/or additional losses compared to the simulation. The material model for polypropylene and the mechanical tolerances of the wires to the connectors are the most likely candidates for an uncertainty like this.

Due to a direct measurement of the coil's inductance, capacitance and loss resistance at the resonant frequency is not feasible, a parameter extraction based on measured data as Touchstone© files and a schematic model was performed.

As a first step, measurements were conducted on the full 4-coil-system with constant matching condition (position of the coupling coils on both sides at 6 cm apart from the resonant coils) and transfer distances from 10 to 50 cm in steps of 2 cm.

The equivalent circuit shown in Fig. 2 has been widely used to characterize WPT systems. Although cross and capacitive couplings between the coils as well as capacitance respectively loss resistance of the coupling coils are neglected, good agreement between measured and simulated responses have been reported in Sample et al. (2010) and Dionigi et al. (2011). Due to this, the model is used to extract the circuit parameters by means of an optimizer run for every transfer distance we measured. Figure 3 shows the measured response and the optimized schematic simulation for four different transfer distances. For the first run (10 cm resonant coil distance, 6 cm coupling coil distance), the schematic values were initially set. The inductance value was taken from the measurements, the capacitance from the simplified calculation using Thomson's formula and the measured resonance frequency. Coupling factors and resistive losses were adapted until a certain fitting was achieved. Then, an optimizer was set up changing capacitance, resistance and coupling factors until the difference between the $S21$-parameter of measured data and schematic simulation was below a threshold of 0.02 (linear scaling). For the following runs, the measured data file was updated to the next higher transfer distance and the same

optimizer was used with the same goal; however, just the coupling factor between the resonant coils was changed (results shown in Fig. 3). The fact that a fit between both curves has been realized shows that the schematic model works fine for describing the system.

Additionally, the extracted parameters tabulated below have proven to be constantly a good approximation of the real physical model. In Table 1, L_2 and L_3 stand for resonant coil inductance, C_1 and C_2 for self-capacitance of the resonant coils, R for loss resistance of the resonant coils, $k_{1,2}$ and $k_{3,4}$ the coupling factor between coupling and resonant coils and L_1 and L_4 for coupling coil inductances.

The extracted coupling factor between the resonant coils, $k_{2,3}$, is shown in Fig. 4, dependent on the transfer distance. Based on the full set of schematic component values, the following section describes a mathematical model used to describe the transfer channel.

2.2 System modelling

Looking again at Fig. 2, which we used to extract the coil's parameters, we can see that there are three pairs of coupled inductors. A pair of coupled inductors, like the ones shown in Fig. 2, can also be represented by means of a T-equivalent circuit. This T-equivalent circuit behaves as an impedance inverter, formed by the three mutual inductances M shown in Fig. 5 (Tosic et al., 2006). An impedance inverter as described below has the following **ABCD** matrix:

$$\mathbf{ABCD} = \begin{bmatrix} 0 & -jK_{1,2} \\ \frac{-j}{K_{1,2}} & 0 \end{bmatrix}. \tag{5}$$

In the above matrix, the inverter constant K is equal to $K_{1,2} = \omega M = \omega k_{1,2}\sqrt{L_1 L_2}$. In this equation, $k_{1,2}$ is the coupling factor between the coils L_1 and L_2, M is the mutual inductance. Replacing the three pairs of coupled inductors by their corresponding impedance inverters and considering the internal resistance and capacitance of every inductor, we end up in a circuit like the one presented in Fig. 6.

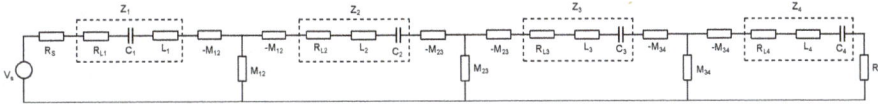

Figure 6. Four coil equivalent circuit.

We can analyse the transfer function of such a circuit by using a cascade of **ABCD** matrices. Thus, we obtain:

$$\mathbf{ABCD} = \begin{bmatrix} 1 & Z_1 \\ 0 & 1 \end{bmatrix} \times \begin{bmatrix} 0 & -jK_{1,2} \\ \frac{-j}{K_{1,2}} & 0 \end{bmatrix} \times \begin{bmatrix} 1 & Z_2 \\ 0 & 1 \end{bmatrix} \qquad (6)$$

$$\times \begin{bmatrix} 0 & -jK_{2,3} \\ \frac{-j}{K_{2,3}} & 0 \end{bmatrix} \times \begin{bmatrix} 1 & Z_3 \\ 0 & 1 \end{bmatrix}$$

$$\times \begin{bmatrix} 0 & -jK_{3,4} \\ \frac{-j}{K_{3,4}} & 0 \end{bmatrix} \times \begin{bmatrix} 1 & Z_4 \\ 0 & 1 \end{bmatrix}$$

A similar analysis has been performed in Dionigi et al. (2011) with equivalent results. If we now consider that the system is symmetric:

$$Z_1 = Z_4$$
$$Z_2 = Z_3$$
$$K_{1,2} = K_{3,4}$$

Then, the **ABCD** matrix gets simplified:

ABCD =

$$\begin{bmatrix} j\dfrac{Z_1 Z_2^2 + K_{1,2}^2 Z_2 + Z_1 K_{2,3}^2}{K_{1,2}^2 K_{2,3}} & \dfrac{Z_1^2 Z_2^2 + K_{2,3}^2 Z_1^2 + 2K_{1,2}^2 Z_1 Z_2 + K_{1,2}^4}{K_{1,2}^2 K_{2,3}} \\[2ex] j\dfrac{Z_2^2 + K_{2,3}^2}{K_{1,2}^2 K_{2,3}} & j\dfrac{Z_1 Z_2^2 + K_{1,2}^2 Z_2 + Z_1 K_{2,3}^2}{K_{1,2}^2 K_{2,3}} \end{bmatrix} \qquad (7)$$

In order to calculate the scattering matrix parameter $S_{2,1}$, we can apply the well-known transformation in Eq. (8) from Hui et al. (2013), where $Z_0 = 50\,\Omega$

$$S_{2,1} = \frac{2(AD - BC)}{A + \frac{B}{Z_0} + CZ_0 + D} \qquad (8)$$

For the sake of verification, we compare the behaviour of the above formula with the schematic simulations. In Fig. 7, we can see the frequency response of a four coils configuration where the resonant coils are 20 cm apart from each other, the coupling coils and the resonant coils are separated 6 cm from each other. For these particular distances, the coupling factors are $k_{1,2} = k_{3,4} = 0.151$ and $k_{2,3} = 0.0156073$.

3   Channel modelling

In the last two sections, we first extracted all parameters for a simplified schematic model describing our prototype system, based on measurements at discrete distances. Afterwards, we

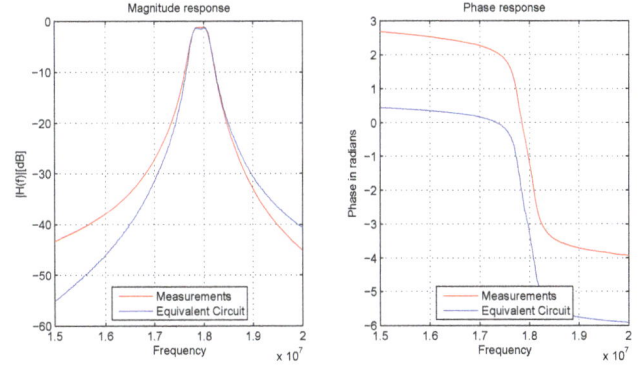

Figure 7. Measurement results vs. equivalent circuit calculation.

found the transfer function shown in Eq. (8) which describes the system behaviour as a simple equation. As this description has proven to be accurate, we are also able to model and predict transfer distances not considered in our measurements.

This development allows us to switch from a physical description in terms of lumped element values to a mathematical model. We use a Finite Impulse Response (FIR) filter to model the transfer in MATLAB© to be able to analyse arbitrary data transfer through the channel.

Since we have already an expression for the $S_{2,1}$ parameter, we can easily obtain the frequency response of such a hypothetical filter by simple substitution of the lumped element values in Eq. (8) by those extracted from the measurements. Furthermore, it is known that the frequency response of a filter is just the Fourier Transform of its impulse response $h[n]$ (Oppenheim et al., 1989). Therefore, using the Inverse Discrete Fourier Transform (IDFT, Eq. 9), we can obtain the impulse response of a channel described by Eq. (7).

$$h[n] = \frac{1}{2\pi} \int_{-\pi}^{\pi} H(e^{j\omega}) e^{j\omega n} d\omega \qquad (9)$$

In case of a FIR filter, the impulse response values are simply the filter's coefficients, as shown in Eq. (10). The IDFT of Eq. (7) values can therefore be directly used as filter coefficients.

$$y[n] = \sum_{k=0}^{N-1} b_k x[n-k] \qquad h[n] = \sum_{k=0}^{N-1} b_k \delta[n-k] \quad (10)$$

Using Eq. (7) and the IDFT, we can now calculate the filter coefficients and compare the transfer functions in Fig. 8.

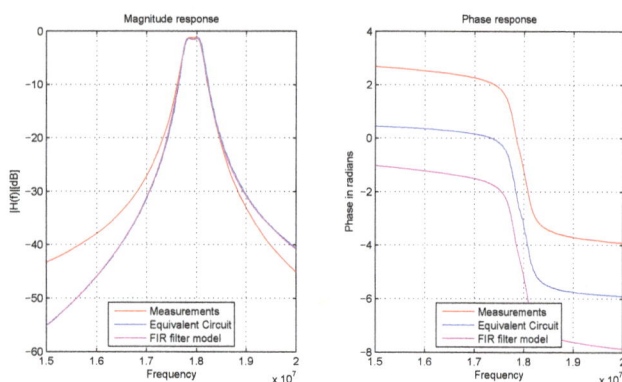

Figure 8. Measurement results vs. equivalent circuit and FIR filter model analysis.

Apart from a constant phase deviation, the FIR filter modelling provides a good approximation for the four coils' frequency response. Having obtained the filter's coefficients will allow us to perform deeper analysis of this kind of communication channels, as well as the evaluation of possible solutions for the challenges it may present. In order to give an idea of what these simulations might look like, we study the following example. We generate a spectrum-shaped pseudo random baseband signal which modulates a 17.935 MHz carrier as Quadrature Phase-Shift Keying (QPSK), as this is the resonant frequency of each single resonator. This results in a 125 KHz RF bandwidth. Then, we send it through a FIR filter with the coefficients we just extracted, and finally we mix down the signal to the baseband and demodulate it. As expected, varying the distance between the coils produces various effects, since they represent different channels.

Assuming that the coupling coils remain in a fixed position with respect to the resonant coils, the first and most obvious effect is resonant frequency splitting, occurring when the resonant coils are over-critically coupled. In this region, the closer the two resonant coils are, the more deviation from the centre appears between the two new resonant frequencies. Moving away the two resonant coils, until a particular distance is reached, causes the two resonant frequencies to merge in a single one (critical coupling). At this point, we obtain the minimum attenuation at the resonant frequency as well as the widest 3 dB bandwidth. Beyond this distance, the attenuation increases while the 3 dB bandwidth may decrease, mainly due to impedance mismatch. Constant adaption of the impedance match can be a first attempt to improve the transmission as highlighted in Sample et al. (2010), and should be applied in addition to the proposal presented in the next section.

In the following figures, a constellation diagram is shown next to the transfer function. It serves as a qualitative measure of the data transmission performance. A totally undistorted, ideal QPSK signal would produce a diagram with one discrete point in every quadrant (absolute value $|I(t)| = |Q(t)| = 1$), representing the four signal states.

Figure 9 shows as an example the transfer function for a distance of 10 cm and the corresponding constellation diagram for the data signal described above. As we can see, mainly the signal's amplitude decreases to a normalised value of around 0.2, corresponding to the attenuation of 11.5 dB. The signal phase is slightly rotated, but the constellation points form quite dense spots.

At an increased distance of 20 cm (Fig. 10), the amplitude is higher than before. This is due to the fact that our signal transmission frequency is now right in the middle of the channel's maximum. The two resonators are no more over-critically coupled but nearly critical and depict an attenuation of 1.5 dB. On the other hand, the constellation points form a less dense cloud, mainly caused by linear phase distortion or in other words a non-constant group delay.

Finally, at 30 cm transfer distance (Fig. 11), both effects described above occur. The constellation diagram shows even wider spread points with a small corresponding amplitude caused by the amplitude and phase distortion induced by the channel.

As we have seen, for data transmission, an untuned system of coupled resonators forms a very dynamic, distance dependent channel. All of the individual data signals described by the constellation plots above can possibly be demodulated by a digital signal processing system. But for a dynamic system, the range of different channels would be too high. In this particular channel, we have the ability to tune the channel for every operating condition. Like this, we can optimize not only the data channel, but also maintain a constant maximum power transfer.

4   Channel adaption

Summing up the last section, by varying the transfer distance, the channel characteristic of the coupled resonators changes considerably in amplitude and phase. To handle this behaviour, we suggest the implementation of an automated, adaptive channel tuning dependent on the actual distance.

Recent research has been carried out by different groups to address the problem of channel dependency on the transfer distance in pure energy transmission systems. Typical experimental implementations to decrease the channel variation make use of additional impedance matching networks attached to the transfer system (Beh et al., 2013; Waters et al., 2012; Sample et al., 2013) or several switchable coupling coils (Kim et al., 2012). Especially (Sample et al., 2013) shows, that this impedance matching network adds additional resonant peaks in between the resonators main peaks, thus providing the possibility to transfer energy at a fixed frequency. Due to the fact that the matching networks consist of switchable elements, it is obvious that the system performance is optimized only for discrete distances dependent

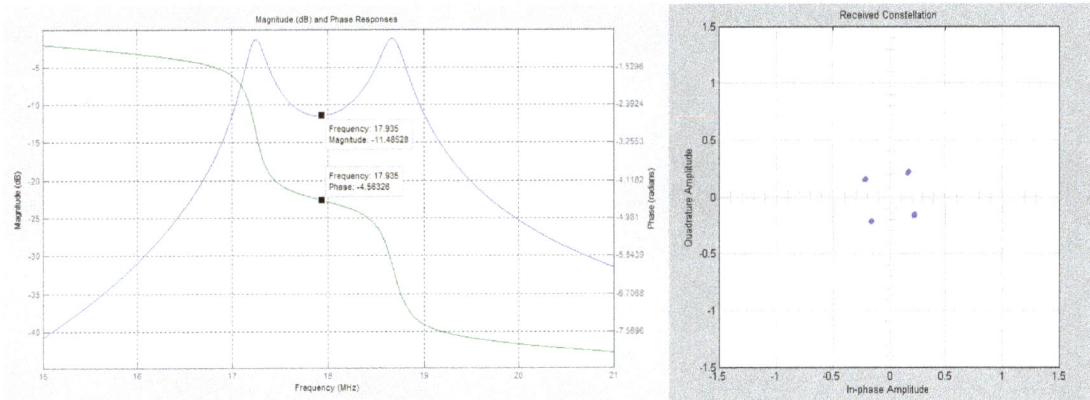

Figure 9. FIR-filter magnitude/phase and constellation plot for 10 cm distance.

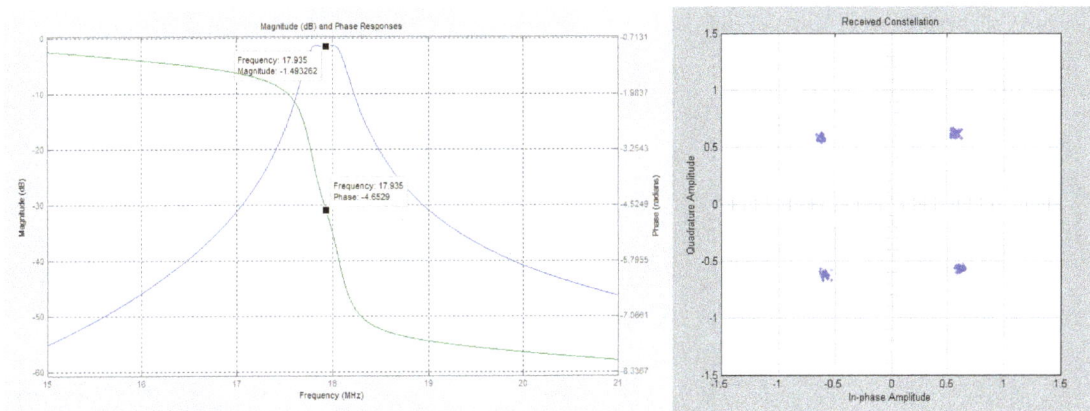

Figure 10. FIR-filter magnitude/phase and constellation plot for 20 cm distance.

on the number of possible switch combinations. Continuous matching networks are presented in (Beh et al., 2010) and (Jang et al., 2012), in which the first system uses motor controlled variable air capacitors, which is certainly limited to large-scale applications like electric vehicles. The second paper reports an impedance match implementation based on varactors. These devices would be also very convenient as tuning elements, unfortunately, their RF voltage handling capability is very limited. Using varactors as tuning elements for the resonators is not useful because of the quite high voltages (tens to hundreds of volts) at the coil's terminals, even for small or mid-range power levels.

Regarding data transmission, the impedance matching network approach introduces again a channel variation due to the different matching states dependant on transfer distance. Additionally, the performance of matching network based implementations degrades compared to a system which shifts the transfer frequency according to the channel response (Sample et al., 2013). Although we stated in the introduction, that frequency shifting is not an option for a data transmission system, it can be used as a reference for other implemen-

tations because it involves the least elements and therefore promises maximum performance.

Based on this fact, tuning the channel response dependent on transfer distance can be considered as counterpart to frequency shifting. A first implementation of such a system has recently been presented in Ricketts et al. (2013). A bank of switchable capacitors is used to shift the resonant frequencies of the coils in a way that the resulting channel shows a transfer maximum always at a fixed frequency. Like this, the channel's amplitude and phase behaviour is a lot less dependent on transfer distance and more stable compared to the impedance matching approach.

To validate the benefits of this approach, we modified our simulation model by increasing the resonators capacitance to locate the lower peak of the transfer function always at 17 MHz which is just below the lower resonant mode for 10 cm distance (see Fig. 9). Now, by adding a capacitance of 0.165 pF, we are able to tune the channel's lower passband to exactly our centre frequency. We did the same for larger distances with appropriately adapted capacitance values. As shown in Fig. 12, again for the case of 10 cm transfer distance, the attenuation has been drastically reduced, while

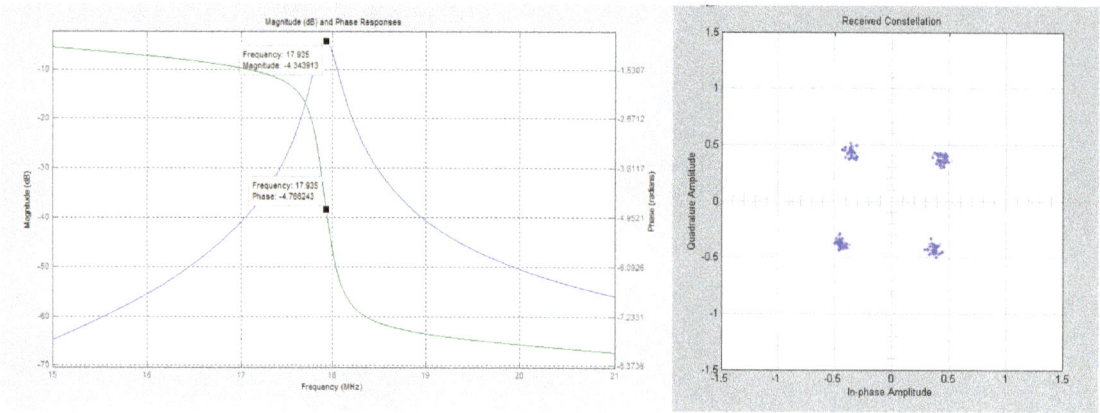

Figure 11. FIR-filter magnitude/phase and constellation plot for 30 cm distance.

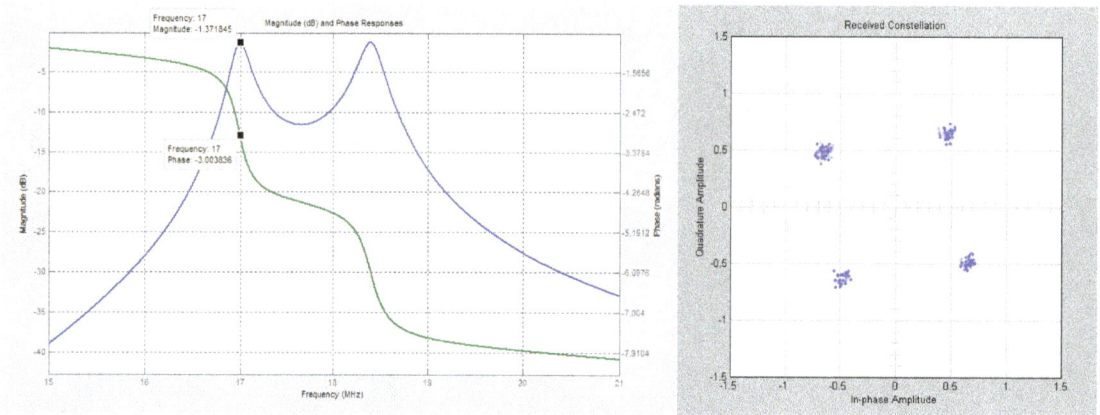

Figure 12. Tuned FIR-filter magnitude/phase response and constellation plot for 10 cm distance.

the linear distortion introduced due to the sharp frequency response around 17 MHz has increased. Anyway, a proper demodulation is still possible.

If this channel adaption scheme is used for all resonant coil distances, the variation of the channel behaviour is minimized. Table 2 shows simulation results, where the error vector magnitude (EVM) has been calculated for different resonant coil separations in case of untuned and capacitively tuned resonant coils. The EVM is defined in different standards as the percentage ratio of the RMS error to the square root of average constellation power.

As can be seen from these results, the channel tuning results in a stable performance with a maximum of 20 % EVM and just a small variation over distance. For distances higher than approximately 20 cm, the system is no longer in the over-critically coupled region (see also Sect. 3). Subsequently, the capacitance value to tune the passband to 17 MHz stays constant. Anyway, the EVM increases because of higher transfer losses and reduced bandwidth in the under-critically coupled regime.

As a result, a couple of tuned coils greatly simplifies the digital signal processing required to recover the transmitted

Table 2. EVM calculation after simulation results.

Resonant coil distance	Untuned coils	Tuned coils	Added capacitance
10 cm	73.33 %	20.18 %	0.165 pF
12 cm	60.86 %	20.12 %	0.319 pF
14 cm	45.82 %	19.90 %	0.420 pF
16 cm	32.65 %	18.90 %	0.472 pF
18 cm	22.04 %	18.15 %	0.524 pF
20 cm	15.72 %	16.70 %	0.605 pF
30 cm	43.21 %	43.15 %	0.605 pF

data. In addition, simultaneous efficient energy transfer is possible, since the transmission is now always at the lower resonant mode's frequency, resulting in increased efficiency. In Fig. 12, S_{21} peak magnitude at 17 MHz is just -1.37 dB in contrast to -11.49 dB in Fig. 9.

To achieve this goal, basic system aspects still have to be examined. First, a method has to be found to perform a fast and reliable channel estimation to gather the information of the current channel status. Second, based on this, the resonators on source and load side have to be tuned. To change the passband's centre to 17 MHz in Fig. 12, a parallel capacitance of 0.165 pF had to be added in the simulation. In a real system, the absolute value of the tuning capacitor can be higher, dependant on the resonator design and the desired transfer frequency. On the other hand, the capacitance values shown in Table 2 highlight that a tuning range of about 500 fF has to be realized with a precision in the range of 10 fF for our prototype system. For comparable system, at least the precision will be in the same range. Such a tuning could for example involve capacitors inductively coupled to the resonators as described in Ricketts et al. (2014). However, further research is required to find a reliable and precise tuning technique.

5 Conclusions

In this paper, we present a system concept for simultaneous energy and data transfer analysis. Based on a prototype system, measurements and parameter extraction using simulations are carried out. An equivalent circuit is validated and embedded in a channel modelling in MATLAB©. The impact of a varying channel on the data and energy transmission is analysed. Regarding a real-life application, a method to tune the channel behaviour is proposed, which promises an optimal, distance-independent transfer in a wide range. Further research has to be done on certain aspects regarding a useful implementation.

Acknowledgements. This paper presents results from the first phase of the research project "ComPad – Smart Modem zur kontaktlosen Sensordaten- und Bildübertragung bei gleichzeitiger Energieaufladung für eine alterssensible Anpassung der hauslichen Infrastruktur". The project is funded by the German Federal Ministry of Education and Research (BMBF), through the "IngenieurNachwuchs" funding stream (young engineers).

References

Awai, I. and Ishida, T.: Design of Resonator Coupled Wireless Power Transfer System by Use of BPF Theory, J. Korean. Inst. Electrom. Eng. Sci., 10, 237–243, 2010.

Beh, T. C., Kato, M., Imura, T., and Hori, Y.: Wireless Power Transfer System via Magnetic Resonant Coupling at Fixed Resonance Frequency-Power Transfer System Based on Impedance Matching, World Electr. Vehicle J., 4, 744–753, 2010.

Beh, T. C., Kato, M., Imura, T., Oh, S., and Hori, Y.: Automated Impedance Matching System for Robust Wireless Power Transfer via Magnetic Resonance Coupling, IEEE T. Ind. Electron., 60, 3689–3698, 2013.

Cheon, S., Kim, Y, Kang, S., Lee, M. L., Lee, J., and Zyung, T.: Circuit-Model-Based Analysis of a Wireless Energy-Transfer System via Coupled Magnetic Resonances, IEEE T. Ind. Electron., 58, 2906–2914, 2011.

Dionigi, M. and Mongiardo, M.: CAD of Efficient Wireless Power Transmission Systems, 2011 IEEE MTT-S, 2011 IEEE MTT-S International Microwave Symposium, Baltimore Convention Center, USA, 5–10 June 2011, 1–4, 2011.

Dionigi, M. and Mongiardo, M.: A novel resonator for simultaneous Wireless Power Transfer and Near Field Magnetic Communications, 2012 IEEE MTT-S, 17–22 June 2012, 1–3, 2012.

Hui, S. Y. R, Zhong, W., and Lee, C. K.: A Critical Review of Recent Progress in Mid-Range Wireless Power Transfer, IEEE T. Power Electr., 29, 4500–4511, 2013.

Jang, B.-J., Lee, S., and Yoon, H.: HF-Band Wireless Power Transfer System: Concept, Issues, and Design, Prog. Electromagn. Res., 124, 211–231, 2012.

Jonah, O., Georgakopoulos, S. V., and Yao, S.: Strongly Coupled Resonance Magnetic for RFID Applications, IEEE Apsursi., 1110–1111, 2013.

Kim, J., Son, H., Kim, D., and Park, Y.: Optimal Design of a Wireless Power Transfer System with Multiple Self-Resonators for an LED TV, IEEE T. Consum. Electr., 58, 775–780, 2012.

Kim, J., Choi, W.-S., and Jeong, J.: Loop Switching Technique for Wireless Power Transfer using Magnetic Resonance Coupling, Prog. Electromagn. Res., 138, 197–209, 2013.

Kurs, A., Karalis, A., Moffatt, R., Joannopoulos, J. D., Fisher, P., and Soljačić, M.: Wireless Power Transfer via Strongly Coupled Magnetic Resonances, Science, 317, 83–86, 2007.

Lee, G., Waters, B. H., Shi, C., Park, W. S., and Smith, J. R.: Design considerations for asymmetric magnetically coupled resonators used in wireless power transfer applications, 2013 IEEE Radio and Wireless Symposium (RWS), 2013 IEEE Radio & Wireless Week, Renaissance Austin, USA, 20–23 January 2013, 328–330, 2013.

Oppenheim, A. V. and Schafer, R. W.: Discrete-Time Signal Processing, International Edition, Prentice Hall, USA, 1989.

Ricketts, David S., Chabalko, Matthew J., and Hillenius, Andrew: Optimization of Wireless Power Transfer for Mobile Receivers Using Automatic Digital Capacitance Tuning, Proceedings of the 43rd European Microwave Conference, 515–518, 2013.

Ricketts, D. S., Chabalko, M. J., and Hillenius, A.: Tri-Loop Impedance and Frequency Matching With High-Q Resonators in Wireless Power Transfer, IEEE Antenn. Wirel. Pr., 13, 341–344, 2014.

Rosa, E. B. and Grover, F. W.: Formulas and Tables for the Calculation of Mutual and Self-Inductance [Revised], Sci. P. Bur. Stand., 169, 116–122, 1916.

Sample, A. P., Meyer, D. A., and Smith, J. R.: Analysis, Experimental Results, and Range Adaptation of Magnetically Coupled Resonators for Wireless Power Transfer, IEEE T Ind. Electron., 58, 544–554, 2010.

Sample, A. P., Waters, B. H., Wisdom, S. T., and Smith, J. R.: Enabling Seamless Wireless Power Delivery in Dynamic Environments, Proceedings of the IEEE, 101, 1343–1358, 2013.

Tosic, D. V. and Potrebic, M.: Symbolic analysis of immittance inverters, 14th Telecommunication Forum, Belgrade (Serbia), 21–23 November 2006, 584–587, 2006.

Waters, B. H., Sample, A. P., and Smith, J. R.: Adaptive Impedance Matching for Magnetically Coupled Resonators, Pr. Electromagn. Res. S., 694–701, 2012

Zhang, Y. and Zhao, Z: Frequency splitting Analysis of Two-Coil Resonant Wireless Power Transfer, IEEE Antenn. Wirel. Pr., 13, 400–402, 2014.

24

Potential of dynamic spectrum allocation in LTE macro networks

H. Hoffmann[1], **P. Ramachandra**[2], **I. Z. Kovács**[3], **L. Jorguseski**[4], **F. Gunnarsson**[2], **and T. Kürner**[1]

[1]Deparmtent of Telecommunication Engineering, TU Braunschweig, Braunschweig, Germany
[2]Ericsson Research, Linköping, Sweden
[3]Nokia, Aalborg, Denmark
[4]TNO, Delft, the Netherlands

Correspondence to: H. Hoffmann (hoffmann@ifn.ing.tu-bs.de)

Abstract. In recent years Mobile Network Operators (MNOs) worldwide are extensively deploying LTE networks in different spectrum bands and utilising different bandwidth configurations. Initially, the deployment is coverage oriented with macro cells using the lower LTE spectrum bands. As the offered traffic (i.e. the requested traffic from the users) increases the LTE deployment evolves with macro cells expanded with additional capacity boosting LTE carriers in higher frequency bands complemented with micro or small cells in traffic hotspot areas. For MNOs it is crucial to use the LTE spectrum assets, as well as the installed network infrastructure, in the most cost efficient way. The dynamic spectrum allocation (DSA) aims at (de)activating the available LTE frequency carriers according to the temporal and spatial traffic variations in order to increase the overall LTE system performance in terms of total network capacity by reducing the interference. This paper evaluates the DSA potential of achieving the envisaged performance improvement and identifying in which system and traffic conditions the DSA should be deployed. A self-optimised network (SON) DSA algorithm is also proposed and evaluated. The evaluations have been carried out in a hexagonal and a realistic site-specific urban macro layout assuming a central traffic hotspot area surrounded with an area of lower traffic with a total size of approximately $8 \times 8\,\mathrm{km}^2$. The results show that up to 47 % and up to 40 % possible DSA gains are achievable with regards to the carried system load (i.e. used resources) for homogenous traffic distribution with hexagonal layout and for realistic site-specific urban macro layout, respectively. The SON DSA algorithm evaluation in a realistic site-specific urban macro cell deployment scenario including realistic non-uniform spatial traffic distribution shows insignificant cell throughput (i.e. served traffic) performance gains. Neverthe-

less, in the SON DSA investigations, a gain of up to 25 % has been observed when analysing the resource utilisation in the non-hotspot cells.

1 Introduction

1.1 Motivation

The fast growing demand of mobile broadband services force mobile network operators (MNOs) to evolve their LTE networks from coverage oriented deployments towards capacity oriented deployment utilising macro cells with multiple LTE frequency carriers (at different LTE bands) complemented with small cells (micro, pico or femto cells) deployment in traffic hotspot areas. The network dimensioning is usually performed based on the estimated load in typical busy hour conditions. Therefore, although the deployed carriers are well utilised during peak hours, e.g. day time in university campus or offices, they become underutilised for the rest of the time, e.g. night time. At the same time the traffic demand may intensify in other areas of the network, e.g. in residential areas in the evening, located outside the busy hour high traffic areas. By means of Dynamic Spectrum Allocation (DSA) the available spectrum resources (LTE carriers) will be allocated according to the spatial and temporal traffic requirements by (autonomously) assigning spectrum to base stations based on the estimated large-scale temporal and spatial offered load (i.e. estimated amount of resources).

Furthermore, in LTE networks the DSA can be complemented by Interference Management (IM) mechanisms, such as the further/evolved Inter-Cell Interference Coordination (feICIC). Typical outcome of these interference management mechanisms is the (semi-)dynamic assignment of the available resources (time, frequency and/or power) between LTE macro and small-cell layers operating on the same carrier frequency.

The objective of our study is to identify the conditions under which a DSA mechanism could be utilised in an LTE macro network deployment in a typical European dense urban area.

1.2 DSA in the academia

A good overview of DSA approaches can be found in Akyildiz et al. (2006). The studies in Leaves et al. (2001, 2002, 2004) focus on spatial and temporal DSA in a multi-radio network consisting of a UMTS and Digital Video Broadcasting-Terrestrial (DVB-T) system. A pre-requisite for the application of temporal DSA is load prediction. Simulation results for perfect and imperfect load prediction are presented showing that DSA has 30 % higher spectrum efficiency compared to fixed channel assignment. The study in Rodriguez et al. (2006) extends the DSA concept with spectrum bidding among different cells from the participating DVB-T and UMTS systems. Further, Kovács and Vidács (2006) and Kovács et al. (2007) propose a spatial and temporal DSA, where the temporal DSA is coordinated by so called Regional Spectrum Brokers (RSB). The RSB considers interference by geographical and radio technology coupling parameters and solves the spectrum assignment problem using integer linear programming approach. The simulation results in Kovács et al. (2007) show that the combined RSB and linear programming optimisation approach achieve gains (e.g. 26 %) in terms of used resources over fixed spectrum assignment.

In Madan et al. (2011) a heuristic distributed algorithm is evaluated for dynamic sub-band partitioning and user associations complemented by transmit power control for LTE heterogeneous networks (macro overlaps with pico, femto and relay nodes). The heuristic algorithm uses light-weight coordination signalling messages between neighbouring (or overlapping) cells and provides significant throughput and delay gains over frequency re-use one.

1.3 Standardisation and deployment status

The related work on DSA within 3GPP can be divided into energy saving via switching on/off cells 3GPP TR36.887 (2014), 3GPP TR36.927 (2014) and inter-cell interference coordination (ICIC). The ICIC activities in 3GPP have started already in Release 8. In particular, for heterogeneous network scenarios the focus has been on the carrier based ICIC (CB-ICIC) for Release 10–11 that addressed scenarios for operational carrier selection, and downlink or uplink interference management for macro-pico as described in Qualcomm Inc. (2011) and 3GPP Overview (2013). Additionally, Release 12 also addresses the signalling needed for robust

solutions for Release-10/11 UEs supporting carrier aggregation, and focus on solutions which do not require tight synchronisation between eNodeBs.

Today's MNOs operating GSM networks in 900 MHz band (for coverage) and in 1800 MHz band (for capacity) deploy DSA to dynamically switch ON and OFF the 1800 MHz capacity cells according to the traffic demand for energy saving purposes, see e.g. Nokia Siemens Networks (2010). The control algorithm monitors the GSM traffic level and controls the activation of 1800 MHz capacity layer by switching off the capacity cells when and where the traffic drops below certain pre-defined thresholds (usually at night hours). If the 1800 MHz capacity layer is already deactivated the algorithm continues to monitor the traffic level and if needed activates back the 1800 MHz capacity cells. Another practical example related to the DSA concept is trialled for wireless cellular communications in the public safety domain. The US public safety market is evolving rapidly with the combined introduction of public and private LTE networks. These networks have available spectrum that is often unused. Therefore, a dynamic spectrum arbitrage solution is trialled RadiSys Corporation (2013) that allows unused spectrum to be easily reallocated across networks to where it is needed most by combining prioritisation of users on the network with a real-time auctioning process.

1.4 Scope of the study

This study investigates the potential gain from DSA for a LTE macro-cellular network covering a geographical area with non-uniform spatial traffic distribution. The remaining of the paper is organised as follows. The system modelling approach and traffic assumptions are illustrated in Sect. 2. The potential DSA performance gains for a homogeneous traffic spatial distribution with hexagonal cell layout as well as with a site-specific urban cell deployment are presented in Sect. 3. Then, Sect. 4 presents the evaluation of the self-optimised DSA algorithm in the site-specific urban deployment scenario. The paper is finalised with the conclusions and recommendations in Sect. 5.

2 System modelling for DSA analysis

For this study, we have selected an initial LTE carrier deployment within the allocated spectrum at a typical European LTE operator having 20 MHz spectrum available in the 1800 MHz band. For investigating the DSA potential two scenarios are considered, namely a hexagonal cell layout and a site-specific urban cell layout as illustrated in Fig. 1a and b, respectively. Both scenarios use a traffic hotspot cell (one macro cell) in the centre of the investigated area illustrated with red colour in Fig. 1. The inter-site distance in the hexagonal layout was chosen such that the resulting cell density is

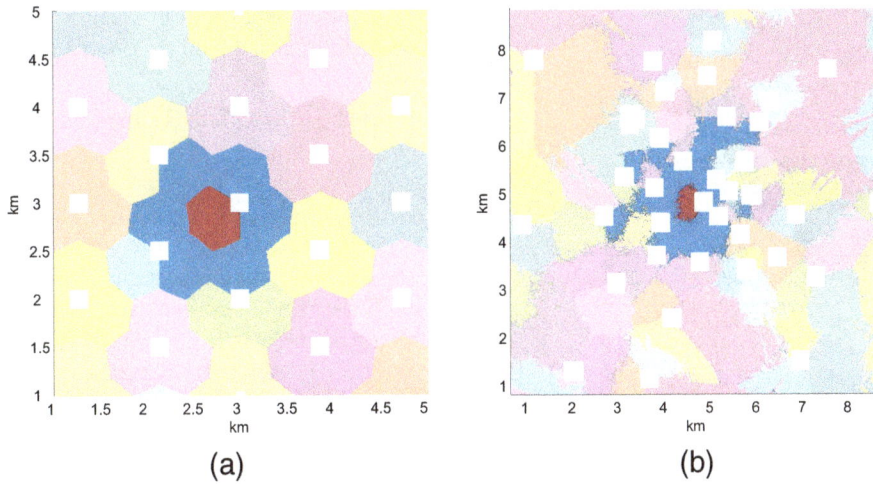

Figure 1. Cell layout (best server areas) with hotspot (red) and its surrounding cells (blue): **(a)** hexagonal layout scenario; **(b)** site-specific layout scenario (Hanover). The white square markers indicate the location of the macro sites.

Table 1. Macro network simulation parameters.

	Hexagonal cell layout (3GPP model)	Site-specific cell layout (Hanover urban network)
Inter-site distance	1000 m	960 m (average)
Pixel size	10 m × 10 m	10 m × 10 m
Scenario area	4 km × 4 km	8 km × 8 km
Macro transmit power	46 dBm per cell	46 dBm per cell
Propagation model	3GPP TS 25.184	Ray-tracing based path loss
Traffic model	Uniform traffic intensity map within each cell; separate scaling for the hotspot cell	Spatially non-uniform distribution within and across cells; separate scaling for the hotspot cell
Considered Cells for SINR computation	20 strongest interfering cells	20 strongest interfering cells
Cell selection	Best signal server	Best signal server

similar to the one in the site-specific network layout in the considered area.

The requested traffic from the users in the network will be referred to as the offered traffic (to the network) in the following. The offered traffic levels in the hotspot cell and the surrounding cells are varied according to spatial traffic intensity maps that are scaled differently within the hotspot cell and the surrounding cells in order to generate different load ratios and interference situations to the area served by the hotspot cell. The traffic intensity maps are assumed to contain the average offered traffic in a certain time interval (e.g. 1 h). Thus, there is no scheduling modelled explicitly and a time averaged resource allocation is simulated as explained in the following section. The main simulation parameters used in the evaluation of the DSA mechanisms are given in Table 1.

2.1 DSA configurations

In order to evaluate the potential DSA gains four different spectrum allocation strategies are investigated for the cells surrounding the hotspot cell, as given in Fig. 2. The bandwidth configuration, *Full Spectrum* is the situation without DSA mechanism activated i.e., all macro cells use the full bandwidth. Correspondingly, *DSA-1*, *DSA-2*, and *Half Spectrum* are configurations where only 17.2 MHz (86%), 14.4 MHz (72%), and 10 MHz (50%) of the 20 MHz spectrum is allowed to be used by the cells surrounding the hotspot cell, respectively. Note that the *DSA-1* and *DSA-2* restrict percentages of spectrum in use with a multiple of 1.4 MHz chunks (i.e. 2.8 MHz restricted or 86% free for DSA-1 and 5.6 MHz restricted or 72% free for DSA-2), which is the minimum LTE bandwidth. As a result a fraction of resources in the hotspot cell does not experience any interference from the surrounding cells. Implicitly, by employing this flexibility in bandwidth allocations a certain degree of interference management is achieved.

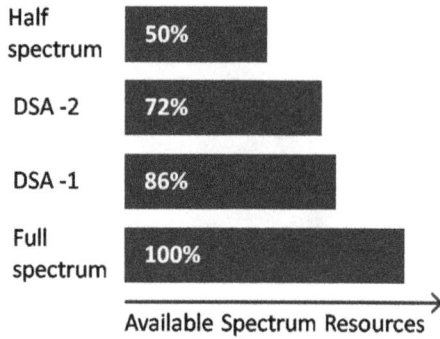

Figure 2. Resource level DSA method used in simulations.

2.2   Traffic and resource allocation

Serving cell selection is based on the best signal strength cell for each pixel of a predicted Reference Symbol Received Power (RSRP) map. A pixel is a square portion of the covered area characterized with its position and pixel area size, which is typically $10 \times 10\,\mathrm{m}^2$. The predicted RSRP map is a collection (or union) of all the pixels in the area under investigation where each pixel is associated with a predicted RSRP transmitted from a given LTE cell.

The amount of bandwidth ($\mathrm{BW}_{\mathrm{p}j}$) required to serve the offered traffic ($\mathrm{OfferedTraffic}_{\mathrm{p}j}$) in a pixel $\mathrm{p}j$ is dependent on the SINR experienced by the pixel ($\mathrm{SINR}_{\mathrm{p}j}$) and is given by:

$$\mathrm{BW}_{\mathrm{p}j} = \begin{cases} 0 & \mathrm{SINR}_{\mathrm{p}1} < \mathrm{SINR}_{\min}, \\[2mm] \dfrac{\mathrm{OfferedTraffic}_{\mathrm{p}j}}{0.6 \cdot \log_2\left(1 + \mathrm{SINR}_{\max}\right)} & \mathrm{SINR}_{\mathrm{p}1} \geq \mathrm{SINR}_{\max}, \\[4mm] \dfrac{\mathrm{OfferedTraffic}_{\mathrm{p}j}}{0.6 \cdot \log_2\left(1 + \mathrm{SINR}_{\mathrm{p}j}\right)} & \text{otherwise,} \end{cases} \tag{1}$$

where $\mathrm{SINR}_{\min} = -6.5\,\mathrm{dB}$ and $\mathrm{SINR}_{\max} = 22.05\,\mathrm{dB}$. These boundary SINR values model the cases where no transmission is possible or the highest modulation and coding scheme might be used, respectively. Note that the approach in this study uses average SINR values i.e. neglecting fast SINR variations over (multiple) 1 ms/TTI periods. The total bandwidth required to serve all the pixels in the cell's coverage area is the sum of bandwidth requirements of each pixel (Eq. 1) in the coverage area and is given by

$$\text{Total required bandwidth for a cell}_k = \sum_{j \in \text{cell}_k} \mathrm{BW}_{\mathrm{p}j}. \tag{2}$$

Based on this value the $\mathrm{CellLoad}_k$ is modelled as:

$$\mathrm{CellLoad}_k = \frac{\text{Total required bandwidth for a cell}_k}{\text{Available bandwidth of a cell}_k}, \tag{3}$$

which is capped at 1 (100 %) for interference computation. It can also be seen as the used resources of a cell. If the to-

tal required bandwidth in a cell is larger than the available bandwidth, then not all the offered traffic can be served, i.e.:

$$\mathrm{Throughput}_k = \mathrm{OfferedTraffic}_k \cdot \min\left(1, \frac{1}{\mathrm{CellLoad}_k}\right). \tag{4}$$

3   Evaluation of the potential DSA gains

The evaluation in this section aims at quantifying the potential DSA gain in terms of total system carried traffic that can be served in the considered cells for the different DSA configurations described in Sect. 2.2.

The offered traffic in the surrounding cells and in the hotspot cell is varied in discrete steps for each of the DSA configurations listed in Fig. 2. The different load situations in the surrounding cells and the hotspot cell cause different interference situations, and hence different system performance. The carried traffic levels in the hotspot cell and the surrounding cells (given in traffic per cell) for the different DSA configurations are illustrated in Fig. 3. The curves present upper limits for the offered traffic such that all of the surrounding cells and the hotspot cell are loaded lower than the maximum cell load threshold of 100 %. Consequently, the offered traffic equals the carried/served traffic in the system. The left most values of the curves represent the maximum traffic that can be served in average in each of the surrounding cells with no or low load (and consequently no or low interference) from the hotspot cell. Note that the traffic in the surrounding cells for the hexagonal cell layout (see Fig. 3a) reaches a maximum of 20, 17, 14 and 10 Mbps per cell for the *Full Spectrum*, *DSA-1*, *DSA-2*, and *Half Spectrum* DSA configurations, respectively. Note that these maximum traffic values scale with the amount of spectrum available in the surrounding cells. Correspondingly, for the site-specific deployment scenario (see Fig. 3b) the traffic in the surrounding cells reaches a maximum at 14, 12, 10, and 6 Mbps per cell. The lower values when compared to the hexagonal cell layout are due to the larger and irregular coverage areas resulting in also different interference conditions. As the offered traffic in the hotspot cell increases, the hotspot cell will cause more interference to the surrounding cells. Consequently, the increase in interference from the hotspot cell will cause the surrounding cells to get overloaded earlier, and their maximum carried traffic will gradually start to decrease. When the hotspot cell traffic crosses a threshold, the hotspot cell itself becomes overloaded and needs help from the surrounding cells to ensure that all the offered traffic is served. This help from the surrounding cells is in terms of reduction in their utilized spectrum so that they interfere less to the hotspot cell.

In Fig. 4 the gain in terms of total offered/carried system traffic values are illustrated for the different DSA configurations when compared with the *Full Spectrum* configuration. We can conclude that:

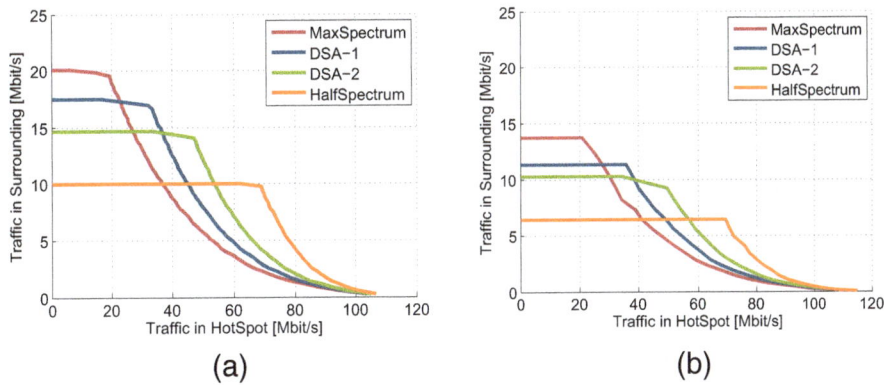

Figure 3. Served traffic in hotspot and the surrounding cells without cell overloading for different DSA configurations: **(a)** hexagonal layout scenario; **(b)** site-specific layout scenario (Hanover).

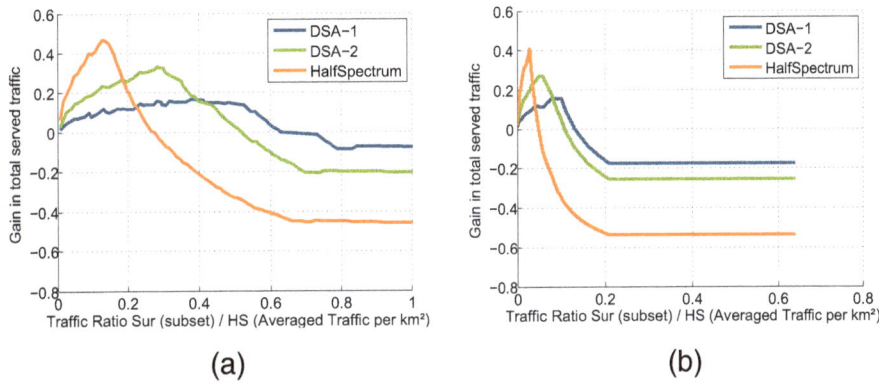

Figure 4. Gain relative to *Full Spectrum* in total carried system traffic for different ratios of surrounding (SUR) vs. hotspot (HS) traffic: **(a)** hexagonal layout scenario; **(b)** site-specific layout scenario (Hanover).

1. For the hexagonal layout (see Fig. 4a) up to 47 % gain in terms of total served traffic can be achieved and the gain is visible in the traffic ratio range up to 0.6. Furthermore, the optimal DSA configuration depends on the traffic ratio between the surrounding cells and the hotspot cell. For traffic ratios up to 0.2 the *Half Spectrum* DSA configuration is best while for the range of traffic ratios between 0.2 and 0.4 the *DSA-2* is outperforming the other DSA configurations. For ratios higher than 0.4 *DSA-1* configuration has the highest gain. An algorithm to allocate spectrum to cells dynamically would therefore take this ratio into account, as illustrated in Sect. 4.

2. For the site-specific urban macro deployment (see Fig. 4b) up to 40 % gain in terms of total served traffic can be achieved and the gain is visible in the traffic ratio range up to 0.16, which is significantly narrower range when compared to the hexagonal cell layout. Again, this is because of the larger and irregular cell area coverage creating different interference conditions.

4 DSA evaluation for realistic layout and self-organised DSA algorithm

Opposite to the uniform spatial traffic distributions used in Sect. 3, in the investigation of the realistic and self-organised DSA algorithm the offered traffic is based on scaled realistic traffic intensity maps for a selected time interval between 8:00 and 18:00 h, as explained in Deliverable 4.1 from SEMAFOUR D4.2 (2013). As the traffic intensity maps do not contain defined hotspots, a time-varying hotspot (per hour) is added to each of the hourly maps. For example, in the time interval between 11:00 and 15:00 h the traffic offered to the hotspot cell is approximately two times higher than the averaged offered traffic to its surrounding cells. The hotspot is placed in the coverage area of only one macro cell, as illustrated in Fig. 1b.

A flowchart of the SON enabled DSA algorithm that is executed in each cell is illustrated in Fig. 5. The DSA algorithm might change the spectrum only stepwise in pre-defined frequency chunks as described in Sect. 2.1, i.e. 20 MHz (*Full Spectrum*), 17.2 MHz (*DSA-1*), 14.4 MHz (*DSA-2*) and 10 MHz (*Half Spectrum*). The DSA algorithm can be acti-

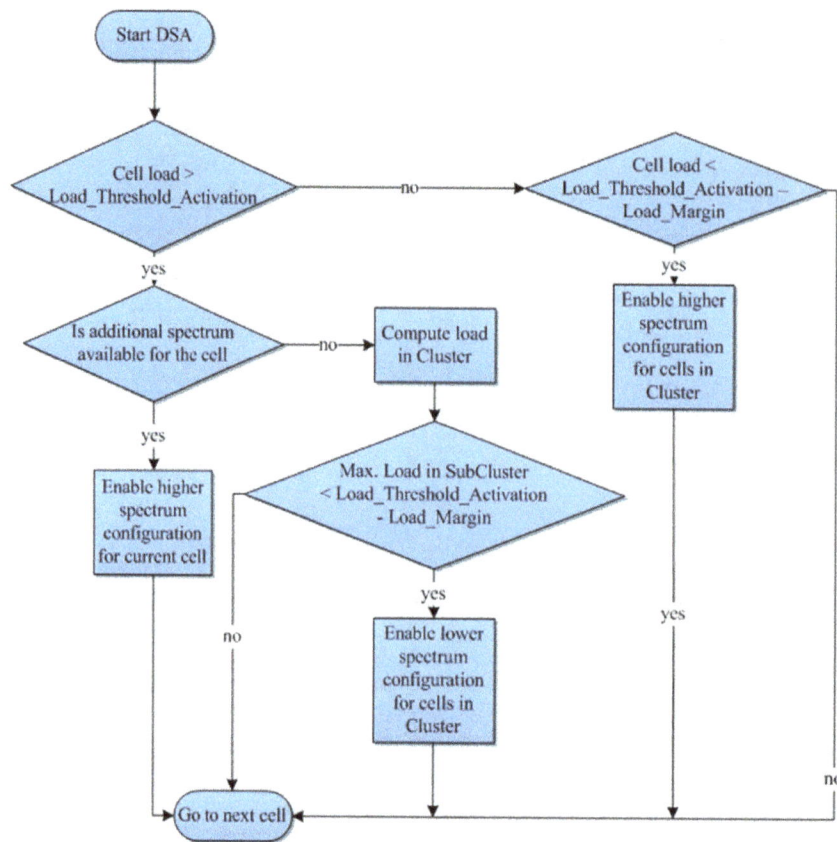

Figure 5. Self-optimised DSA algorithm flow chart.

vated in any cell and is triggered by a high load condition, as explained below. It is assumed that each cell has a pre-configured *Cluster of interfering cells.* In this study the top 20 interfering cells (surrounding the hotspot cell) are considered for the cluster.

The different load thresholds used in the flow chart in Fig. 5 can be explained from a cell and from the corresponding *cluster neighbour* perspective, as follows:

1. *Load_Threshold_Activation:* It is the load level in the cell above which the cell either requests for higher bandwidth configurations or requests its *cluster neighbours* to reduce their spectrum (not guaranteed of reduction yet as the actual reduction depends on the load levels in the neighbours).

2. *Load_Threshold_Activation – Load_Margin:*

   a. *From a cell perspective:* When the load in the cell goes below this threshold, the cell allows its *cluster neighbours* to enable higher bandwidth configurations.

   b. *From the cluster cells' perspective:* When the maximum load among the cell in the defined *cluster neighbours* goes below this threshold, they will ac-

cept any request for further reduction in their bandwidth configurations.

The SON-enabled DSA algorithm evaluates every 20 min (i.e. three times per hour) the loads per cell and takes actions as defined in the flow chart in Figure 5. The values for the *Load_Threshold_Activation* and *Load_Margin* are set to 0.7 and 0.15, respectively. In practice these parameters would be part of the overall DSA configuration in the network, and would be set by the MNO depending on the expected traffic variations, available spectrum, etc.

The performance results for the SON-enabled DSA algorithm are presented in Fig. 6. In the investigation the centre hotspot cell is always using 20 MHz (full) spectrum. As it can be observed the DSA algorithm is triggered between 11:00 and 15:00 h, when the load in the hotspot cell increases above 0.7. The DSA algorithm then reduces the spectrum of the surrounding (top 20) cluster cells as long as the maximum load among the cells is below *Load_Threshold_Activation* = 0.7.

It can be seen in Fig. 6 that cluster throughput curves for the simulation with DSA enabled and with DSA disabled are the same. The absence of DSA gain in terms of cluster throughput, when compared to the hexagonal layout and uniform traffic intensity per pixel, can be explained as follows:

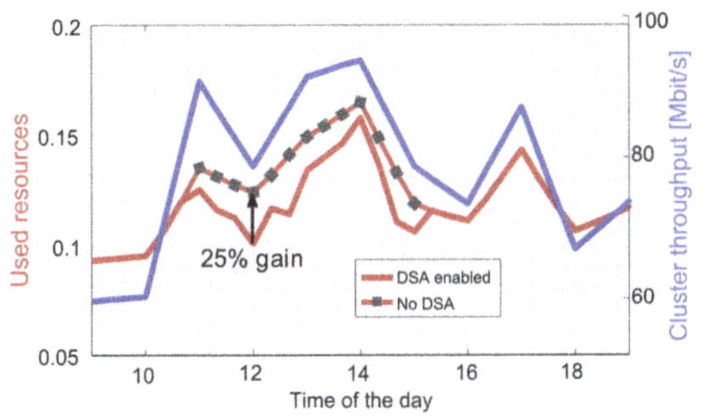

Figure 6. Performance of the self-optimised DSA algorithm in the site-specific layout scenario.

1. The non-uniform traffic intensity per pixel based on the realistic traffic intensity maps does not allow for very asymmetric traffic ratios between the hotspot cell and its corresponding cluster. Note that the high potential DSA gains for the site-specific deployment scenario (see Fig. 4b) were observed only in the traffic ratio range below 0.16.

2. The SON enabled DSA algorithm cannot reduce the spectrum of the surrounding cluster cells down to half-spectrum configuration i.e. only DSA-1 and DSA-2 configurations are used, due to the relatively high traffic demand in the surrounding (top 20) cluster cells.

3. The non-uniform cell coverage areas i.e., largely different cell sizes when compared to the hexagonal layout, result in significantly different interference and cell load conditions when compared to the regular hexagonal layout.

In addition to the cluster throughput the corresponding (spectrum) resource utilisation is also plotted in Fig. 6. It can clearly be seen that the activation of the SON DSA algorithm achieves approximately the same cluster throughput while using significantly less resources. At the maximum (at 12 o'clock) there are 25 % more resources used without DSA. These resource savings directly relate to the optimised interference situation in the network. Due to the restricted spectrum use in some of the cells pixels assigned to the remaining cells experience a higher SINR and thus require fewer resources.

5 Conclusions

This study investigated the potential performance gain from using dynamic spectrum allocation (DSA) in an LTE macro-cellular network. The DSA evaluation has been carried out in a hexagonal cell layout with uniform spatial traffic distribution, as well as in a realistic site-specific cell deployment with non-uniform spatial traffic distribution scenarios. The centre cell was assumed to be a high traffic cell (hotspot) and always to utilise the total available spectrum of 20 MHz while the spectrum allocation in surrounding cells was varied. Four different spectrum allocation strategies were used: *Full Spectrum* (20 MHz), *DSA-1* (17.2 MHz), *DSA-2* (14.4 MHz) and *Half Spectrum* DSA (10 MHz).

The results show that DSA can provide significant gains only in spatially highly non-uniform traffic distributions. Under highly asymmetric traffic ratio between the surrounding cells and the hotspot cell, 0.5 or 0.16 for the hexagonal or realistic scenario, respectively, DSA gains up to 47 % can be achieved. However, the same network configuration also lead to losses of about 50 % when insufficient spectrum is allocated to the surrounding cells.

In the site-specific cell deployment the self-optimised DSA algorithm did not provide significant gain in the overall cell cluster throughput. This is because in the realistic network deployments the required highly asymmetric traffic between the hotspot and surrounding cells, when the full potential of DSA could be exploited, are difficult to achieve.

However, even though there is hardly any gain in cluster or cell throughput for the realistic network deployments, there is a significant gain of up to 25 % in terms of resource utilisation. The free resources can be used for two purposes depending on the operator's policies. They can either be used for energy saving purposes, e.g. carriers can be switched off if a sufficient number of resources is unused or these free resources can be used as input for Multi-RAT DSA algorithms, i.e. they can be re-assigned to other RATs for capacity optimisation.

As a follow-up of this study the DSA gains for inter-RAT scenarios will be investigated for the case of LTE macro cells deployed in the same area with GSM macro cells. It is expected that DSA will be beneficial in dynamically re-farming parts of the allocated GSM spectrum as allowed by the ongoing GSM traffic and acceptable level of interference from the LTE cells re-using the same spectrum.

Acknowledgements. The research leading to these results has been carried out within the FP7 SEMAFOUR project and has received funding from the European Union Seventh Framework Programme (FP7/2007-2013) under grant agreement no. 316384.

References

Akyildiz, I. F., Lee, W. Y., Vuran, M. C., and Mohanty, S.: NeXt generation/dynamic spectrum access/cognitive radio wireless networks: a survey, Comput. Netw., 50, 2127–2159, 2006.

Kovács, L. and Vidács, A.: Spatio-temporal spectrum management model for dynamic spectrum access networks, in: Proceedings of the first international workshop on Technology and Policy for Accessing Spectrum (TAPAS 2006), ACM, New York, NY, USA, 5 August, 2006.

Kovács, L., Vidács, A., and Tapolcai, J.: Spatio-Temporal Dynamic Spectrum Allocation with Interference Handling, IEEE Conf. Comm., Glasgow, UK, 24–28 June 2007, 5575–5580, 2007.

Laselva, D.,Altman, Z., Balan, I., Bergström, A., Djapic, R., Hoffmann, H., Jorguseski, L., Kovács, I. Z., Michaelsen, P. H., Naudts, D., Ramachandra, P., Sartori, C., Sas, B., Spaey, K., Trichias, K., and Wang, Y.: SON functions for multi-layer LTE and multi-RAT networks (first results), INFSO-ICT-316384 SEMAFOUR, Report, 141 pp., 2013.

Leaves, P., Ghaheri-Niri, S., Tafazolli, R., Christodoulides, L., Sammut, T., Staht, W., and Huschke, J.: Dynamic spectrum allocation in a multi-radio environment: concept and algorithm, 2nd International Conference on 3G Mobile Communication Technologies, London, UK, 26–28 March 2001, Conf. Publ. No. 477, 53–57, 2001.

Leaves, P., Huschke, J., and Tafazolli, R.: A summary of dynamic spectrum allocation results from DRiVE, IST Mobile and Wireless Telecommunications Summit, Thessaloniki, Greece, 17–19 June 2002, 245–250, 2002.

Leaves, P., Moessner, K., Tafazolli, R., Grandblaise, D., Bourse, D., Tönjes, R., and Breveglieri, M.: Dynamic spectrum allocation in composite reconfigurable wireless networks, IEEE Commun. Mag., 42, 72–81, 2004.

Madan, R., Borran, J., Sampath, A., Bhushan, N., Khandekar, A., and Tingfang, J.: Cell Association and Interference Coordination in Heterogeneous LTE-A Cellular Networks, IEEE J. Sel. Area Comm., 28, 1479–1489, 2010.

Nokia Siemens Networks: Smart Energy Control cuts CO_2 footprint and saves up to 1.2 GWh of power annually, available at: http://nsn.com/portfolio/customer-successes/success-stories/smart-energy-control-cuts-co2-footprint-and-saves-up-to (last access: 19 May 2015), 2010.

RadiSys Corporation: Radisys LTE Solutions Enable Rivada Networks' Technology to Dynamically Allocate Excess Spectrum for Public Safety Networks, available at: http://www.radisys.com/2013/ (last access: 19 May 2015), 2013.

Rodriguez, V., Moessner, K. K., and Tafazolli, R.: Market driven dynamic spectrum allocation over space and time among radio-access networks: DVB-T and B3G CDMA with heterogeneous terminals, Mobile Netw. Appl., 11, 847–860, 2006.

Qualcomm Inc.: R3-112609, Carrier-based HetNet ICIC use cases and solutions, The 3rd Generation Partnership Project (3GPP), 2011.

The 3rd Generation Partnership (3GPP): Overview of 3GPP Release 12 V0.1.0, 3GPP, available at: http://www.3gpp.org/specifications/releases/68-release-12 (last access: 19 May 2015), 2013.

25

A test method for analysing disturbed ethernet data streams

M. Kreitlow[1]**, F. Sabath**[1]**, and H. Garbe**[2]

[1]Bundeswehr Research Institute for Protective Technologies and NBC Protection, Munster, 29633 Germany
[2]Institute of Electrical Engineering and Measurement Technology, Leibniz University Hannover, Hannover, 30167 Germany

Correspondence to: M. Kreitlow (matthiaskreitlow@bundeswehr.org)

Abstract. Ethernet connections, which are widely used in many computer networks, can suffer from electromagnetic interference. Typically, a degradation of the data transmission rate can be perceived as electromagnetic disturbances lead to corruption of data frames on the network media. In this paper a software-based measuring method is presented, which allows a direct assessment of the effects on the link layer. The results can directly be linked to the physical interaction without the influence of software related effects on higher protocol layers. This gives a simple tool for a quantitative analysis of the disturbance of an Ethernet connection based on time domain data. An example is shown, how the data can be used for further investigation of mechanisms and detection of intentional electromagnetic attacks.

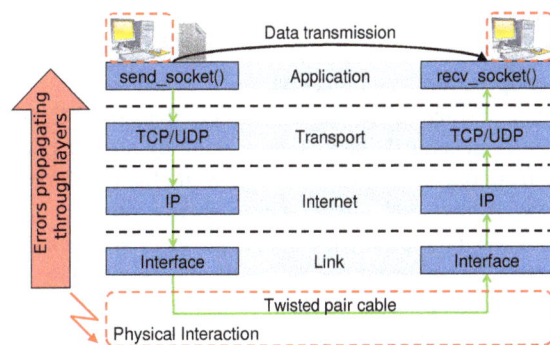

Figure 1. Communication with the TCP/IP reference model.

1 Introduction

Ethernet following the IEEE 802.3 standard (IEEE Standard Association, 2012) is a technique, which is used in many network environments. At least the lowest level in big network installations is usually realised with CAT5e copper twisted pair lines for signal transmission. This enables for data transmission using the variant 1000BASE-T, which is also known as Gigabit Ethernet. Twisted pair lines of the category CAT5e do not require a metallic shielding. Therefore, these lines are more susceptible against radiated electromagnetic interferences. Especially intentional interferences have been subject of many investigations over the past years (Mojert et al., 2001; Jeffrey et al., 2004; Parfenov et al., 2008; Brauer, 2010). Amongst others, it was shown that interference from high-power electromagnetics (HPEM) can have a dramatic impact on data transmissions over Ethernet networks.

Typically the effects can be perceived on the application layer with regard to their criticality, e.g. when the data trans-

mission rate of a file transfer drops significantly or an interactive network application like voice-over-IP shows big lags (Sabath, 2008). In practise, it is very hard to find out what exactly is causing such a behaviour as the effects on application layer are normally not directly linked to the physical interaction. Figure 1 illustrates the process of logical communication within modern IP networks and physical interaction on the basis of the TCP/IP reference model (Postel, 1981a). This model can be considered as simpolified breakdown of the more general OSI layer model (ITU, 1994).

Errors within the communication can be caused by a hardware problem, a bottle-neck in the network infrastructure due to high utilization or even by a intentional electromagnetic interference (IEMI). Observing and detecting the last-mentioned is a real challenge. Reliable detection with field monitors is still subject of ongoing research and in big network infrastructures an effective field monitoring will probably be a high cost factor (Adami et al., 2014).

Therefore a new approach shall be discussed which allows for observation of disturbed data transmissions over an Ethernet network. Software-related effects on higher protocol layers, like control algorithms reacting to data errors, have to bypassed. It will be shown that a practical software implementation of a new test method gives a finely granulated time resolution of Ethernet frames and disturbances on the physical layer. This data can be used for further analysis to detect IEMI. Additionally this new method allows for testing network equipment with comparable results independently from user applications and specific software environments.

2 Error mechanisms in classical TCP based communication

Many network applications, such as file transfers, require a reliable data transmission. Altering or loss of data is unacceptable. Therefore typical techniques, like the file transfer protocol (FTP), utilize the transmission control protocol (TCP) on the transport layer, as this protocol offers integrated security measures. TCP operates on top of the IP protocol (Postel, 1981b). Whereas IP is only responsible for the logical routing of data packets from one destination to another through the network topology, TCP opens the communication endpoint for an application.

When testing a network for effects due to EMI, traffic has to be generated. This is often done by simply copying a file from one computer to another while the interference is present (Adami et al., 2012). This approach presents a problem, as the TCP protocol's integrated security mechanisms will react to occurring packet losses and delays. Therefore, the perceived effects are not linked to the physical processes on the network cables or within the computers. In van Leersum (2013) and Kreitlow et al. (2014) it was shown that an observed drop of the transmission speed can be a result of the control algorithms, although the physical degradation of the network is not very strong.

To conclude this, it can be said that investigating an IT network under the influence of EMI with TCP based applications will only give information on this specific network application in a specific network environment. This data is hardly comparable, as the TCP implementation will act as a "black box". To overcome these issues s new method shall be implemented.

3 Design of a new test method

The basic idea for a new measuring method is not to use a "black box" software like a FTP server/client to generate load on the network, which can be observed using network sniffing tools like Wireshark. Rather the data stream itself shall be generated in a well defined way without the influence of software based control algorithms. To achieve this, the user datagram protocol (UDP) will be used. UDP is also a part of the network stack of every modern operating system and operates on top of IP like TCP does (Postel, 1980).

Compared to this, UDP is a stateless protocol that provides no security and integrity checks on the transport layer. If a data packet gets lost or the data gets corrupted within an UDP connection, there will be no way to detect and handle this on protocol level as opposed to TCP. This makes UDP a very simple protocol with less overhead, but for this reason the application itself has to handle segmenting, flow-control and especially loss or corruption of data. Exactly this behaviour of UDP will be used to detect interferences without having influences of unknown control algorithms.

3.1 Testing procedure

As stated before the new method is based on UDP. It consists of a client, which requests a data stream from the server addressed by its IP address. The request itself is sent as an UDP datagram and defines all necessary testing parameters. In particular these are the test duration in seconds and the payload, which is the same for each packet to be sent. The payload size also defines the actual size of the packet. Additionally, there is a possibility to define delay-times for throttling the data transmission by inserting a delay after each UDP datagram. This is usually set to zero to achieve the maximum speed.

After sending the request to the server, the client immediately starts listening for incoming UDP datagrams originating from the servers IP address. The server will process the request and start sending UDP datagrams back to the clients IP address. The first four bytes of each datagram represent an 32-bit integer number, which acts as a sequence number. For each packet this number is increased consecutively, then the packet is filled up with the given payload up to the defined size. This process is running in a loop, that stops, when the server has been sending data for the specified duration.

On the other side of the connection the client will wait for incoming data. If a UDP datagram from the sever arrives, the client will check first check whether the payload specified in the request has been altered. This can be handled as a corruption. In practical tests this will actually never happen as data corruption due to bit flips will cause the frame check sequence of this Ethernet frame to fail. If this happens, the whole frame will be discarded. If an UDP datagram with the correct payload arrives, the sequence number will be read from the first 4 bytes of the datagram. This sequence number in combination with a time stamp is consecutively logged to a file. The client will stop its receiving and logging process, when it receives no datagrams for a specified time. After each test run, the logged data can be analysed off-line. This file will have the following format, where the first column is the time stamp in microseconds and the second column is the sequence number: *[time stamp];[SEQ#];*.

The whole process is shown in Fig. 2 as a flow chart. The structure of each resulting Ethernet frame is illustrated in in Fig. 3. It is obvious that each Ethernet frame corresponds to exactly one UDP datagram and therefore to one unique sequence number. If an Ethernet frame is discarded or lost,

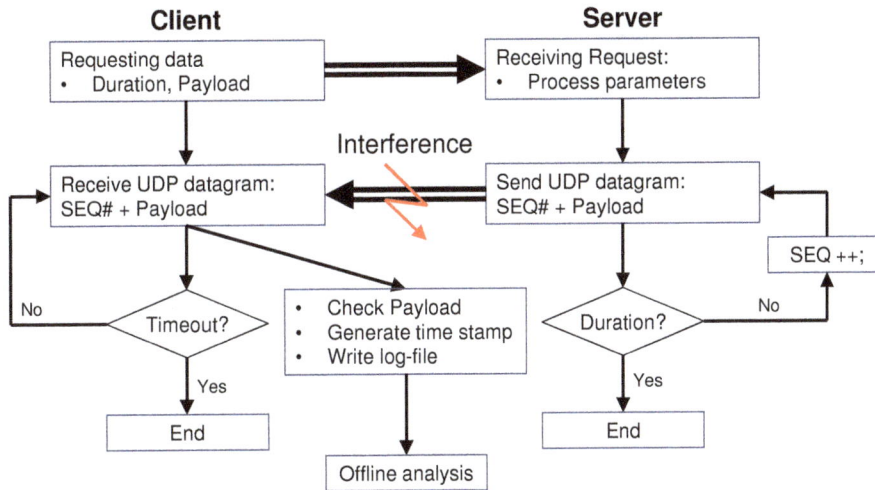

Figure 2. Flowchart for the test procedure.

8 bytes	14 bytes	20 bytes	8 bytes	18 – 1472 bytes	4 bytes	12 bytes
Preamble + start of frame delimiter	Ethernet header	IP header	UDP header	SEQ# + payload	Frame check sequence	Interframe gap

Figure 3. Frame structure.

this can be directly seen in the log file as the sequence number in the corresponding UDP datagram will not occur, because UDP uses no techniques like retransmission or error correcting for securing the data transmission. Also all following Ethernet frames will not be affected by a single event due to the absence of flow control algorithms. By using this behaviour of UDP and the implementation above it is possible to detect interferences with the data transmission with a resolution down to single Ethernet frames.

3.2 Implementation

For a of concept the method was implemented in C as a Win32 application. The software has a user interface as shown in Fig. 4. An off-line analysis of the log file was done in MATLAB. However, there are some limitations in this implementation. As every UDP datagram is unique and must be generated in real-time, the maximum transmission speed is currently limited to $7.5\,\mathrm{Mbyte\,s^{-1}}$ at the moment. This is equivalent to approximately 5000 full-sized Ethernet frames ($1538\,\mathrm{bytes\,frame^{-1}}$) per second. Using high-resolution timers from Microsoft Windows, the time resolution is basically in the range of microseconds. Actually the absolute precision of this timers differs up to 10 % from the exact time in this testing set-up. This is a consequence from the fact, that a high resolution and a good absolute accuracy with negligible drift, while summing up each small time step, is not possible at the same time using the software defined clock of an operating system.

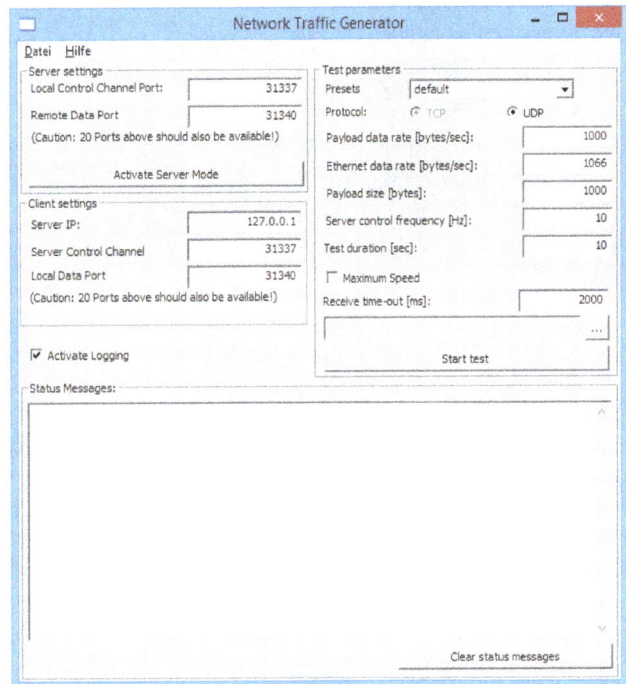

Figure 4. User interface of the test software.

4 Test results

The new method was used to perform some test of a IT network under the influence of IEMI. A simple topology as shown in Fig. 5 was set up. Two computers PC1 and PC2 are

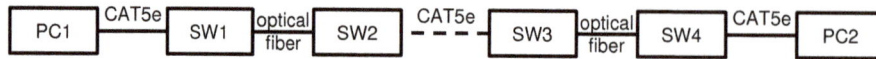

Figure 5. Topology of the test network.

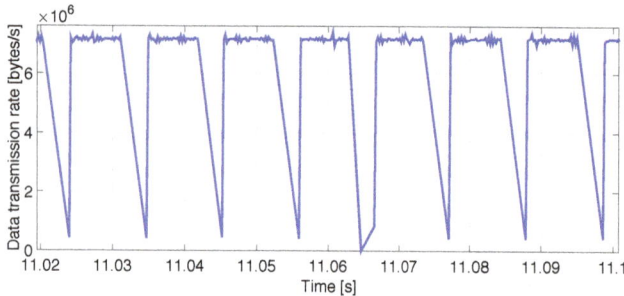

Figure 6. Data transmission rate over time.

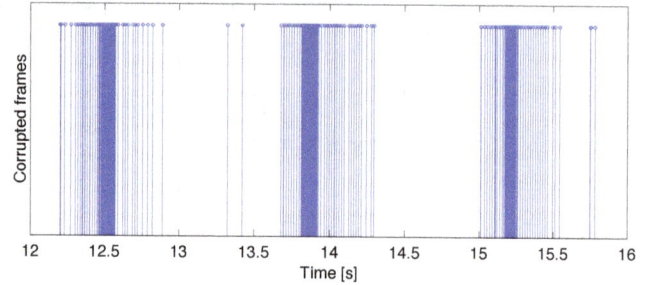

Figure 7. Corrupted Ehernet frames over time.

connected over a line of switches SW1 to SW4. The computers are connected to the switches using CAT5e twisted pair cables. The switches SW2 and SW3 are also connected by twisted pair cables. SW1 and SW2 as well as SW3 and SW4 are linked with an optical fibre. Using an optical connection between these switches ensures that electromagnetic disturbances can not spread out over the whole network when only one component is exposed to interferences.

4.1 Measuring data transmission rate

Each line in the log file represents one UDP packet and therefore one Ethernet frame. The size of the frame is defined within in the test software. This information can be used to calculate the current data transmission rate R on the lowest layer as

$$R_n = \frac{S}{t_n - t_{n-1}}, \tag{1}$$

where S is the size of the complete Ethernet frame, t_n is the time when the actual frame arrived and t_{n-1} is the time when the previous frame arrived.

Figure 6 shows the result for an experiment where PC2 was exposed to ultra wide band (UWB) pulses with a a fieldstrength of $7.5\,\mathrm{kV\,m^{-1}}$ and a pulse repetition frequency of 100 Hz, while PC2 was sending data to PC1. The data shows that the data transmission rate drops with every UWB pulse, but almost no data gets corrupted. In this detailed view only at 11.065 s, where the data rate drops to zero, a packet loss actually happens. From this data it can be deduced, that the data transmission itself is not directly affected most of the time, but the PC delays sending new data for a short time, when it is affected by an UWB pulse.

This time domain data can be used to detect intentional attacks on computers as the pulse repetition frequency is reflected in the data transmission rate measured with the new method. Otherwise it would be hard to detect due to the ab-

sence of exorbitant packet loss, although the transmission speed within applications will slow down.

4.2 Measuring of packet losses

Corrupted Ethernet frames can be detected by enumerating the sequence numbers in the log file. Without disturbances they should increase by one with each line. If the difference between two sequence numbers of two subsequent lines in the log file is higher than one, this indicates the number of missing Ethernet frames. The time stamps give the time window for the data corruption.

Another experiment was done, where an error current was directly coupled into the twisted pair cable between SW2 and SW3. The current also had the shape of an UWB pulse and the pulse repetition frequency was set to 20 kHz. Figure 7 shows at which time Ethernet frames get corrupted due to the interference. Each line in the diagram represents a corrupted frame. Although the disturbance source was permanently activated, the errors occur in bursts. The reason for this is a beat effect between the the packet rate ($\sim 5000\,\mathrm{Hz}$) and the repetition frequency of the source.

The data show that it is possible to catch single disturbance for a data stream with a high resolution in time. However due to the limited packet rate not every single event on the link layer can be caught as the software is not able to utilize the network line by 100 %. But if an effect occurs while the test software occupies the network line, it will be detected by this method. This makes the approach behind this method suitable especially for detecting repetitive interferences.

5 Conclusions

In this paper the challenges, when assessing the disturbance in a real world Ethernet set-ups, have been discussed. It has to be distinguished between the hardware effects and software-related effects to get comparable results. To overcome this issue, a new test method was developed and implemented as a simple user application using the stateless user datagram protocol. In contrast to observing normal data transfers from e.g. FTP servers/clients, the effects on application layer perceived with this new method are directly linked to the disturbances on the physical layer. The results of the method are presented as time-domain data within a resolution in the range of approximately a few hundred microseconds. This data can be used for further investigations using known techniques of signal analysis to detect intentional interferences.

The implementation of the test-software is a proof of concept and still under development status. The source code can be requested from the corresponding author.

References

Adami, C., Braun, C., Clemens, P., Jöster, M., Suhrke, M., and Taenzer, H.-J.: High Power Microwave Tests of Media Converters, International Symposium on Electromagnetic Compatibility 2012 (EMC Europe), Rome, Italy, 17–21 September, 2012.

Adami, C., Braun, C., Clemens, P., Jöster, M., Suhrke, M., Schmidt, H.-U., and Tänzer, H.-J.: HPM-Detektionssystem mit Frequenzbestimmung, EMV Konferenz 2014, Düsseldorf, Germany, 11–13 March, 2014.

Brauer, F., ter Haseborg, J. L., Potthast, S.: Protection Circuits for IT Equipment under HPEM Conditions, AMEREM 2010, Ottawa, Canada, 2010.

IEEE Standard Association: IEEE 802.3-2012 – IEEE Standard for Ethernet, Section 3, 2012.

International Telecommunication Union (ITU): Telecommunication Standardization Section Sector of ITU (ITU-T), Information technology – Open Systems Interconnection – Basic Reference Model: The basic model, Recommendation X.200, 1994.

Jeffrey, I., Gilmore, C., Siemens, G., and LoVetri, J.: Hardware invariant protocol disruptive interference for 100BaseTX Ethernet communications, IEEE Transactions on Electromagnetic Compatibility, 46, 412–422, 2004.

Kreitlow, M., Garbe, H., and Sabath, F.: Influence of Software Effects on the Susceptibility of Ethernet Connections, IEEE International Symposium on Electromagnetic Compatibility EMC, Raleigh, NC, USA, 3–8 August, 2014.

Mojert, C., Nitsch, D., Friedhoff, H., Maack, J., Sabath, F., Camp, M., and Garbe, H.: UWB and EMP Susceptibility of Microprocessors and Networks, 14th International Zürich Symposium & Technical Exhibition on Electromagnetic Compatibility, Zurich, Switzerland, 20–22 Februar, 2001.

Parfenov, Y. V., Kohlberg, I., Radasky, W. A., Titov, B. A., and Zdoukhov, L. N.: The Probabilistic Analysis of Immunity of a Data Transmission Channel to the Influence of Periodically Repeating Voltage Pulses, Asia-Pacific Symposium on Electromagnetic Compatibility & 19th International Zurich Symposium on Electromagnetic Compatibility, Singapore, 283–286, 2008.

Postel, J.: User Datagram Protocol, RFC 768, USC/Information Sciences Institute, available at: http://tools.ietf.org/rfc/rfc768.txt, 1980.

Postel, J. (Ed.): Internet Protocol, DARPA Internet Program – Protocol Specification, RFC: 791, Defense Advanced Research Projects Agency (DARPA), Arlington, Virginia, USA, available at: http://tools.ietf.org/rfc/rfc791.txt, 1981a.

Postel, J. (Ed.): Transmission Control Protocol, DARPA Internet Program – Protocol Specification, RFC: 793, Defense Advanced Research Projects Agency (DARPA), Arlington, Virginia, USA, available at: http://tools.ietf.org/rfc/rfc793.txt, 1981b.

Sabath, F.: Classification of electromagnetic effects at system level, Proceedings of the 2008 International Symposium on Electromagnetic Compatibility (EMC Europe), Wroclaw, Poland, 18–22 August 2008, 1–5, 2008.

van Leersum, B. J. A. M., Buesink, F. J. K., Bergsma, J. G., and Leferink, F. B. J.: Ethernet susceptibility to electric fast transients, Proceedings of the 2013 International Symposium on Electromagnetic Compatibility (EMC Europe), Brugge, Belgium, 2–6 September 2013, 29–33, 2013.

26

Validation of the radiation pattern of the VHF MST radar MAARSY by scattering off a sounding rocket's payload

T. Renkwitz, C. Schult, R. Latteck, and G. Stober

Leibniz-Institute of Atmospheric Physics at the Rostock University, Schloss-Str. 6, 18225 Kühlungsborn, Germany

Correspondence to: T. Renkwitz (renkwitz@iap-kborn.de)

Abstract. The Middle Atmosphere Alomar Radar System (MAARSY) is a monostatic radar with an active phased array antenna designed for studies of phenomena in the mesosphere and lower thermosphere. Its design, in particular the flexible beam forming and steering capability, makes it a powerful instrument to perform observations with high angular and temporal resolution. For the configuration and analysis of experiments carried out with the radar it is essential to have knowledge of the actual radiation pattern. Therefore, during the time since the radar was put into operation various active and passive experiments have been performed to gain knowledge of the radiation pattern. With these experiments the beam pointing accuracy, the beam width and phase distribution of the antenna array were investigated. Here, the use of a sounding rocket and its payload as a radar target is described which was launched in the proximity of the radar. The analysis of these observations allows the detailed investigation of the two-way radiation pattern for different antenna array sizes and beam pointing positions.

1 Introduction

The Middle Atmosphere Alomar Radar System (MAARSY) was built in 2009/2010 by the Leibniz-Institute of Atmospheric Physics (IAP) on the island Andøya in Northern Norway (69.30° N, 16.04° E). The MAARSY radar facilitates studies of the arctic atmosphere with high spatial and temporal resolution within the mesosphere/lower thermosphere and the troposphere/lower stratosphere and is therefore categorized as a MST radar. The main active phased array antenna consists of 433 individual 3-element Yagi antennas (see Fig. 1) optimized for approximately 5 MHz bandwidth at the operating frequency of 53.5 MHz. Each individual antenna of

the main array is connected to its own transceiver allowing independent phase and amplitude settings. This arrangement facilitates very flexible beam forming and steering properties of the radar. A detailed technical description of the radar is given by Latteck et al. (2012). Furthermore, additional external antenna groups or individual antennas may be connected to the receivers allowing e.g. interferometric observations with antenna spacings and baseline lengths divergent to the main array and its subarrays. Extensive simulations of the radiation pattern of the main antenna array and the additional groups have been performed during the design phase and since the radar was put into operation. These simulations were performed with the Numerical Electromagnetic Code (version 4.1) investigating the radiation properties like the antenna gain, beam width and shape, the side lobe suppression, the influence of the soil properties and mutual coupling for various beam pointing directions.

In the beginning passive experiment like the observation of cosmic radio emissions were used to validate the simulation results. By the observation of galactic radio emissions the beam pointing accuracy, the beam width as well as a rough estimation of the side lobe attenuation (Renkwitz et al., 2012, 2013) and the antenna-receiver phase distribution were investigated (Chau et al., 2014). Therein, the passive observations of cosmic radio sources were compared to a highly accurate reference model by de Oliveira-Costa et al. (2008). Additional active two-way radar experiments with MAARSY using targets like satellites, Earth's moon as well as the piecewise comparison to meteor head echo events, already described in the previously given references, augmented the findings. However, the observation of targets within or beyond the Earth's ionosphere are challenging using linearly polarized antennas and needs to be carefully analyzed. Another method of validating the one-way radiation pattern of

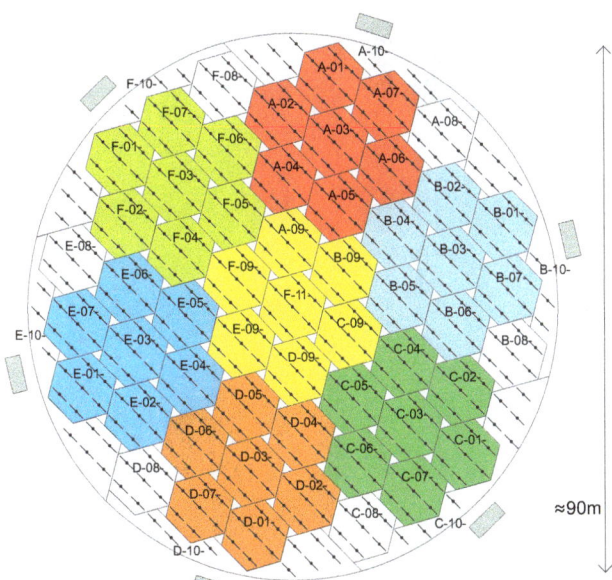

Figure 1. Sketch of the MAARSY VHF radar antenna array with the nomenclature of the individual subarrays. The colored subgroups mark the MAARSY343 subarray.

Table 1. Experiment settings used to observe the WADIS payload.

Pulse repetition frequency	1200 Hz
Sampling range	≈ 75 to 119 km
Code	16 bit complementary
Pulse length	$2\,\mu s$
Maximum peak power	nom. $433 \times 2\,kW$
Number of beams	18

a VHF antenna array is the placement of e.g. a transmitter onboard a satellite. This has to be carried out carefully as the wave's propagation through the ionosphere is subject to Faraday rotation and is therefore often limited to circularly polarized antenna arrays (see e.g. Fukao et al., 1985; Sato et al., 1989). Alternatively, a receiver may be placed on an airplane or helicopter to sample directly the radar's emissions, advantageously in the radiation far field.

In this paper we present an active radar experiment performed with MAARSY, which allows the determination of the beam pointing accuracy, the effective beam width and side lobe attenuation. For this purpose we use the backscatter of a sounding rocket's payload that was launched in the proximity of the radar. In the subsequent section we will describe the experiment we conducted to evaluate the radiation pattern of the MAARSY radar by the use of the sounding rocket's payload as a radar target. Subsequently we will present and discuss the results of this experiment and compare these to the simulation findings. Finally, conclusions and an outlook will be presented.

2 Experiment description

In the end of June 2013 the WADIS sounding rocket campaign (WAve propagation and DISsipation in the middle atmosphere) was conducted at the Andøya Space Center located in the proximity of MAARSY. During this campaign several meteorological and one instrumented rocket were launched. These rocket launches were accompanied by simultaneous operations of various ground based radar, lidar

and balloon facilities. Herein, MAARSY was the leading radar facility to detect and observe polar mesospheric summer echoes (PMSE) with multiple beam directions to propose favorable launch conditions and to derive detailed information about the PMSE structure during the flight. For this purpose MAARSY was operated with a dedicated experiment configuration during the flight to sequentially observe 16 different beam directions along the predicted payload's trajectory.

Besides the initial objectives of this scan configuration, the subsequent analysis of the spatially and temporally spread PMSE and the in-situ observations of the rocket's payload, additional backscatter was observed. The analysis of the radar data set turned out to contain strong backscatter from the rocket, the payload and likely the rocket's nose cone. Favorably, the rocket motor and the payload had a length of approximately half the radar wavelength allowing strong backscatter. The beam positions MAARSY was sequentially scanning using the entire antenna array for transmission and the GPS trajectory[1] of the payload with color-coded height are depicted in the left panel of Fig. 2. The GPS trajectory of the payload has been validated by interferometric means of MAARSY, using various combinations of subarrays carefully calibrated in range and phase. Due to the large radar cross-section of the payload we detected signatures of it in basically all beam positions. Thus, within 75 and 119 km range the flight of the payload has been observed almost continuously as shown in an power integrated plot in the right panel of Fig. 2. The prominent data gaps are caused by the selected experiment configuration, mainly the chosen sample rate and the available data transfer speed of the acquisition system existent at that time. The experiment settings are shown in Table 1.

The partially strong backscatter of at most 50 dB signal-to-noise ratio allows the detection of the radar's main lobe and its first side lobe and is therefore exceptionally valuable for radiation pattern comparisons. In the subsequently presented analysis of the experiment's data different sizes of the antennas array are used for reception, where the use of the entire antenna array (MAARSY433) should provide the maximum gain of all configurations in the selected beam pointing direction and therefore the minimum available beam width. Addi-

---

[1]Courtesy of M. Hörschgen-Eggers/DLR-MORABA.

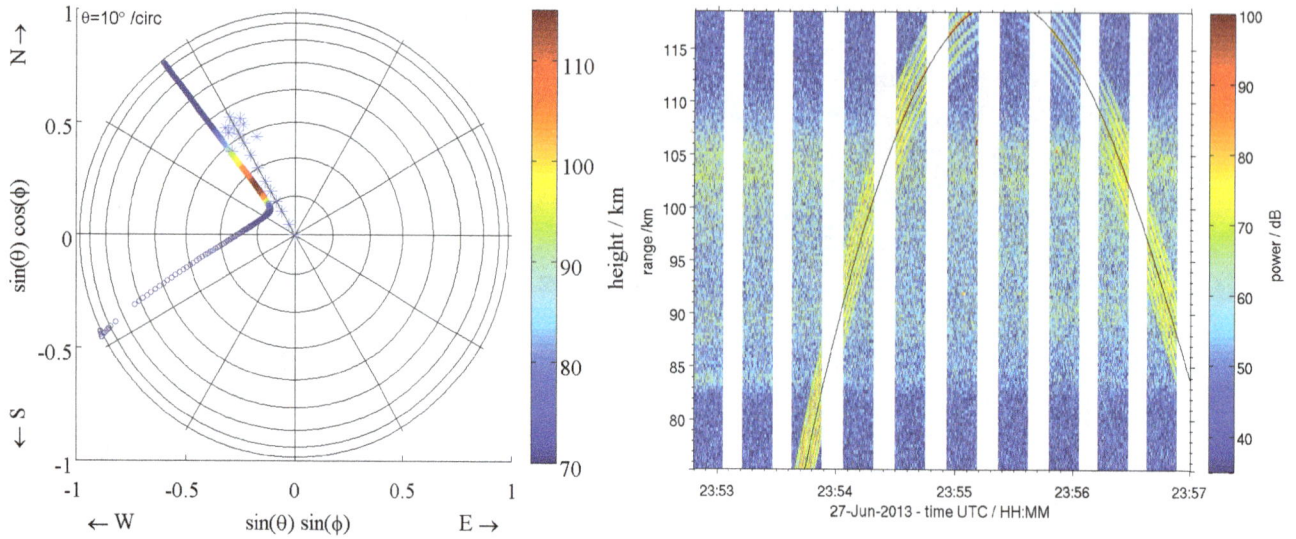

Figure 2. Left panel: trajectory of the WADIS rocket payload in reference to MAARSY, depicted in spherical coordinates, overlaid by the beam positions used in the WADIS scan experiment, marked by asterisks. The height of the payload during its flight is depicted color coded. Right panel: integrated power of all beams in the scan (median removed) depicted over range and time. The thin black line marks the range calculated from the GPS coordinates, which agrees exceptionally well to the observed range during the entire flight. The enhanced intensity between approximately 82 and 106 km range origins PMSE observed with different beam positions.

tional to this completely hardware combined channel, 15 additional receiving channels have been connected to subarray groups of 49 or 7 antennas each (Anemones and Hexagons, respectively). With the knowledge of the absolute phases of these individual subarray groups larger subarrays may be formed in software by integrating the signals of e.g. seven Anemones resulting in MAARSY343 (marked in color in Fig. 1). This software combined subarray is exceptionally interesting due to its large aperture and thus good angular resolution. The use of different sizes of the antenna array allows the verification of the beam pointing accuracy, beam width and side lobe attenuation for multiple beam directions for every single configuration. Furthermore the use of small subarrays in this experiment is exceptionally interesting as those broad radiation patterns do not contribute significantly to the two-way pattern. As an extreme case the use of an omnidirectional antenna for reception result in a two-way pattern that is identical to the pattern of the one-way transmitting antenna as the two-way pattern is the product of both individual radiation patterns.

The radiation pattern of the corresponding subarray groups and the entire antenna array have been simulated for the respective beam pointing positions for the entire hemisphere with 1° resolution in both azimuth and zenith angle. Within these simulations the antenna array have been modeled with best accuracy of the antenna structure itself and the soil properties. Thus, the most realistic pattern and so the side lobes are taken into account for the subsequent analysis. The resulting two-way pattern for the different receiving subarrays were computed afterwards to obtain simulated intensi-

ties along the payload's trajectory. The simulated intensities were then corrected for the changing radar cross-section of the target during its flight.

For this purpose we assume a metallic mainly pure cylindrical surface and a simplified spherical bottom side of appropriate dimensions. The most significant approximation used here refers to both ends of the payload, where various instruments are placed. These instruments represent diverse surface structures significantly smaller than the used wavelength which makes it intricate to derive a comprehensive expression and therefore a half sphere is assumed. This imperfection is, however, relieved as the total effective area is dominated by the payload's cylindrical body. Considering the length of the target of approximately half the radar wavelength Rayleigh scattering appears most appropriate for the purpose of this analysis and the desired accuracy.

Applying these approximations of the geometric shape the radar cross-sections σ_s and σ_b for the cylindrical side and the bottom surface may be expressed by the approximations given by Fuhs (1983), respectively.

$$\sigma_s^o = 9/4\pi h^2 (2\pi r/\lambda)^4 \tag{1}$$

$$\sigma_s^p = \frac{\pi \cdot h^2}{(\pi/2)^2 + (\ln(\lambda/1.78\pi r))^2} \tag{2}$$

$$\sigma_s = \left(\sigma_s^o + \sigma_s^p\right)/2 \cdot \sin\theta_a \tag{3}$$

$$\sigma_b = 7.11\pi r^2 \cdot (2\pi/\lambda)^4 \cdot \cos\theta_a \tag{4}$$

The height and radius of the payload are denoted by h and r, which are 2.8 m and 0.35 m respectively. The angle of attack from the radar to the payload is represented by θ_a,

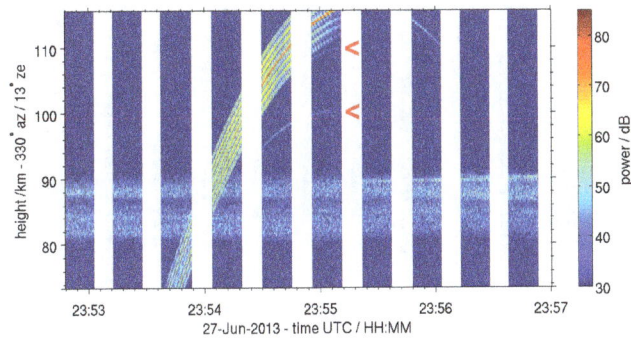

Figure 3. Detected intensities for the beam direction $\phi = 330°$ and $\theta = 13°$ using MAARSY433 on reception. The backscatter at around 85 km height originates PMSE. The red angles mark backscatter likely caused by the motor and the rocket's nose cone.

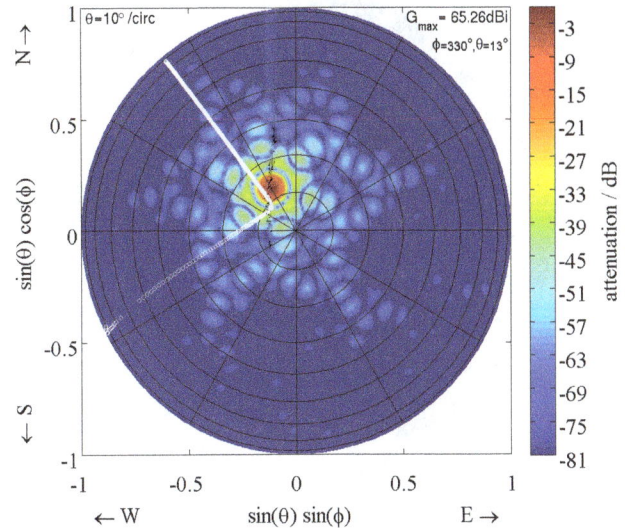

Figure 4. Simulated two-way radiation pattern pointing to $\phi = 330°$ and $\theta = 13°$ by the superposition of MAARSY433 and MAARSY343 for transmission and reception, respectively. Equivalently to Fig. 2 left, the trajectory of the payload is marked by white circles.

which varies between 10 to 30° during the evaluated period. In Eqs. (1) to (3) σ_s^o and σ_s^p describe the orthogonal and parallel components of the total radar cross section σ_s of the cylindrical body, where the first dominates significantly. As the MAARSY antenna array was mainly circularly polarized during the experiments, no consideration of the orientation angle of the payload and the polarization angle are necessary as it would have been for linear polarization near the resonance scattering. Besides this angular independency of circular polarization for the mentioned target it, however, leads to a reduction of effective area by a factor of 2 (in Eq. 3), comparable to the polarization loss between linear and circular polarization. Alternatively, equations and approximations given by e.g. van Vleck et al. (1947), Mailloux (1994) and Skolnik (2008) yield to results of equivalent cross sections. It has to be noted, diffraction and creeping waves around the backside of the body are neglected in the shown approximations. The radar cross-section was computed for every point of the trajectory considering the pitch angle of the rocket (and its payload) and the geometric angle of attack from the radar to the target. The radar cross-section thus varies for the evaluated beam direction by a factor of 3 dB at most accounting for the previously mentioned approximations. This mainly affects the simulated intensity of the outward side lobe and to a lesser extent the simulated beam width. This factor is mainly caused by the variable effective area of the dominating cylindrical surface with the minimum at approximately 80 km height during the upleg and the maximum during the downleg due to the increasing zenith angle and thus angle of attack and effective area.

3   Analysis and discussion

In this section the analysis of the experiment's data for few beam positions and their interpretation are discussed.

Figure 3 shows the detected intensities for the beam pointing direction of $\phi = 330°$ and $\theta = 13°$. Here, besides the sig-

nature the of PMSE at approximately 81 to 90 km height, the major backscatter is detected from the rocket's payload almost throughout its maximum range. For the periods of very strong backscatter the observed intensities are marked by the properties of the used code and thus also the code's side lobes smearing ±4 km around the maximum. Since 23:54 UT the maximum of the payload's detected intensity per time agrees well to the range and height calculated from the GPS trajectory. Prior to 23:54 UT the backscatter intensity does not maximize in the middle of the signature. This indicates imperfect decoding of the signal due to the large Doppler shift caused by the high departure speed in respect to the radar's position. Unfortunately no un-decoded raw data have been stored as the experiment's outcome was not foreseen and thus its potential needs were unknown during the configuration process. Interestingly, below the strong payload backscatter two additional weaker signatures can be spotted at approximately 100 km and 109 km apogee respectively, which are most likely caused by the rocket's motor and the nose cone, marked by red angles in Fig. 3. The reason of seeing the payload's backscatter during the entire upleg with this beam position can be seen in Fig. 4, where MAARSY's two-way radiation pattern is shown overlaid by the trajectory of the payload. The payload essentially stays within the main lobe and the first side lobe ring for the beam pointing of $\phi = 330°$ and $\theta = 13°$. The beginning downleg can be seen e.g. for the beam pointing of $\phi = 330°$ and $\theta = 20°$ (see Fig. 2, respectively). Here, the payload traverses a part of the side lobe ring of MAARSY and thus allows short but strong backscatter (40 dB signal-to-noise ratio). Furthermore, only weak traces can be seen during the upleg period as for this beam direction

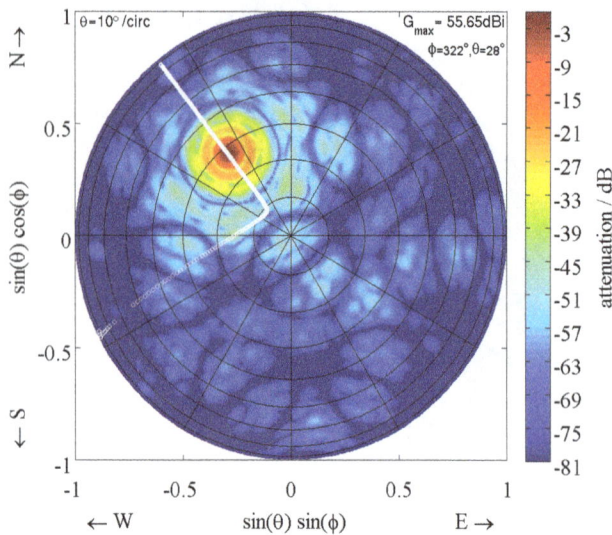

Figure 5. Simulated two-way radiation pattern pointing to $\phi = 322°$ and $\theta = 28°$ by the superposition of MAARSY433 and an Anemone group of 49 antennas for transmission and reception, respectively. Equivalently to Fig. 2 left, the trajectory of the payload is marked by white circles.

the payload was far outside the main and the first side lobe resulting in at least 40 dB less gain in these directions. More interesting are the observations for the earlier highlighted beam pointing direction of $\phi = 322°$ and $\theta = 28°$ (see Fig. 5).

Here, the payload almost exactly passes through the radar's beam pointing direction and thus facilitates strong backscatter from both, the main lobe and the first side lobe ring (see Fig. 6). Therefore this beam pointing direction was used for the detailed analysis and comparison with the simulated radiation pattern. For this purpose the maximum detected intensities close to the GPS trajectory are used, as marked with crosses in Fig. 7 for both MAARSY433 (blue) and MAARSY343 (red) on reception. The simulated intensities, computed for the actual state of the antenna array during this experiment for reception and transmission, are depicted as solid lines in the respective color for both array sizes.

Similarly to the detected power in Fig. 6, the intensities have been simulated for each point of the GPS trajectory and thus the direction from the radar to the target. It can be seen that the relative intensities and the beam shape for both array sizes are approximately the same. MAARSY's beam pointing appears to be shifted very slightly towards zenith, which is however hardly to see due to the existing data gap at the simulated maximum intensity. Nevertheless the misalignment can be approximated to be in the order of less than 0.5°. The observed beam width for MAARSY433 on reception appears to be slightly broader than the width of MAARSY343. This is just opposite to the simulation and the general approximation since the aperture of the MAARSY433 supersedes MAARSY343 and thus the theoretical relative beam width is inversely proportional to their aperture ratio. Furthermore,

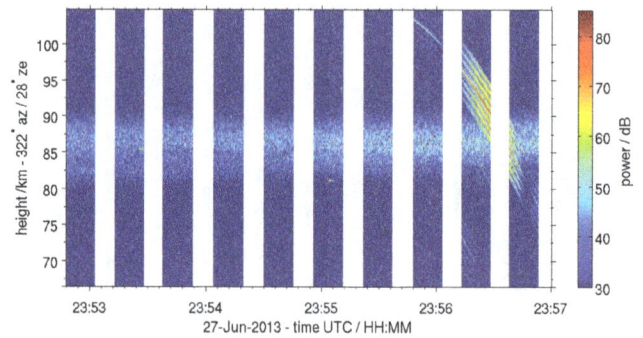

Figure 6. Detected intensities for the beam direction $\phi = 322°$ and $\theta = 28°$ using MAARSY343 on reception. The backscatter at around 85 km height originates PMSE.

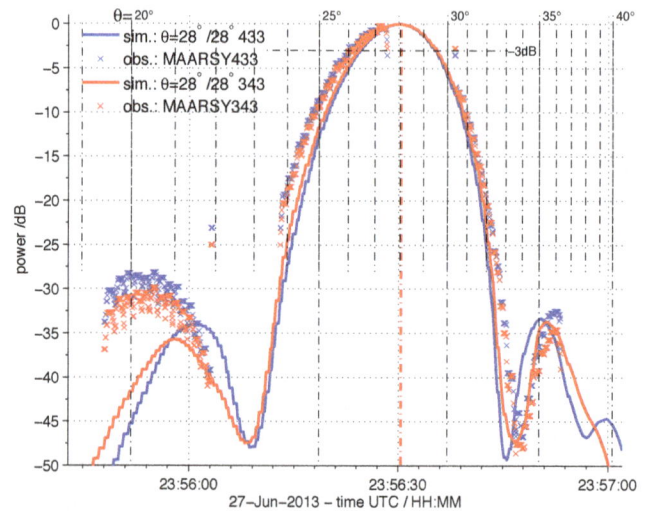

Figure 7. Detected and simulated intensities (marked by crosses and solid line, respectively) for the beam direction $\phi = 322°$ and $\theta = 28°$ using MAARSY433 and MAARSY343 (blue and red, respectively). Black vertical broken lines mark the zenith angle labeled on top of the figure. Colored vertical broken line accentuates the maximum of simulated intensity.

the beam width of the main lobe of both arrays seem to be broadened by 0.5° at most, which is consistent to the outward shifted positions of the minima between the main lobe and the first side lobe ring. The side lobe attenuation agrees fairly well to the simulations with about 2 dB increased intensity towards zenith.

It has to be noted that the MAARSY antenna array was mixed-polarized during this experiment. Since May 2013 the antennas of the seven Anemones (A to F and the Middle, see Fig. 1) were completely converted from linear to circular polarization, while the outer antennas were still linearly polarized. Additionally, during that time in total 21 transmitter modules and thus antennas were non-operational on transmission, which disturbs the radiation pattern mainly affect-

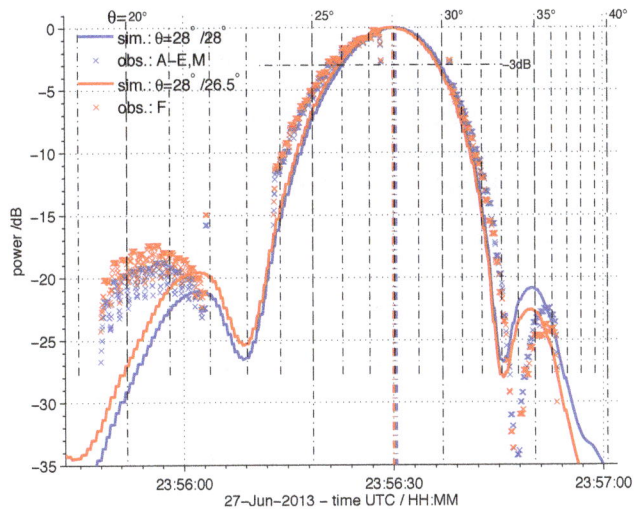

Figure 8. Equivalent to Fig. 7. Detected intensities for the beam direction $\phi = 322°$ and $\theta = 28°$ using Anemones A–E, M (blue) and Anemone F (red) marked by crosses. The corresponding simulated intensities are shown for $\phi = 322°$ and $\theta = 28°$ (blue) and $\phi = 322°$ and $\theta = 26.5°$ (red).

Figure 9. Detected intensities for the beam direction $\phi = 330°$ and $\theta = 13°$ using a Hexagon group of seven antennas.

ing the side lobe level. Both issues have already been incorporated in the simulations used for this comparison.

To investigate the reason of the broadened beam width smaller subarray groups have been used in a similar manner as before. In Fig. 8 the simulated and observed intensities for Anemone subarrays (49 antennas each) are shown. It was found that six out of the seven available Anemones show equivalent detected intensities, while Anemone F slightly differed. The main lobe of Anemone F seems to be shifted towards zenith, which is also underlined by an intensified side lobe towards zenith and vice versa with the outer side lobe. The best agreement to the observations was found for the simulation of pointing the Anemone at most 1.5° off the nominal beam direction. This misalignment however should not result in broadening the beam width of the entire antenna array in such an extent as seen before. Equivalent to the previous example using (almost) the entire antenna array, the two-way beam width using the Anemone subarrays on reception appears to be broadened by 0.5°. Consistently, the minima between main and side lobe are shifted outwards by approximately 0.6°.

As quoted in the prior section a rather broad, in an extreme case omnidirectional, antenna does not change the two-way radiation pattern significantly. During this experiment, the smallest available antenna groups were Hexagon subarrays, composed of seven antennas each. In total eight Hexagons spread over the antenna array have been used and analyzed (see Fig. 9). Besides mediocre variation of the detected intensities via the side lobes (e.g. B-08), all Hexagons show an equivalent shape and width of the main and side lobe. However, the beam width appears to be still enlarged by approx-

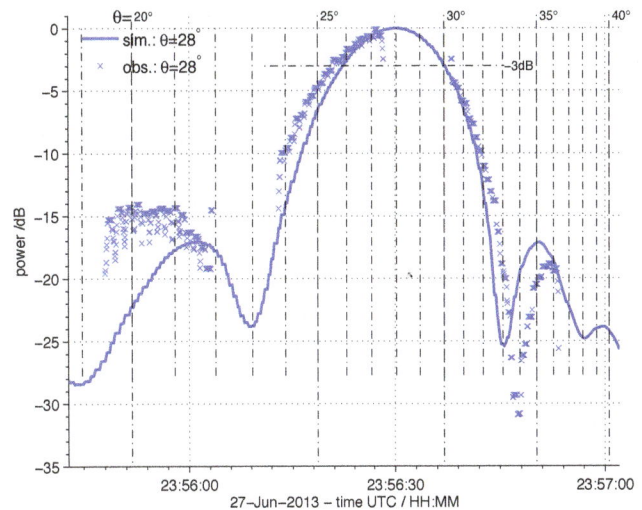

imately 0.7°. Due to the wide beam width of the Hexagons (31°) the effective width of the two-way radiation pattern is dominated by the transmission pattern of the entire antenna array. Even though the width of the Hexagons could also be enlarged to some extent it appears to be more likely the beam width of the entire antenna array is broadened.

A generally smaller aperture of the antenna array and thus broader beam width as planned can be excluded as the array has been built with best precision. Furthermore passive experiments observing cosmic radio sources and active experiments observing meteor head echoes indicated a better agreement with simulations (see e.g. Renkwitz et al., 2012, 2013; Chau et al., 2014).

Possible reasons of the observed broadening are: (a) additional ineffective antennas at the rim of the entire antenna array, (b) unintentional amplitude taper, or (c) imperfect simulation. The outermost antennas of the entire array have not been exceptionally peculiar in neither earlier experiments nor the automatically measured impedances and reflected output power for each experiment run for the considered time. Unintentional random amplitude taper (± 1 dB) and phase variations ($\pm 10°$) for the individual antennas have been simulated and are shown in Fig. 10. The perfectly phased array MAARSY 343 with uniform amplitude has a minimum beam width of 4°, while for the simulation with random phase and amplitude variations a broadening of 0.1° can be seen in one cross-section. An imperfect simulation may be caused by the mixed polarization within the antenna array, leading to e.g. divergent mutual coupling, or the non-operational transmitter modules and thus individual antennas (21 in total). Simulations of the entire antenna array considering these missing antennas have shown a beam width broadening approximately of 0.2°, which already has been incorporated in the simulations of the previously shown comparisons to the

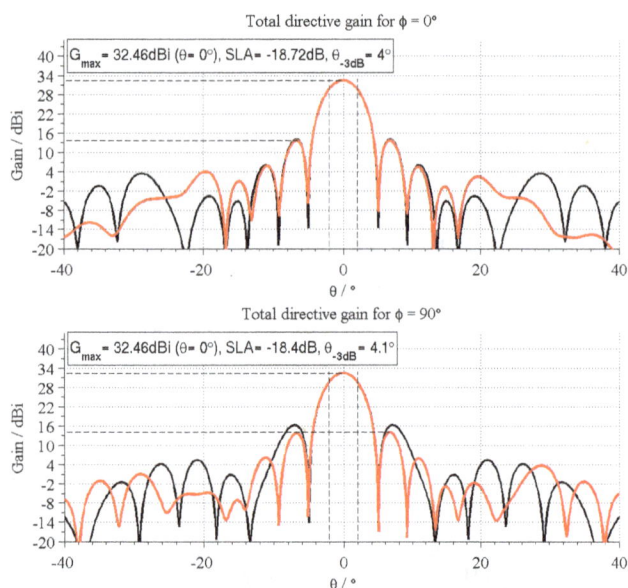

Total directive gain for φ = 0°

Total directive gain for φ = 90°

Figure 10. Comparison of simulations for the ideally phased and uniform amplitude MAARSY343 (black) to the case of random fluctuations of every array element of ±1 dB and ±10° for two cross-sections ($\phi = 0°$ and $\phi = 90°$).

observations. Finally, we assume a superposition of primarily the points (b) and (c) causing the observed beam width broadening of approximately 0.5°.

4   Conclusions and outlook

In this paper we presented an active experiment to validate the radiation pattern of MAARSY. This actively scanning radar experiment was initially planned to derive background parameters, like PMSE strength and wind components for the trajectory of the rocket's payload launched during the WADIS campaign. Additionally to these objectives strong backscatter from the payload were detected which facilitates the comparison to the simulated two-way radiation pattern of the radar. During this experiment the payload's flight could be seen in all 18 scheduled beam directions along the predicted trajectory. This is primarily caused by the payload's size (almost exactly half wave length of the radar) and its cylindrical shape that exhibit a large radar cross-section. Though MAARSY reaches a reasonably good side lobe suppression, the payload's backscatter could often also be seen in the first side lobe of the radar's radiation pattern for the entire antenna array and smaller subarrays on reception. The two-way radiation pattern for the combination of the entire antenna array on transmission and various arrays sizes for reception were simulated for the individual points of the payload's trajectory. These simulations were subsequently corrected for the varying radar cross-section due to the orientation of the payload and thus the angle of attack.

Doing so, we were able to find very good agreement of the simulated radiation pattern and the observed backscatter intensities. Though, we also found a generally enlarged beam width, which could be verified by the use of smaller subarrays on reception, seven antennas at the least. Therefore the transmit pattern incorporating the entire antenna array already has to be broadened. The enlarged beam width is most likely caused by the superposition of circularly polarization emitted by approximately 80 % of the antenna array in its center and the still linearly polarized antennas at the rim of the array. Furthermore, with the simulation we could demonstrate that the subarray Anemone F appeared to squint to an extent of maximum 1.5° towards zenith during this experiment. This is assumed to be caused by a flaw in the receiver phases, which however could not be verified during maintenance measures. Overall, we are impressed by the excellent opportunity to validate the radiation pattern by the observation of a sounding rocket, which is of significant benefit to the radar operators. Therefore, we are looking forward to the next equivalent rocket campaign in the proximity of the radar as MAARSY's antenna array is completely converted to circular polarization since autumn 2013. Such rocket campaign could be used to validate the beam width of MAARSY's radiation pattern for the entirely circularly polarized antenna array. For the first half-year of 2015 a second rocket launch within the WADIS campaign is now planned with an equivalent payload. The authors are looking forward to this excellent opportunity which may solve some of the uncertainties illustrated here.

Acknowledgements. The authors explicitly acknowledge the kindly provided GPS trajectory by M. Hörschgen-Eggers/DLR-MORABA and the PI of IAP's sounding rocket group Dr. Boris Strelnikov. The WADIS project was funded by the German Space Agency (DLR) under grant 50 OE 1001. Furthermore, we like to express our gratitude to the Andøya Space Center for their permanent support for the operation and maintenance of the MAARSY radar. The radar development was supported by the german grant 01 LP 0802A of Bundesministerium für Bildung und Forschung.

References

Chau, J. L., Renkwitz, T., Stober, G., and Latteck, R.: MAARSY multiple receiver phase calibration using radio sources, J. Atmos. Sol.-Terr. Phy., doi:10.1016/j.jastp.2013.04.004, 2014.

de Oliveira-Costa, A., Tegmark, M., Gaensler, B. M., Jonas, J., Landecker, T. L., and Reich, P.: A Model of Diffuse Galactic Radio Emission from 10 MHz to 100 GHz, Mon. Not. Roy. Astron. Soc., 338, 247–260, doi:10.1111/j.1365-2966.2008.13376.x, 2008.

Fuhs, A. E.: Radar cross section lectures, Naval postgraduate school Monterey, Dep. of Aeronautics, California, 1983.

Fukao, S., Sato, T., and Kato, S.: Monitoring of the MU Radar Antenna Pattern by Satellite OHZORA (EXOS-C), J. Geomag. Geoelectr., 37, 431–441, 1985.

Latteck, R., Singer, W., Rapp, M., Vandepeer, B., Renkwitz, T., Zecha, M., and Stober, G.: The new MST radar on Andøya: System description and first results, Radio Science, 47, RS1006, doi:10.1029/2011RS004775, 2012.

Mailloux, R. J.: Phased Array Antenna Handbook, vol. 1, 1st Edn., Artech House, London, 1994.

Renkwitz, T., Singer, W., Latteck, R., Stober, G., and Rapp, M.: Validation of the radiation pattern of the Middle Atmosphere Alomar Radar System (MAARSY), Adv. Radio Sci., 10, 245–253, doi:10.5194/ars-10-245-2012, 2012.

Renkwitz, T., Stober, G., Latteck, R., Singer, W., and Rapp: New experiments to validate the radiation pattern of the Middle Atmosphere Alomar Radar System (MAARSY), Adv. Radio Sci., 11, 283–289, doi:10.5194/ars-11-283-2013, 2013.

Sato, T., Inooka, Y., Fukao, S., and Kato, S.: Multi-Beam Pattern Measurement of the MU Radar Antenna by Satellite OHZORA, J. Geomag. Geoelectr., 41, 743–752, 1989.

Skolnik, M. I.: Radar handbook, 3rd Edn., McGraw-Hill, New York, 2008.

van Vleck, J. H., Bloch, F., and Hamermesh, M.: Theory of radar reflection from wires or thin metallic stripes, J. Appl. Phys., 18, 274–294, doi:10.1063/1.1697649, 1947.

27

Remote sensing and modeling of energetic electron precipitation into the lower ionosphere using VLF/LF radio waves and field aligned current data

E. D. Schmitter[1],[†]

[1]University of Applied Sciences Osnabrueck, 49076 Osnabrueck, Germany
[†]deceased, 25 March 2015

Correspondence to: M. Förster (mfo@gfz-potsdam.de)

Abstract. A model for the development of electron density height profiles based on space time distributed ionization sources and reaction rates in the lower ionosphere is described. Special attention is payed to the definition of an auroral oval distribution function for energetic electron energy input into the lower ionosphere based on a Maxwellian energy spectrum. The distribution function is controlled by an activity parameter which is defined proportional to radio signal amplitude disturbances of a VLF/LF transmitter. Adjusting the proportionality constant allows to model precipitation caused VLF/LF signal disturbances using radio wave propagation calculations and to scale the distribution function. Field aligned current (FAC) data from the new Swarm satellite mission are used to constrain the spatial extent of the distribution function. As an example electron precipitation bursts during a moderate substorm on the 12 April 2014 (midnight–dawn) are modeled along the subauroral propagation path from the NFR/TFK transmitter (37.5 kHz, Iceland) to a midlatitude site.

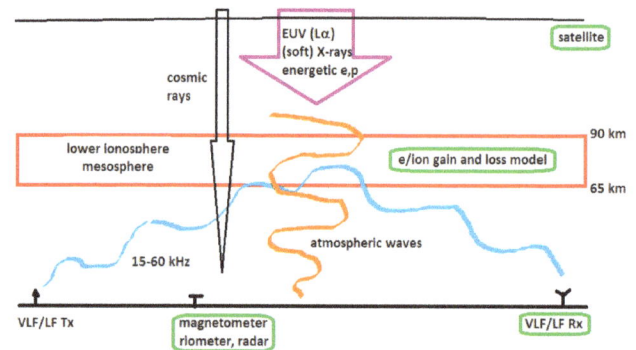

Figure 1. Forcing of the lower ionosphere/mesosphere and remote sensing instrumentation.

1 Introduction

The lower ionosphere (60–85 km height) is forced from above by extreme UV, especially Lyman α, solar flare X-rays and energetic particle precipitation (electrons $\gtrsim 30$ keV, protons $\gtrsim 4$ MeV, lower energetic particles do not penetrate so deeply). Forcing from below takes place via tides, gravity- and planetary waves, but also lightning electromagnetic pulses and upward discharges (Fig. 1). Ionization by particles not only directly enhances the local electron/ion density but also leads to the formation of HOx and NOx molecules. The latter family is rather long living and by vertical transport also reaches the stratosphere where it catalytically affects the ozone budget, especially in the polar region (Clilverd et al., 2007; Salmi et al., 2011). This is one example of the strong coupling of the different ionospheric and atmospheric layers. Very low frequency and low frequency (VLF/LF) electromagnetic radiation from man made transmitters (15–60 kHz), but also from lightning, propagates mainly between ground and the lower ionosphere. The propagation conditions are strongly affected by the conductivity of the lower ionosphere which is proportional to the quotient of the electron density and collision frequency height profiles. Remote VLF/LF sensing therefore since decades proves as an inexpensive and reliable way to assess forcing processes in this layer, also with regard to auroral pre-

Figure 2. The monitored propagation path (red line; transmitter: NRK, receiver: Rx). Also indicated: magnetometers LEIRV: Leirvogur, LER: Lerwick; riometers: ABI, SOD, OULU: Abisko, Sodankyla, Oulu.

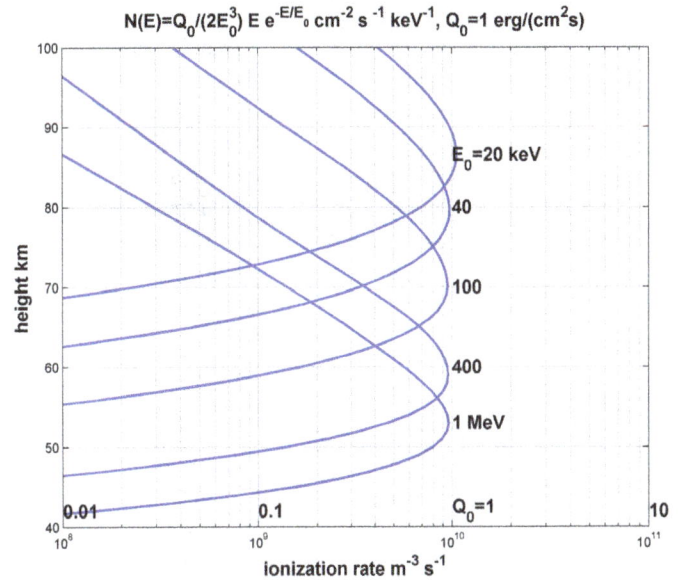

Figure 3. Ionization rate vs. height profile for Maxwellian electrons of different folding energies E_0 (normalized to the energy flux rate $Q_0 = 1\,\mathrm{mW\,m^{-2}} = 1\,\mathrm{erg\,cm^{-2}\,s^{-1}} = 10^{-7}\,\mathrm{J\,cm^{-2}\,s^{-1}} = 6.25\cdot10^8\,\mathrm{keV\,cm^{-2}\,s^{-1}}$). Atmospheric model: COSPAR International Reference Atmosphere (CIRA86, June, 50° N).

cipitation activity (Cummer et al., 1996, 1998). During the last years we have developed a modeling scheme that allows to assess ionospheric forcing parameters from VLF/LF amplitude and phase recordings of distant Minimum Shift Keying (MSK) transmitters (Schmitter, 2010, 2011, 2012, 2013, 2014). As the amplitude of these transmissions is constant, any observed variations are caused along the propagation path. The monitored propagation path from the VLF/LF transmitter with the call sign NRK/TFK (37.5 kHz, 63.9° N, 22.5° W, Iceland) to our receiver site 52° N, 8° E (NW Germany, 2210 km) starts in the subauroral domain and extends to midlatitudes (Fig. 2). In this paper we extend our model to include forcing by energetic electrons. The penetration depths of energetic electrons are shown in Fig. 3. Assuming a Maxwellian electron energy distribution electrons with a folding energy exceeding 30 keV have their ionization maximum in the lower ionosphere ($\lesssim 85$ km) strongly affecting VLF/LF propagation. Riometers monitor cosmic noise absorption at about 90–100 km height and in this respect they are also sensitive to electron precipitation, however with regard to lower energies (~ 10 keV, i.e. auroral electrons). Magnetometer recordings reflect the current variations caused by geomagnetic storm/substorm conditions which often are accompanied by particle precipitation. Large scale energetic electron precipitation modify the auroral electrojet and lead to variations of the AL index. In this way ground based observations by riometers and magnetometers at proper sites yield important additional information with regard to the identification of energetic electron precipitation along the VLF/LF propagation path. The sources of the pre-

cipitation are the radiation belts (Thorne, 1974; Carson et al., 2012; Thorne et al., 2013). So, for a detailed characterization of precipitation, satellite data are indispensable, even if a specific domain (e.g. around the VLF/LF propagation path) cannot be monitored continuously.

Mapping down electric and magnetic field measurements and derived data (field aligned currents, FACs, Ritter et al., 2013) from the Swarm satellites yields important information for the assessment of the energy input into the ionosphere, particularly with regard to the path and spatial extent of particle precipitation as well as the acceleration mechanisms. At least a part of the energetic electron precipitation is coupled to the midnight–dawn region 2 (equatorward) currents and the dusk–midnight region 1 (poleward) currents (Ohtani et al., 2010). So it may be expected, that the subauroral VLF propagation path is mainly affected by energetic electron precipitation coupled to region 2 (equatorward) field aligned currents, however compare the discussion of measured FAC data with regard to this simple picture in Sect. 5. In this paper we describe the VLF remote sensing procedure followed by the discussion of our model for the calculation of the electron density profiles, especially with regard to forcing events like electron precipitation. After the description of the VLF radio wave propagation calculations we discuss an example event: electron precipitation bursts during a moderate substorm on the 12 April 2014 (midnight–dawn) and the derivation of the precipitation energy input along a VLF propagation path in space and time from VLF data as well as using field aligned current data from the new Swarm satellite mission (launch:

22 November 2013). This mission was designed to measure the magnetic signals from Earth's core, mantle, crust, oceans, ionosphere and magnetosphere. It consists of a constellation of three identical satellites in low polar orbits: two of them orbit side-by-side, descending from an initial altitude of 460–300 km over 4 years; the third maintains an altitude of about 530 km. Each satellite carries a vector field magnetometer, an absolute scalar magnetometer and an electric field instrument (for a complete description see www.esa.int). In our paper the field aligned current (FAC) data, a Level 2 product generated from the magnetic field data, are used.

2 Remote sensing

Remote sensing of lower ionosphere conditions (bottomside sounding) by monitoring low and very low frequency radio signal propagation has been a well known method for several decades. MSK (Minimum Shift Keying) transmitters prove useful in this respect because of their constant amplitude emissions. We have analyzed the signal amplitude and phase variations of the NRK/TFK transmitter (37.5 kHz, 63.9° N, 22.5° W, Iceland) received at a midlatitude site (52° N, 8° E) with a great circle distance of 2210 km. The receiver has been set up with a ferrite coil oriented for maximum signal amplitude of the horizontal magnetic field. After preamplification a stereo sound card computer interface with 192 kbit sampling rate is used. The second channel is fed with the 1-s pulse of a GPS receiver. Our software reads a 170 ms signal train each second and extracts within the narrow MSK bandwidth (200 Hz with NRK) amplitude and phase with regard to the rising GPS-pulse flank yielding a time synchronization better than 100 ns, corresponding to phase detection errors of 1.4° at 37.5 kHz. For the amplitude the signal to noise ratio (SNR) is recorded. SNR = 0 dB is defined by the averaged signal level received during transmitter maintenance drop outs. The time stability of both transmitters proves to be sufficient for continuous day and night monitoring not only of the amplitude but also of the phase. Our phase detection algorithm for MSK signals records phases between −90 and +90°.

The NRK propagation path proceeds most of its way through the subauroral domain to the midlatitude receiver site and proves well suited to study lower ionosphere forcing from above by particle precipitation (Schmitter, 2010) and solar flares (Schmitter, 2013) as well as forcing from below by planetary wave activity (Schmitter, 2011, 2012). In this paper we concentrate on forcing by energetic electron precipitation.

3 Modeling electron density profiles

A well known parametrization for the conductivity of the lower ionosphere (about 60–95 km height) is the two parameter model of Wait and Spies (1964) with effective height h'

(km) and profile steepness β (1 / km):

$$\sigma(h) = \sigma_0 e^{\beta(h-h')}, \tag{1}$$

with $\sigma_0 = \sigma(h = h') = 2.22 \cdot 10^{-6}\,\mathrm{S\,m^{-1}}$.

From Eq. (1) together with the collision frequency profile $f_c(h) = f_0 e^{-h/H}$, $f_0 = 1.816 \cdot 10^{11}$ Hz and the relation between conductivity and electron density appropriate for the very low frequency range $\sigma = \epsilon_0 \frac{\omega_p^2}{f_c} = \frac{e^2 n_e}{m_e f_c}$ (ω_p: plasma frequency) we get the classic Wait and Spies (1964) electron density parametrization of the lower ionosphere:

$$n_e = n_0 e^{-h/H} e^{\beta(h-h')}, \tag{2}$$

with $n_0 = 1.43 \cdot 10^{13}\,\mathrm{m^{-3}}$ and scale height $H = 1/0.15 = 6.67$ km corresponding to an isothermal atmosphere with $T = 230$ K.

While this loglinear electron density–height relation often is a reasonable first order approximation and intuitive with regard to the interpretation of its two parameters, it does not explain anything about the physical and chemical processes generating the profiles. We therefore want to go a step further in this respect and use reaction rate equations describing the development of the density profiles of electrons and ions. Chemical reaction based modeling has been done in the past at various levels of specification. A very detailed model of the complex D-layer chemistry is the Sodankyla Ion Chemistry (SIC) model which takes into account several hundred reactions and external forcing due to solar radiation (1–422.5 nm wavelength) as well as particle precipitation (Verronen et al., 2002). Several approaches have been made to reduce complexity to most relevant reaction types – at the cost of introducing effective reaction parameters lumping together different processes. For example: the $14-\text{ion}+\text{electron}$ model of Torkar and Friedrich (1983); the $6-\text{ion}+\text{electron}$ Mitra–Rowe model (Mitra and Rowe, 1972; Mitra, 1975); the $3-\text{ion}+\text{electron}$ Glukhov–Pasko–Inan model (Glukhov et al., 1992); the electron "only" model (Rodger et al., 1998, 2007).

With regard to VLF/LF remote sensing the electron density profiles are of main interest. For this reason and also to keep the effort for the propagation calculations tolerable we use an extended version of the electron "only" model. The model is extended by taking into account electron detachment reactions. They provide an additional electron source which gains special importance around dusk and dawn, when Lyman α radiation is weak. This leads to the following rate equations for the electron-, positive ion- and negative ion density distributions, n_e, n_-, n_+ (m^{-3}):

$$\frac{\partial n_e}{\partial t} = q - \alpha_D n_+ n_e - \beta_n n_e + (\gamma_n + \rho)n_-, \tag{3}$$

$$\frac{\partial n_-}{\partial t} = -\alpha_i n_+ n_- + \beta_n n_e - (\gamma_n + \rho)n_-. \tag{4}$$

Average neutrality is assumed: $n_+ = n_e + n_-$ and spatial diffusion is neglected.

- q: ion-pair production ($m^{-3} s^{-1}$), during daytime caused by Lyman α radiation and solar X-rays (from flares), during nighttime mainly by scattered Lyman α radiation and cosmic rays. Particle precipitation can cause additional ionization at any time. See the next subsection for a detailed discussion.

- $\alpha_{D,i}$: electron (D) and negative ion (i) recombination co-efficients ($m^3 s^{-1}$)

- β_n: three body attachment coefficient (s^{-1})

- γ_n, ρ: coefficients for electron detachment from negative ions by collision with neutrals or by photons respectively.

The reaction rates depend on the density of the reaction partners, i.e. the pressure as well as the temperature T (below 100 km height we can assume local thermodynamic equilibrium: $T = T_{neutrals} = T_{ions} = T_{electrons}$). This, as well as relative abundances of ion species with different recombination coefficients, means, that the reaction rates vary with height. We use reaction rates as provided by Brekke (1997), with the exception of the attachment coefficient β_n, where we use the more detailed relations provided in Rodger et al. (1998, 2007).

All numerical integrations are done using the classic Runge-Kutta algorithm with variable time step size ($0.1, \ldots, 2$ s).

We now describe the ionization processes in the lower ionosphere which are included in our model.

3.1 Forcing processes characterizing undisturbed conditions

Before considering disturbance events the processes generating the background ionization have to be modeled. In the lower ionosphere we have continuous forcing from cosmic radiation and solar Lyman α extreme UV:

- cosmic radiation: the ionization rate is proportional to the neutral density n_n absorbing the radiation:

$$q_{cosmic} = c_0 n_n, \qquad (5)$$

where c_0 increases with geomagnetic latitude ($1.2 \cdot 10^{-18} \ldots 1.2 \cdot 10^{-17} s^{-1}$ between 0 and 60° latitude) (Nicolet and Aikin, 1960). With undisturbed conditions cosmic radiation is the main forcing process below 80 km height during the night. During daytime this transition height drops to 65–70 km.

- solar Lyman α radiation (121.57 nm) on the day side and night side via scattering:

The Lyman α photon energy of 10.2 eV is high enough to ionize NO but not the major species O_2, N_2, so, with $I_{L_\alpha,\infty}$ as the Lyman α intensity incident at the top of the atmosphere:

$$q_{L_\alpha} = \sigma_{NO} \cdot n_{NO} \cdot I_{L_\alpha,\infty} e^{-\tau}. \qquad (6)$$

$I_{L_\alpha,\infty}$ varies with solar activity in the range $2.5, \ldots, 5 \cdot 10^{15}$ photons $s^{-1} m^{-2}$. The optical depth integrated along the ray path with O_2 as the main absorbing species is:

$$\tau = \int \sigma_{O_2} \cdot n_{O_2} ds. \qquad (7)$$

With Nicolet and Aikin (1960) we use $\sigma_{NO} = 2 \cdot 10^{-22} m^2$ for the ionization cross section of NO and $\sigma_{O_2} = 10^{-24} m^2$ for the absorption cross section of O_2 at 121.57 nm. n_{O_2} is 0.21 of the total number density at mesospheric heights. n_{NO} varies considerably: as the measurements of the Halogen Occultation Experiment (HALOE) of the Upper Atmosphere Research Satellite (UARS) have shown, the NO number density between 60 and 90 km height at undisturbed conditions during the same day can vary at least between 10^{11} and $10^{13} m^{-3}$. We adopt $5 \cdot 10^{12} m^{-3}$ as an average undisturbed value. The concentration of NO is significantly enhanced during particle precipitation and for hours afterwards (Barth et al., 2001; Saetre et al., 2004) – which has to be taken into account for the description of precipitation events during daylight.

3.2 Energetic electron precipitation

With regard to forcing processes leading to disturbed conditions we focus on electron precipitation in this paper. Electrons with energies $\gtrsim 30$ keV (Fig. 3) and protons with energies $\gtrsim 4$ MeV penetrate down into the lower ionosphere ($\lesssim 85$) km. Proton precipitation are mostly confined to the polar domain, which is not crossed by the propagation path used in this investigation (Iceland – NW Germany). Different approaches to predict the auroral oval were summarized by Sigernes et al. (2011).

With Fang et al. (2008) we model the electron impact ionization rate as

$$q_{precip} = Q_0 \frac{f(E_0, \rho H)}{2 \Delta \epsilon H}, \qquad (8)$$

with Q_0 (keV $cm^{-3} s^{-1}$): total vertically incident electron flux at the top of the atmosphere.

The fraction $\frac{f(E_0, \rho H)}{2 \Delta \epsilon H}$ stands for the ionization efficiency of an electron of (folding) energy E_0 at a specific height and temperature:

ρ (g cm^{-3}): atmospheric mass density, $H = \frac{kT}{Mg}$ the scale height in cm at temperature T and with mean atmospheric molecular mass M, and gravitational acceleration g (to ease the comparison with literature we use keV and cm units here). $\Delta\epsilon = 35 \cdot 10^{-3}$ keV: mean energy loss per ion pair production, E_0 (keV): the folding energy of a Maxwellian electron energy spectrum:

$$N(E) = \frac{Q_0}{2E_0^3} E e^{-E/E_0} \tag{9}$$

$N(E)$ is the differential hemispherical electron number flux (keV cm^{-2} s^{-1}).

The numerical evaluation of the energy deposition function $f(E_0, \rho H)$ is described in detail in Fang et al. (2008).

We model the incident electron flux Q_0 (keV cm^{-2} s^{-1}) as a function of the local auroral activity a, geographic latitude, longitude and universal time UT according to Schmitter (2010):

$$Q_0(a, \text{lat}, \text{lon}, \text{UT}) = c \, a \, w^2 e^{-\frac{w}{2\sigma^2}}, \tag{10}$$

with $c = 7.5 \cdot 10^9$ (keV cm^{-2} s^{-1}). w is a dimensionless distance parameter with regard to the magnetic pole defined as follows: with the projected coordinates $x = r \cos(\text{long})$, $y = r \sin(\text{long})$, $r = r_e \cos(\text{lat})$, $r_e = 6371$ km as components of a vector $\boldsymbol{r}^T = (x, y)$ we define the vector $\boldsymbol{r}_n = R_{\text{UT}}(\boldsymbol{r} - \boldsymbol{r}_{\text{mag pole}})$ which centers the coordinates at the magnetic pole and rotates according to the universal time UT using the usual 2 dimensional rotation matrix with UT expressed as rotation angle. $w = \boldsymbol{r_n}^T \mathbf{M} \, \boldsymbol{r_n}$ is the bilinear form of $\boldsymbol{r}_n$ scaling these coordinates to an elliptic figure with semi axes $a_s = (1590 + 130a)$ km and $b_s = (1270 + 100a)$ km with the scaling matrix $\mathbf{M} = (1/a_s^2 \; 0; 0 \; 1/b_s^2)$. With this procedure the auroral oval is fixed relative to the longitude of the sun with the more intensive part at local midnight. As an example of this type of distribution function see Fig. 4. Our approach for modeling the auroral oval distribution can be compared to that described in Semeniuk et al. (2008). They use poleward and equatorward boundaries to parametrize the auroral oval. Our formulation was inspired by the POES auroral oval data which made use of an auroral activity parameter (this product is replaced by the Ovation–Prime–model, www.swpc.noaa.gov/models since 14 February 2014). In contrast to the POES global (hemispheric) activity our activity parameter quantifies the local precipitation intensity Q_0.

The constant $c = 7.5 \cdot 10^9$ (keV cm^{-2} s^{-1}) is defined such that the elliptic shapes according to the NOAA POES statistical auroral activity maps (www.swpc.noaa.gov/pmap/) are reproduced and by integrating Q_0 over the hemispheric area the total hemispheric power input P at auroral activity index $a = 0, \dots, 10$ results in accordance with the NOAA POES data (www.swpc.noaa.gov/ftpdir/lists/hpi/). It is roughly $P = e^{a/2}$ GW (GigaWatt).

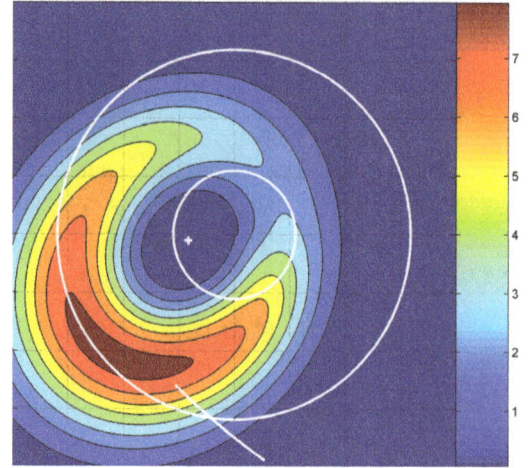

Figure 4. The auroral oval distribution function for 03:00 UT and POES activity level 10. The two rings mark 60 and 80° geographic latitude. The cross left from the North Pole (center of the rings) is the magnetic south pole. Colorbar: erg cm^{-2} s^{-1} = mW m^{-2}. The VLF propagation path from Iceland to NW Germany (2210 km distance) is also marked.

The local auroral activity a at time t is parametrized proportional to the VLF/LF amplitude dip:

$$a(t) = c_{\text{dB}}(\text{amp}_{\text{undisturbed}} - \text{amp}_{\text{disturbed}}(t)) \tag{11}$$

The proportionality constant is measured in auroral activity units per signal amplitude drop (dB).

3.3 Constraining the precipitation distribution function using Swarm FAC data

Because the definition of the electron energy input distribution function Q_0 is a main result of this paper we discuss it in some more detail. For a better understanding we approximate the oval semi axes parameters a_s and b_s by the circular average $\gamma = 1380 + 115a$ km and get

$$Q_0 \cong c \, a \left(\frac{d}{\gamma}\right)^4 e^{-\left(\frac{d}{\gamma}\right)^2 / 2\sigma^2}, \tag{12}$$

where d is the distance from the magnetic pole for the geographic point in question. At at $d_{\max} = 2\gamma\sigma$ the function attains its maximum $Q_0(d_{\max}) = c \, a (2\sigma)^4 / e^2$ ($e = \exp(1)$). The dimensionless width σ is parametrized by $\sigma = \sigma_0 + \sigma_1 a (1 + 0.25 \cos(\phi_{\text{UT}}))$ with $\phi_{\text{UT}} = 0$ at local midnight.

Assuming that electron precipitation is confined to the range of significant field aligned currents we constrain Q_0 using Swarm FAC data, compare Figs. 5 and 6. Figure 5 shows the FAC distribution as recorded by the Swarm A,B,C satellites during their orbits between 0:00 and 4:00 UT on 12 April 2014. Figure 6 shows the FACs along a nearly meridional cut through the auroral oval at 5° W longitude between 50 and 85° N latitude. Displayed in this figure are the results

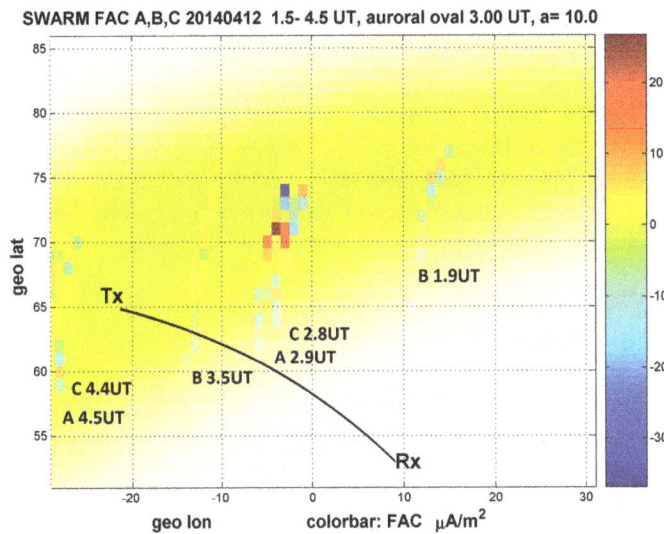

Figure 5. Field aligned currents from Swarm satellites A, B, C on day 12 April 2014 from orbits between 1.9 and 4.5 h together with the VLF propagation path Iceland-NW Germany. Underlying (green) the auroral oval distribution function according to Fig. 4 is shown.

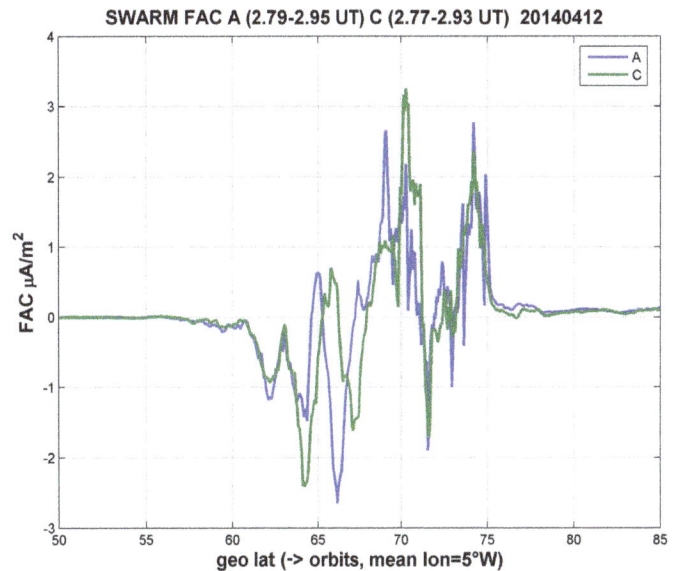

Figure 6. Field aligned currents from Swarm satellites A and C flying side by side, 12 April 2014, 2.8–2.9 h at about 5° W longitude (Atlantic, north of Scotland), 20 s low pass filtered. The data correspond to the (A, C) orbits near the center of Fig. 5. This figure and Fig. 5 are based on Swarm Level 2 data.

from the satellites A and C which are flying side by side and cross the indicated latitudes in about 12 min. Their separation according to the specifications is 1.4° in geographic longitudes corresponding to 78 km at 60° N. A time delay of max. 10 s along their orbit corresponds to max. 70 km distance in latitude. At the time of preparation of this paper the combined FAC data evaluation from both satellites was not available. The currents are significant between 61° N and 76° N at a common longitude of $\cong$ 5° W, which corresponds to a local auroral oval width of 1800 km. On the average FACs are negative in the equatorward half of the auroral oval and positive on the poleward half (by definition the FACs are parallel to the magnetic field lines and point inwards at northern latitudes). In detail however the 20 s low passed data exhibit 5 periods of different FAC domains across the auroral oval, each of a width of about 360 km. With regard to the FAC evaluation method see Ritter et al. (2013).

Now let d_p, d_e be the distances from the magnetic pole of the poleward and equatorward FAC limit points (76° N, 5° W; 61° N, 5° W) for the parallel overflight of Swarm satellites A and C between 2.77 and 2.93 h on the day 12 April 2014. Requiring $Q_0(d_p) = Q_0(d_e) \cong 0$ we numerically find for an activity $a = 10\sigma_0 = 0.2$ and $\sigma_1 = 0.02$ as consistent parameters for the width $\sigma = \sigma_0 + \sigma_1 a (1 + 0.25 \cos(\phi_{UT}))$, UT = 2.85 h. For the resulting shape of the Q_0-function see Figs. 4 and 5. An investigation of other precipitation events is necessary to test the range of validity of the parametrization of the electron input distribution function.

With all the ionization sources and reaction rates defined by the rate equations (Eqs. 3 and 4) can be integrated. The

resulting electron density n_e as function of time, height and position along the radio wave propagation path is the essential input for the propagation calculation. The c_{dB} parameter, Eq. (11), is adjusted to yield the best possible fit of the calculated signal amplitude to the disturbed recorded signal amplitude.

4 Propagation calculations

For frequencies below the plasma frequency $\sqrt{\frac{e^2 n_e}{\epsilon_0 m_e}}$ the space between the conducting Earth ground and the ionosphere behaves like a leaky waveguide. For the VLF/LF range the diffuse upper waveguide boundary is formed by the lower ionosphere (60–90 km height). Propagation calculation in our case is the task to calculate at a ground based receiver position the signal field amplitude and phase of a ground based transmitter (vertical electric and horizontal magnetic field components). Besides transmitter radiated power and antenna characteristics the essential input to the calculation are the conductivities of the waveguide along the great circle propagation path. We use the LWPC (Long Wave Propagation Capability) code (Ferguson, 1998) for this purpose. It includes a world wide map of the ground conductivities. For the ionospheric conductivity ($\sigma = \frac{e^2 n_e}{m_e f_c}$) the collision frequency and electron density height profiles along the path have to be provided. The collision frequency height profile is usually assumed to be constant along the path: $f_c(h) = f_0 e^{-h/H}$, $f_0 = 1.816 \cdot 10^{11}$ Hz, $H = 6.67$ km. It may how-

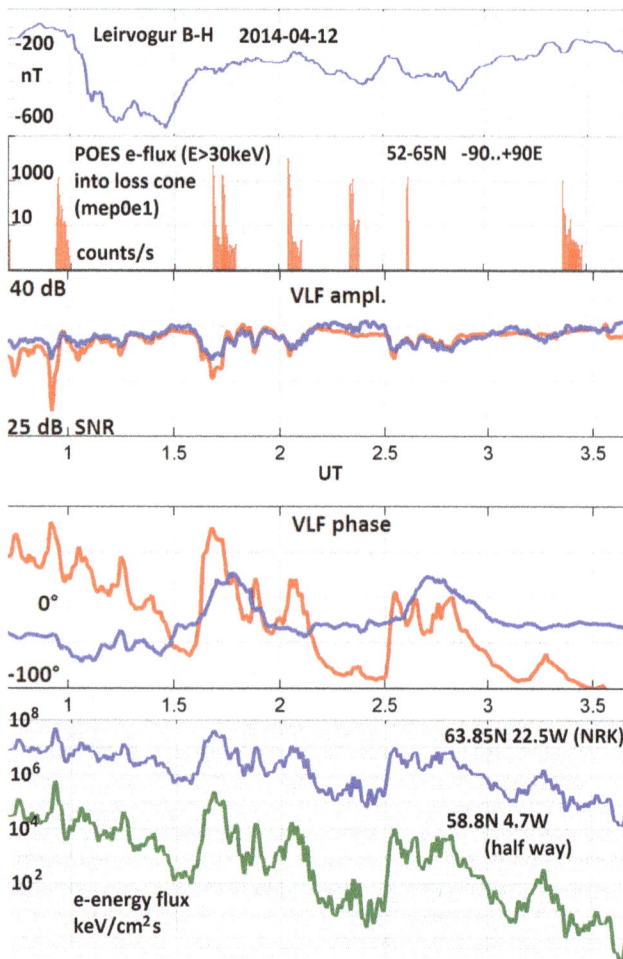

Figure 7. 12 April 2014: From top to bottom: Leirvogur magnetometer data (horizontal magnetic field component); energetic electron precipitation data from the mep0e1-sensors of 5 POES/Metop satellites that had overflights at the time (UT hours) and within the area indicated; recorded (blue) and modeled (red) VLF/LF amplitude and phase data during pulsed electron precipitation along the propagation path; electron kinetic energy input flux Q_0 (Eq. 10) between 0.7 and 3.7 h on 12 April 2014 derived from VLF propagation modeling exemplary for two segments along the propagation path (at the transmitter and half way to the receiver) using a Maxwellian electron energy spectrum with a folding energy of 40 keV. The Leirvogur magnetometer and VLF/LF data display a zoomed part of Fig. 8.

ever be modulated by gravity or planetary waves (Schmitter, 2012). The electron density $n_e(h)$ at height h changes along the path with sun zenith angle, activity of the sun (e.g. solar flares, Schmitter, 2013) and because of local forcing, e.g. particle precipitation.

The propagation path is divided in 50 km segments. Each segment is provided with the proper electron density height profile (integrated from the rate equations, Sect. 3) for the time step and the location in question. Propagation calculations are done in one minute intervals (however the integra-

tion of the rate equations yielding the electron density profiles has to use much smaller step sizes, typically 0.1–2 s).

LWPC then does a full wave calculation resulting in the vertical electric field amplitude and phase at the receiver site, the horizontal magnetic field data being directly proportional. To allow comparisons with our recorded data amplitude results are displayed in dB above the noise level, phases in degree, compare Fig. 7, panels 3 and 4 from top.

5 Results

In Fig. 7 as an example we see the course of the VLF/LF amplitude and phase (blue) during pulsed electron precipitation (panels 3 and 4 from top, zoomed portion of Fig. 8, bottom panel) together with our model calculation (red) where the ionization by energetic electrons is accounted for by a Maxwellian spectrum with 40 keV folding energy with impact ionization parametrization from Fang et al. (2008) and an intensity and space distribution according to the auroral oval model function Q_0, Eq. (10), which accounts for the precipitation intensity distribution in space and time, as described in Sects. 3.2 and 3.3. For comparison the energetic electron ($E > 30$ keV) flux into the loss cone as observed by the low orbit polar orbiters POES15,16,18,19 and Metop02 satellites during their overflights (panel 2 from top, mep0e1 data) and the Leirvogur magnetometer data (top panel) are shown. Within 52–65° N latitude and −90 to +90° E longitude encompassing the propagation path of the POES/Metop satellite data confirm sporadically energetic electron precipitation during the time of the VLF events (there was no enhanced proton level at that time). The magnetometer data confirm a synchronous ionospheric current activity in the vicinity of the NRK transmitter (Leirvogur is situated about 50 km NW of the NRK transmitter). In the non zoomed Fig. 8, top panel, the difference to the undisturbed situation is more clearly to be seen.

There are two parameters in the model that are fitted to the data: the proportionality constant c_{dB} (auroral activity per signal amplitude drop (dB)) yielding the activity parameter $a(t) = c_{dB}(amp_{undisturbed} - amp_{disturbed}(t))$, and the folding energy E_0 of the electron energy spectrum (40 keV in our example case). c_{dB} has been adjusted for best fit of the modeled signal amplitude to the disturbed recorded signal amplitude. Changing the folding energy by more than −5 or +10 keV degenerates the fit quality significantly.

Level 2 data from the Swarm mission are used to identify the spatial extension of field aligned current densities (FACs). With this information the precipitation boundaries are assessed and the electron input distribution function Q_0 has been constrained.

The FAC constrained spatial distribution of the precipitation along the propagation path together with the VLF/LF amplitude variations yields $c_{dB} = 0.2$ (auroral activity units per dB signal amplitude drop) at 37.5 kHz and supports the

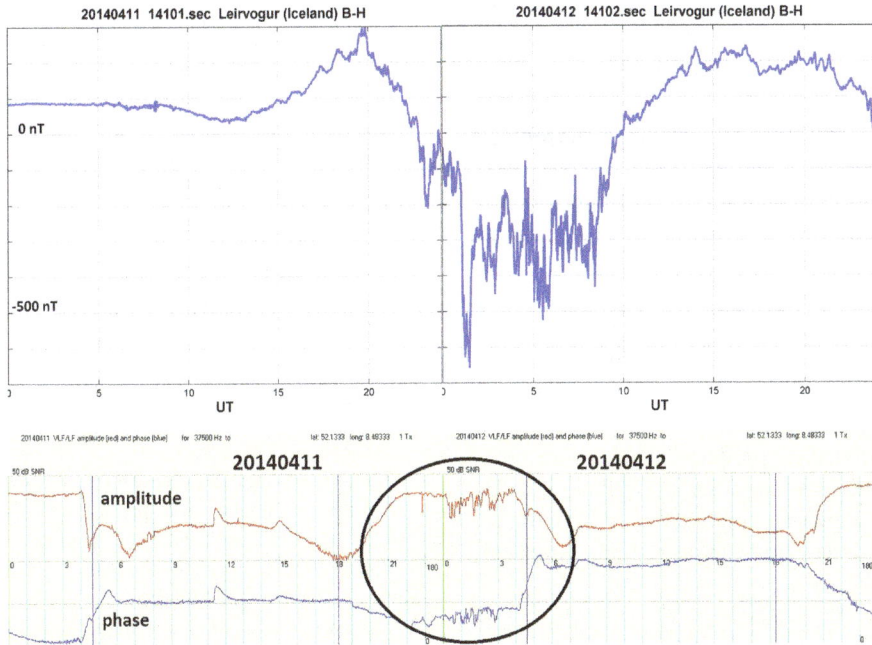

Figure 8. Days 11 and 12 of April 2014: effect of a series of energetic electron precipitation bursts with regard to VLF/LF signals. Lower panel: VLF amplitude (dB SNR, red) and phase (deg, blue). The top panel displays the LEIRV magnetometer response reflecting ionospheric current fluctuations during the substorm condition starting in the evening hours of 11 April 2014. (LEIRV: Leirvogur, Iceland, 50 km NW of the NRK transmitter). The humps in the VLF/LF signal during daytime of 11 April 2014 are solar flare responses. Sunrise and sunset are marked by vertical lines.

assumption of an $E_0 = 40$ keV Maxwellian electron spectrum in the model calculations. c_{dB} is expected to be constant in our modeling scheme, so that after calibration an application with regard to other electron precipitation events can concentrate on the determination of E_0, which can be a function of time.

As the main modeling result the energy input from electron precipitation as a function of time is available for each of the 50 km segments along the VLF propagation path (Q_0 distribution function). For two segments (near the transmitter and half ways along the propagation path to the receiver) the electron input energy flux results are shown in Fig. 7, bottom panel. The maximum energy flux rate at the transmitter coordinates 64° N 22.5° W is $5 \cdot 10^7$ keV cm^{-2} s^{-1} = 0.08 mW m^{-2}. Half way to the receiver coordinates (58.8° N, 4.7° W) the flux is reduced by two orders of magnitude. The ionization rate vs. height characteristics for the reported fluxes and $E_0 = 40$ keV can be read from Fig. 3 (maximum: $8 \cdot 10^8$ m^{-3} s^{-1} at 78 km height).

The variation of the Q_0 function (Fig. 7, bottom panel) reflects the course of the disturbed VLF/LF signal amplitude because we define the activity parameter $a(t)$ (which controls the electron input) proportional to the signal amplitude, Eq. (11). The constant c_{dB} is chosen for a good fit of recorded and modeled VLF/LF signal amplitude data. The modeled phase also has to reflect the $a(t)$ course (here in an inverted way) because it is an input into the propagation calculation,

however there is only little correlation to the recorded phase. This can be understood if we take into account that signal phases in contrast to amplitudes are sensitive to all kinds of effects along the propagation path in a cumulative way.

With regard to the spatial resolution of the derived flux values we recall that the main part of the electromagnetic energy that is transferred from a transmitter to a receiver is contained in the first Fresnel zone, an ellipsoid encompassing the propagation path with an extension orthogonal to the propagation path of the order of one wavelength ($\lambda = 8$ km for 37.5 kHz) near the transmitter and the receiver and a maximum extension of $0.5\sqrt{\lambda d} = 66$ km at half way (Tx–Rx distance $d = 2210$ km). So with a segment length of 50 km along the propagation path the spatial resolution of our VLF/LF remote sensing model, i.e. the averaging area, varies between $8 \cdot 50$ km^2 near to the transmitter and near to the receiver and $66 \cdot 50$ km^2 at half way.

6 Conclusions

The main objective of the work described in this paper is to advance our VLF/LF propagation model by including the effect of electron particle precipitation. The extended model allows for the characterization of Maxwellian electron energy spectra along the VLF/LF propagation path by using data from continuous VLF/LF remote sensing and field aligned current density data from the new Swarm mission. The current status of the modeling efforts has been demonstrated with regard to electron precipitation bursts during a moderate substorm condition on 12 April 2014 (Kp = 5, min. DST = −80 nT, min. AL = −700 nT). Signal amplitude disturbances along a 37.5 kHz radio propagation path from Iceland (63.9° N, 22.5° W) to a midlatitude site (52° N, 8° E) have been modeled successfully and an electron energy input distribution function has been derived. Future work intends to model many different precipitation events and to further validate the parametrization of the electron input distribution function. For this purpose our software has to be optimized with regard to speed to allow for faster modeling cycles. The space time characterization of electron energy spectra can help to quantify the effectiveness of energetic electron precipitation with regard to NO_x production and ozone depletion at mesospheric heights, underlining the importance of energetic electron precipitation as a part of solar influence on the atmosphere and the climate system (Andersson et al., 2014).

Acknowledgements. Swarm Level 2 data have been provided by the European Space Agency and the POES/Metop data have been provided by the US NOAA National Geophysical Data Center.

References

Andersson, M. E., Verronen, P. T., Rodger, C. J., Clilverd, M. A., and Seppaelae, A.: Missing driver in the SunEarth connection from energetic electron precipitation impacts mesospheric ozone, Nat. Comm., 5, 5197, doi:10.1038/ncomms6197, 2014.

Barth, C. A., Baker, D. N., Mankoff, K. D., and Bailey, S. M.: The northern auroral region as observed in nitric oxide, Geophys. Res. Lett., 8, 1463–1466, doi:10.1029/2000GL012649, 2001.

Brekke, A.: Physics of the upper polar atmosphere, John Wiley & Sons, New York, USA, 1997.

Carson, B. R., Rodger, C. J., and Clilverd, M. A.: POES satellite observations of EMIC-wave driven relativistic electron precipitation during 1998–2010, J. Geophys. Res.-Space, 118, 232–243, doi:10.1029/2012JA017998, 2012.

Clilverd, M. A., Seppaelae, A., Rodger, C. J., Thomson, N. R., Lichtenberger, J., and Steinbach, P.: Temporal variability of the descent of high-altitude NOX inferred from ionospheric data, J. Geophys. Res., 112, A09307, doi:10.1029/2006JA012085, 2007.

Cummer, S. A., Bell, T. F., and Inan, U. S.: VLF remote sensing of the auroral electrojet, J. Geophys. Res., 101, 5381–5389, 1996.

Cummer, S. A., Inan, U. S., and Bell, T. F.: Ionospheric D region remote sensing using VLF radio atmospherics, Radio Sci., 33, 1781–1792, doi:10.1029/98RS02381, 1998.

Fang, X., Randal, C. E., Lummerzheim, I. D., Solomon, S. C., Mills, M. J., Marsh, D. R., Jackman, C. H., Wang, W., and Lu, G.: Electron impact ionization: A new parameterization for 100 eV to 1 MeV electrons, J. Geophys. Res., 113, A09311, doi:10.1029/2008JA013384, 2008.

Ferguson, J. A.: Computer Programs for Assessments of Long-Wavelength Radio Communications, Version 2.0, Technical Document, SPAWAR Systems Center, San Diego, USA, available at: http://www.dtic.mil/dtic/tr/fulltext/u2/a350375.pdf (last access: 12 May 2015), 1998.

Glukhov, V. S., Pasko, V. P., and Inan, U. S.: Relaxation of transient lower ionospheric disturbances caused by lightning-whistler-induced electron precipitation bursts, J. Geophys. Res., 97, 16971–16979, 1992.

Mitra, A. P. and Rowe, J. N.: Ionospheric effects of solar flares – VI. Changes in D-region ion chemistry during solar flares, J. Atmos. Sol.-Terr. Phy., 34, 795–806, 1972.

Mitra, A. P.: D-region in disturbed conditions, including flares and energetic particles, J. Atmos. Sol.-Terr. Phy., 37, 895–913, 1975.

Nicolet, M. and Aikin, A. C.: The Formation of the D Region of the Ionosphere, J. Geophys. Res., 65, 1469–1483, 1960.

Ohtani, S., Wing, S., Newell, P. T., and Higuchi, T.: Locations of nights' side precipitation boundaries relative to R2 and R1 currents, J. Geophys. Res., 115, A10233, doi:10.1029/2010JA015444, 2010.

Ritter, P., Lühr, H., and Rauberg, J.: Determining field-aligned currents with the Swarm constellation mission, Earth Planets Space, 65, 1285–1294, 2013.

Rodger, C. J., Molchanov, O. A., and Thomson, N. R.: Relaxation of transient ionization in the lower ionosphere, J. Geophys. Res., 103, 6969–6975, 1998.

Rodger, C. R., Clilverd, M. A., Thomson, N. R., Gamble, R. J., Seppaelae, A., Turunen, E., Meredith, N. P., Parrot, M., Sauvaud, J. A., Berthelier, J. J.: Radiation belt electron precipitation into the atmosphere: Recovery from a geomagnetic storm, J. Geophys. Res., 112, A11307, doi:10.1029/2007JA012383, 2007.

Saetre, C., Stadsnes, J., Nesse, H., Aksnes, A., Petrinec, S. M., Barth, C. A., Baker, D. N., Vondrak, R. R., and Ostgaard, N.: Energetic electron precipitation and the NO abundance in the upper atmosphere: A direct comparison during a geomagnetic storm, J. Geophys. Res., 109, A09302, doi:10.1029/2004JA010485, 2004.

Salmi, S.-M., Verronen, P. T., Thölix, L., Kyrölä, E., Backman, L., Karpechko, A. Yu., and Seppälä, A.: Mesosphere-to-stratosphere descent of odd nitrogen in February-March 2009 after sudden stratospheric warming, Atmos. Chem. Phys., 11, 4645–4655, doi:10.5194/acp-11-4645-2011, 2011.

Semeniuk, K., McConnell, J. C. , Jin, J. J. Jarosz, J. R., Boone, C. D., and Bernath, P. F.: N_2O production by high energy auroral electron precipitation, J. Geophys. Res., 113, D16302, doi:10.1029/2007JD009690, 2008.

Schmitter, E. D.: Remote auroral activity detection and modeling using low frequency transmitter signal reception at a midlatitude site, Ann. Geophys., 28, 1807–1811, doi:10.5194/angeo-28-1807-2010, 2010.

Schmitter, E. D.: Remote sensing planetary waves in the midlatitude mesosphere using low frequency transmitter signals, Ann. Geophys., 29, 1287–1293, doi:10.5194/angeo-29-1287-2011, 2011.

Schmitter, E. D.: Data analysis of low frequency transmitter signals received at a midlatitude site with regard to planetary wave activity, Adv. Radio Sci., 10, 279–284, doi:10.5194/ars-10-279-2012, 2012.

Schmitter, E. D.: Modeling solar flare induced lower ionosphere changes using VLF/LF transmitter amplitude and phase observations at a midlatitude site, Ann. Geophys., 31, 765–773, doi:10.5194/angeo-31-765-2013, 2013.

Schmitter, E. D.: Remote sensing and modeling of lightning caused long recovery events within the lower ionosphere using VLF/LF radio wave propagation, Adv. Radio Sci., 12, 241-250, doi:10.5194/ars-12-241-2014, 2014.

Sigernes, F., Dyrland, M., Brekke, P., Chernouss, S., Lorentzen, D. A., Oksavik, K., and Deehr, C. S.: Two methods to forecast auroral displays, J. Space Weather, 1, A03, doi:10.1051/swsc/2011003, 2011.

Thorne, R. M.: A cause of dayside relativistic electron possible precipitation events, J. Atmos. Sol.-Terr. Phy., 36, 635–645, doi:10.1016/0021-9169(74)90087-7, 1974.

Thorne, R. M., Li, W., Ni, B., Ma, Q., Bortnik, J., Chen, L., Baker, D. N., Spence, H. E., Reeves, G. D., Henderson, M. G., Kletzing, C. A., Kurth, W. S., Hospodarsky, G. B., Blake, J. B., Fennell, J. F., Claudepierre, S. G., and Kanekal, S. G.: Rapid local acceleration of relativistic radiation-belt electrons by magnetospheric chorus, Nature, 504, 411–414, doi:10.1038/nature12889, 2013.

Torkar, K. and Friedrich, M.: Tests of an Ion-Chemical Model of the D- and Lower E-Region, J. Atmos. Terr. Phys., 45, 369–385, 1983.

Verronen, P. T., Turunen, E., Ulich, T., and Kyrola, E.: Modelling the effects of the October 1989 solar proton event on mesospheric odd nitrogen using a detailed ion and neutral chemistry model, Ann. Geophys., 20, 1967–1976, 2002, http://www.ann-geophys.net/20/1967/2002/.

Wait, J. R. and Spies, K. P.: Characteristics of the Earth-ionosphere waveguide for VLF radio waves, NBS Tech. Note 300, available at: http://nova.stanford.edu/~vlf/IHY_Test/Tutorials/SfericsAndTweaks/Papers/Wait1964f.pdf (last access: 12 May 2015), 1964.

28

Multi-view point cloud fusion for LiDAR based cooperative environment detection

B. Jaehn, P. Lindner, and G. Wanielik

Professorship of Communications Engineering, Chemnitz University of Technology, Chemnitz, Germany

Correspondence to: B. Jaehn (bjae@hrz.tu-chemnitz.de)

Abstract. A key component for automated driving is 360° environment detection. The recognition capabilities of modern sensors are always limited to their direct field of view. In urban areas a lot of objects occlude important areas of interest. The information captured by another sensor from another perspective could solve such occluded situations. Furthermore, the capabilities to detect and classify various objects in the surrounding can be improved by taking multiple views into account.

In order to combine the data of two sensors into one coordinate system, a rigid transformation matrix has to be derived. The accuracy of modern e.g. satellite based relative pose estimation systems is not sufficient to guarantee a suitable alignment. Therefore, a registration based approach is used in this work which aligns the captured environment data of two sensors from different positions. Thus their relative pose estimation obtained by traditional methods is improved and the data can be fused.

To support this we present an approach which utilizes the uncertainty information of modern tracking systems to determine the possible field of view of the other sensor. Furthermore, it is estimated which parts of the captured data is directly visible to both, taking occlusion and shadowing effects into account. Afterwards a registration method, based on the iterative closest point (ICP) algorithm, is applied to that data in order to get an accurate alignment.

The contribution of the presented approch to the achievable accuracy is shown with the help of ground truth data from a LiDAR simulation within a 3-D crossroad model. Results show that a two dimensional position and heading estimation is sufficient to initialize a successful 3-D registration process. Furthermore it is shown which initial spatial alignment is necessary to obtain suitable registration results.

1 Introduction

Humans were able to use indirect signals over mirrors or through the windows of other vehicles to observe relevant areas e.g. at difficult crossroad situations. To achieve a similar understanding of the surrounding with depth sensors, one important aspect for future autonomous systems will be the cooperative exchange of environment information denoted as car to car (C2C) communication. Together with the use of real time cloud or local infrastructure based services this will improve the recognition and reaction capabilities towards objects located outside the direct field-of-view (FoV) of the own sensors.

E.g. the green car of Fig. 1 could provide useful information about the blue truck for the red car. The foundation of such a dense data fusion is the availability of an highly accurate relative pose estimation which can be transferred into a homogeneous transformation matrix. This is hard to derive by traditional satellite based relative localization methods but their estimate could be improved using the gathered environment data of both cars. A method for achieving this is the content of this paper which is mainly based on the master thesis by Jähn (2014). For the rest of this paper we will refer to the red car as target and to the green car as source and we will use the coordinate definitions of the right picture from Fig. 1.

To support the understanding of the reader the present paper is organized as follows: Sect. 2 provides relevant background information's which are necessary to ensure comprehensibility. Afterwards the general outline and the novel ideas of the presented method are shown in Sect. 3. An evaluation of the presented approach is then given in Sect. 4. The conclusion in Sect. 5 contains a summary of all findings, it's limitations as well as improvements and future work.

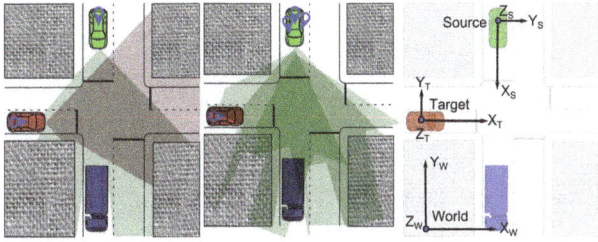

Figure 1. The figure shows two cars which could exchange their environment information if an accurate enough pose estimation would be available (left picture). If the pose estimation is inaccurate this would yield an alignment error (center picture). The right picture shows the coordinate system definitions which where used in the paper.

2 Background

2.1 Depth sensors

Typical depth sensors, like LiDARs and stereo cameras, measures the surface of the surrounding environment. Using the projective geometry their range images can be converted into a point cloud which represents the surface shape of the surrounding. During this conversion the pixel wise neighborhood information is maintained which is useful for further processing steps like normal feature estimation. In the case that two of these sensors measures the same surface area from different perspectives, this data can be used to align them against each other if their relative pose estimation is inaccurate. The process of doing this is called registration or shape matching which is part of Sect. 2.3.

2.2 Normal feature estimation

To gain information about the underlying surface, which the point cloud data represents, it is very common in 3-D data processing to determine normal features[1]. We apply here the standard weighted normal averaging scheme which uses triangle combinations incorporating the neighbourhood points (Klasing et al., 2009). In this work we use a weightning factor w_j which is chosen as the reciprocal product of the distance between the neighbours $q_{i,j}$ of p_i according to Eq. (1). We refer to this method as distance weighted cross product (DWCP).

$$w_j = \frac{1}{|q_{i,j} - p_i| \cdot |q_{i,j+1} - p_i|} \tag{1}$$

2.3 Registration of point clouds

Registration, or shape matching, is the process to align two or more surfaces against each other in such a way that a

mathematical metric is minimized. Although other registration methods exists (compare Pottmann et al., 2002) the most popular one is the so called ICP algorithm introduced by Besl and McKay (1992).

The general idea is to find at first point correspondences between two slightly misaligned data sets. This is done usually by searching the currently closest point in the other data set. Secondly a locally optimal rigid transformation matrix is calculated which minimizes a certain distance metric in order to align them against each other. Using the result from the previous iteration this procedure is repeated. New correspondences where determined and the alignment is optimized again until some stopping criteria is fulfilled.

The original formulation of the algorithm by Besl and McKay (1992) delivers bad results if the incorporated data sets does include outliers[2]. This happens especially if the two surfaces do not overlap completely or if there are holes due to occlusion effects. As a consequence a still growing number of different variations has been proposed. The main goals are to speed up the convergence rate and improve the robustness against outliers. Therefore, several heuristics and modifications are applied and additional features, like normals, are incorporated. A comparison study of several methods can be found at Pomerleau et al. (2013).

In this work we utilize a point to plane (Chen and Medioni, 1991) based ICP method which utilizes additional rejection methods (Rusinkiewicz and Levoy, 2001). First all correspondence where the angle between the corresponding normals exceeds 30° are ignored and secondly we remove the worst 10 % due to the Euclidean distance (Pulli, 1999).

3 Method

The primary goal of this work is to include the captured point cloud data of a sensor platform, like the green car (compare Fig. 1), into the coordinate system of another, e.g. the red car. Each of them has its own coordinate system indicated with the superscript prefix S and T respectively. Thus, an arbitrary point $^S p = [x_S, y_S, z_S]^T$ in the source coordinate system can be transformed in the target coordinate system, using $\mathbf{T}^{TS}$. It follows that the corresponding point $^T p = [x_T, y_T, z_T]^T$ from $^S p$ can be calculated with the following equation.

$$\begin{bmatrix} ^T p \\ 1 \end{bmatrix} = \mathbf{T}^{TS} \cdot \begin{bmatrix} ^S p \\ 1 \end{bmatrix} \tag{2}$$

Unfortunately, the exact transformation matrix $\mathbf{T}^{TS}$ is in general not known and has to be estimated by $\hat{\mathbf{T}}^{TS}$. In order to achieve an initial guess of this matrix, it is further assumed that the target platform estimates the relative position and orientation of the source with a suitable, e.g. satellite based, relative localization system which delivers a state

---

[1] Vectors assigned to each point in space representing a local plane patch of the approximated surface

[2] E.g. points which do not have a corresponding partner representing the same physical object surface within the other data set.

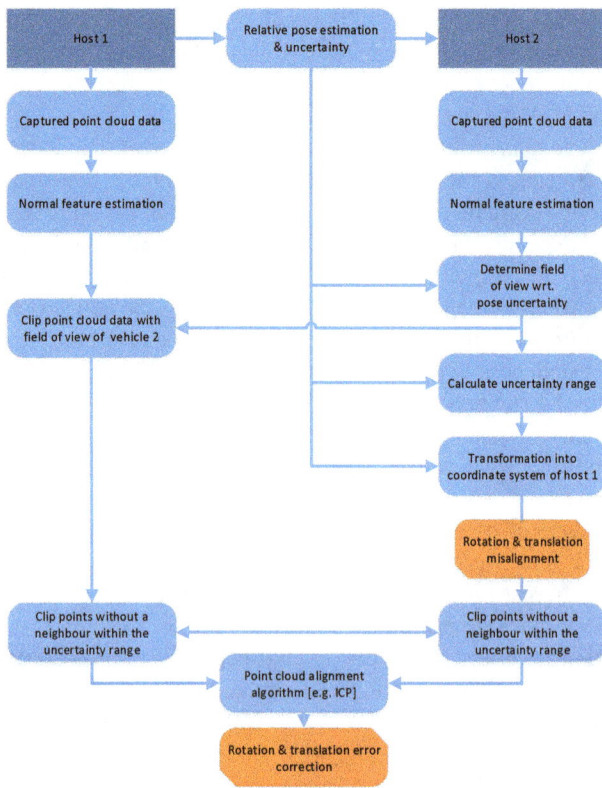

Figure 2. Overview of the processing steps within the presented approach.

$x = \begin{bmatrix} x, y, z, \phi, \theta, \psi \end{bmatrix}^T$ and its uncertainty, as covariance matrix Σ_{Cov}. The accuracy of this transformation can be improved, using a registration algorithm with environment information captured by both sensor platforms simultaneously.

To achieve this the environment data, which was captured by both sensor platforms simultaneously, is determined. Therefore the state and the corresponding uncertainty information, namely the standard deviation values $\text{Diag}(\Sigma_{\text{Cov}}) = \begin{bmatrix} \sigma_x^2, \sigma_y^2, \sigma_z^2, \sigma_\phi^2, \sigma_\theta^2, \sigma_\psi^2 \end{bmatrix}$, were used to estimate the theoretical field of view (FoV) of the source sensor within the targets coordinate system in order to determine the data within the overlapping FoV. Further on occlusion effects were resolved to achieve the relevant points for the registration process within the target point cloud (denoted in the following as **T**) and the source point cloud (**S**).

Once an accurate transformation is determined it can be applied to the whole source point cloud. Both data sets are now fused within the target coordinate systems and available for further processing. An overview of the previously explained steps is given in Fig. 2 and some steps are explained in detail in the following.

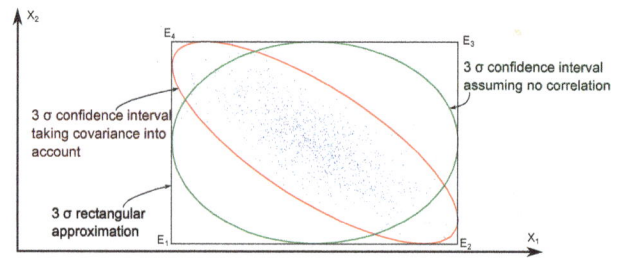

Figure 3. Within this chart three different 3σ ($\gamma = 3$) confidence intervals of the same two dimensional correlated normal distributed sample set are shown. The red one uses the full covariance matrix whereas the green one just neglects the covariance values. The rectangular confidence interval includes both ellipses and thus acts as an upper bound.

3.1 Overlapping FoV estimation

The FoV is the geometrical area, which is theoretically visible for a sensor, without taking occlusion effects into account. For the most typical range sensors, like automotive LiDAR, this can be approximated as a pyramid. All sensor data have to lie within this field, thus it can be used as a rough geometrical boundary. Such a pyramid can be defined for the source and target sensor. In order to determine the overlapping part it is necessary to transform them into the same coordinate system.

3.1.1 Uncertainty field of view

Problematic in this context is the accuracy of $\hat{\mathbf{T}}^{TS}$. The exact origin and orientation of the source FoV is thus not known. To overcome this problem the track uncertainty is used to increase the source FoV such that it includes the error space, which is considered as normal distributed, up to a certain confidence interval. How this can be achieved efficiently is dependent on its original shape. In this work a pyramid model is used and therefore the basic idea to derive a proper uncertainty field approximation is described in the following. The full derivation can be found at Jähn (2014).

As mentioned already the uncertainty space is assumed to be normal distributed with an expectation value of **0**. Consequently a surrounding confidence interval $\gamma \cdot \sigma$ would result in a six-dimensional elliptical error space. If this is used to transform the pyramid model into the uncertainty space this would result in an extremely complex mathematical shape. Therefore just the diagonal values of Σ_{Cov} are used to approximated this with a rectangular upper boundary space. With this simplification it doesn't matter if one would take cross correlation values into account or not. To support the understanding of the reader a two-dimensional example is shown in Fig. 3.

Using this rectification, the confidence interval can be described just with the maximum divergence from the state space in positive and negative direction for each dimension.

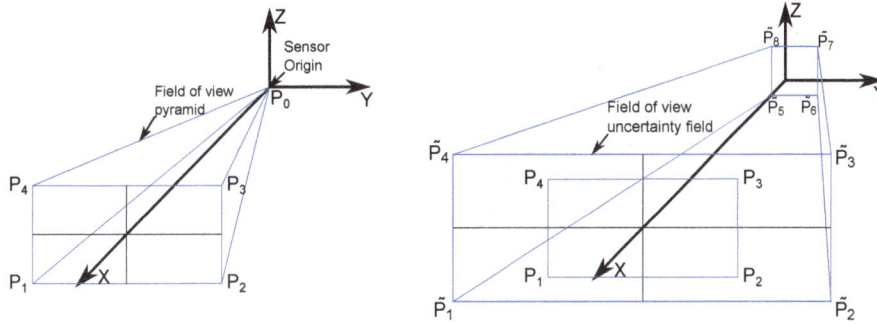

Figure 4. The two pictures show the coordinate system definitions and the nomenclature of the original field of view (left) and the increased uncertainty field of view (right).

Consequently there are $2^6 = 64$ different maximum error variations. If all of these error variations are applied to the FoV pyramid (compare Fig. 4, left panel) the resulting shape can be approximated by a frustum pyramid as it is shown in Fig. 4 (right panel).

This approximation of the uncertainty FoV can be used to clip the points of the target sensor which could have no corresponding partner within the source data set.

3.1.2 Point uncertainty ranges

In this section it is estimated which parts of both point clouds could be visible to both sensors taking occlusion effects into account. E.g. if an object is located within both fields of view but it is occluded through another object from just one perspective. To figure this out the uncertainty of the initial transformation matrix $\hat{\mathbf{T}}^{TS}$ have to be taken into account again. Therefore the same assumptions, namely the rectification of the uncertainty space, from the previous section are applied. This time the uncertainty space is determined for each point of the source point cloud separately and the resulting space is approximated by a sphere around the point $^S p_i \in \mathbf{S}$ with a radius r_i according to Eq. (3).

$$r_i = \max\left\{r_{\min}, \max\left\{r_{i,j} \forall j\right\} + \|t_{\mathrm{err}}\|_\infty\right\} \quad (3)$$

$$j \in \{1, 2, \ldots, |\Sigma_{\mathrm{Rot}}|\} \quad t_{\mathrm{err}} \in \Sigma_{\mathrm{Transl}} \quad (4)$$

$$r_{i,j} = |\boldsymbol{p}_{S,i} - \mathrm{Rot}\{\boldsymbol{r}_j\} \cdot \boldsymbol{p}_{S,i}| \quad \boldsymbol{r}_j \in \Sigma_{\mathrm{Rot}} \quad (5)$$

$$\Sigma_{\mathrm{Transl}} = \{\pm\sigma_x\} \times \{\pm\sigma_y\} \times \{\pm\sigma_z\} \quad (6)$$

$$\Sigma_{\mathrm{Rot}} = \{\pm\sigma_\phi\} \times \{\pm\sigma_\theta\} \times \{\pm\sigma_\psi\} \quad (7)$$

Transferred into the target coordinate system using $\hat{\mathbf{T}}^{TS}$ these spheres represents the area where $\begin{bmatrix} \boldsymbol{p}_{T,i} \\ 1 \end{bmatrix} = \mathbf{T}^{TS} \cdot$

$\begin{bmatrix} \boldsymbol{p}_{S,i} \\ 1 \end{bmatrix}$ could be if $\mathbf{T}^{TS}$ would be known exactly. Further it represents the area where corresponding points of $\mathbf{T}$ could be which were captured from the same object. Thus each source point, which has no target point within its range defined by r_i, is clipped. The other way around each target point is clipped

if it is not within at least one of the source uncertainty spheres (compare Fig. 5).

4 Evaluation

The objective of the following evaluation is to show the applicability of the presented approach for cooperative environment recognition applications. Compared to the master thesis (Jähn, 2014), where this paper is based on, the data which was used for the evaluation was re-evaluated to fit the limited scope of this paper. For a more exhaustive evaluation please refer to Jähn (2014).

4.1 Evaluation strategy

To show the effects of different steps of the presented approach a detailed cross road 3-D model was built with SketchUp®[3]. This was used to simulate range data by two virtual LiDAR depth sensors, which were moved through the static model such that they overlap partially and enough non-parallel surface details for registration were included. The recorded data includes ground truth range data as well as position and orientation of both sensors. This was then superimposed by range and pose (position and orientation) noise. These point clouds were afterwards aligned and the output was compared with the applied pose error vector. This procedure was than repeated 21 times with different data set combinations each with 100 different range and tracking noise samples. Thus the quality of the registration process can be statistically appraised.

4.1.1 Range data generation

The range data was captured using a simple ray tracing approach. Each laser beam of the virtual LiDAR has been modelled as a single ray. In contrast to the simulation applied by Gabriel (2010) the complete sensor optic simulation and the 3 dimensional beam structure was not taken into account.

---

[3]A 3-D modeling software by Trimble Navigation

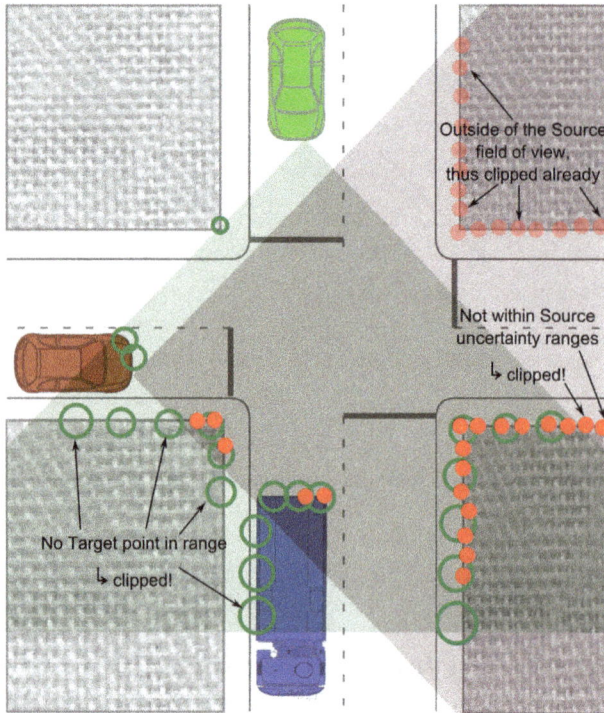

Figure 5. To each source point an uncertainty range is assigned. Within each sphere at least one target point has to be present and each target point have to lie within at least on sphere.

Table 1. Tracking noise setups.

	$\sigma_\phi, \sigma_\theta, \sigma_\psi$	$\sigma_x, \sigma_y, \sigma_z$
TN1	1°	30 cm
TN2	2°	60 cm
TN3	3°	120 cm
TN4	4°	240 cm

Table 2. Percentiles from $\mathcal{N}(0, \sigma)$ intervals for direct comparison with non-normal distributed error spaces.

$\mathcal{N}(0, \sigma)$		Percentile	Name
$\|x\| < 0.5\sigma$	$\hat{=}$	38.29 %	p_{38}
$\|x\| < 1.0\sigma$	$\hat{=}$	68.27 %	p_{68}
$\|x\| < 1.5\sigma$	$\hat{=}$	86.64 %	p_{86}
$\|x\| < 2.0\sigma$	$\hat{=}$	95.45 %	p_{95}
$\|x\| < 2.5\sigma$	$\hat{=}$	98.76 %	p_{98}

4.1.2 Analysis of the results

Within each test cycle the resulting alignment vector x_{Alg} is subtracted from the previously applied tracking noise vector x_{Err} which results in the remaining error vector x_{Rem} (compare Eq. 9). Over all, in total 2100, test cycles the expectation value of this is assumed to be $\mathbf{0}$. To allow an easy comparison between the input error space, which is normal distributed, and the output error space, which is in general not normal distributed, the equivalent percentiles are used according to Table 2. These percentiles are calculated for each component of x_{Rem}. If the resulting value of the percentile is smaller than the equivalent interval from the input error space normal distribution the alignment was improved during the registration.

$$x_{\text{Err}} - x_{\text{Alg}} = x_{\text{Rem}} \qquad (9)$$

4.2 Contribution of clipping & confidence interval

To figure out the contribution of the FoV and uncertainty range clipping from Sects. 3.1 and 3.1.2 the influence of the applied confidence interval will be investigated now. Therefore the confidence interval parameter γ is modified and an additional test was done where both clipping methods were deactivated.

According to Table 3 the smallest remaining error is for $\gamma = 1.75$. With $\gamma = 1.75$ and without clipping at all the error increases. This effect decreases with higher tracking noise due to the huge uncertainty ranges where occlusion and shadowing effects cannot be detected. Obviously the rejection step works very efficiently such that bad correspondences are detected reliably. Consequently this pre-processing does affect the alignment accuracy just slightly but noticeable. Thus we conclude that the FoV and uncertainty range clipping stabilize the registration process. Furthermore it has an influence on the number of points, which have to be aligned and

This approach allows a much faster simulation but it does not reflect the special properties of LiDAR sensors, e.g. just the first echo is used. Anyhow this simple sensor model can be applied also to other range sensors, like stereo and time of flight cameras, hence it is used here. For the evaluation in this work we used a LiDAR model with 40 layers each with 451 beams covering 72° horizontally and 10° vertically.

The range image taken directly from the simulation is noiseless and therefore not very realistic. That's why it is superimposed by Gaussian noise. According to the manual (Ibeo Automotive Systems GmbH, 2008) of the Ibeo LUX laser scanner family, a typical standard deviation value σ_r is 10 cm, which is used here.

To simulate tracking noise n random error sample from a normal distributed error space is superimposed to x according to Eq. (8). The uncertainty standard deviation values $\sigma_x, \sigma_y, \sigma_z, \sigma_\phi, \sigma_\theta, \sigma_\psi$ were taken from Table 1 to simulate different tracking conditions and quality. As described in Sect. 3.1 this will influence how the overlapping area of the sensor data is estimated.

$$x = x + x_{\text{Err}} \quad x_{\text{Err}} \in \mathcal{N}\left(0, \left[\sigma_x, \sigma_y, \sigma_z, \sigma_\phi, \sigma_\theta, \sigma_\psi\right]^T\right) \qquad (8)$$

Table 3. The table shows the results of the remaining error percentiles for the point to plane metric with different clipping confidence intervals γ. The light green colour indicates nearly perfect alignment results whereas dark red stands for unsuitable results above $1°$ or $20\,cm$. The achieved accuracy and the necessary number of iterations is quite similar in all cases. Especially for higher tracking noise levels where the uncertainty ranges were huge.

		Point to Plane [DWCP] @ γ = 1,75							Point to Plane [DWCP] @ γ = 2,5							Point to Plane [DWCP] without Clipping						
		φ [°]	θ [°]	ψ [°]	x [cm]	y [cm]	z [cm]	Iter	φ [°]	θ [°]	ψ [°]	x [cm]	y [cm]	z [cm]	Iter	φ [°]	θ [°]	ψ [°]	x [cm]	y [cm]	z [cm]	Iter
TN1	p_{38}	0,03	0,02	0,04	2,34	1,41	0,82	4	0,02	0,03	0,05	2,87	1,79	0,81	5	0,05	0,04	0,09	4,47	3,03	1,14	5
	p_{68}	0,06	0,08	0,10	5,35	3,49	1,51	5	0,05	0,10	0,12	7,07	4,47	1,50	6	0,09	0,13	0,23	10,66	7,57	2,15	6
	p_{86}	0,10	0,12	0,21	11,04	6,53	2,57	6	0,10	0,15	0,28	14,52	9,15	2,44	7	0,19	0,20	0,41	21,27	13,48	3,53	7
	p_{95}	0,13	0,15	0,32	26,89	10,15	3,41	7	0,15	0,17	0,52	25,44	16,16	3,42	9	0,47	0,27	0,70	37,67	20,86	6,88	9
	p_{98}	0,16	0,16	0,42	41,16	13,65	4,97	8	0,19	0,19	0,68	41,35	22,33	4,75	11	0,57	0,44	1,46	54,86	36,95	9,60	12
TN2	p_{38}	0,04	0,03	0,06	3,52	2,32	1,07	5	0,04	0,04	0,07	4,34	3,00	1,08	6	0,05	0,04	0,08	4,32	2,99	1,20	6
	p_{68}	0,08	0,11	0,16	8,62	5,84	2,01	6	0,07	0,12	0,18	10,17	7,20	2,04	8	0,09	0,13	0,22	10,71	8,12	2,24	7
	p_{86}	0,11	0,17	0,34	16,38	11,47	3,29	7	0,12	0,17	0,42	20,66	14,65	3,57	10	0,19	0,20	0,43	22,80	14,24	3,68	9
	p_{95}	0,21	0,21	0,59	34,52	18,33	5,27	9	0,22	0,22	0,91	42,03	28,96	5,74	13	0,46	0,27	0,87	40,93	23,80	7,41	11
	p_{98}	0,25	0,22	0,87	47,67	26,11	8,57	11	0,28	0,24	2,75	80,49	112,58	8,37	17	0,56	0,43	2,64	83,89	109,92	9,42	15
TN3	p_{38}	0,05	0,04	0,10	4,86	3,62	1,37	6	0,05	0,04	0,09	4,37	3,16	1,27	6	0,05	0,05	0,10	4,90	3,38	1,31	7
	p_{68}	0,10	0,14	0,24	12,28	9,37	2,61	8	0,09	0,13	0,23	11,79	8,04	2,43	8	0,10	0,14	0,26	13,38	9,49	2,48	9
	p_{86}	0,18	0,21	0,55	27,43	18,69	4,12	10	0,19	0,20	0,48	24,98	16,49	4,26	10	0,24	0,21	0,65	33,29	20,15	4,93	11
	p_{95}	0,30	0,26	2,44	79,39	104,60	8,31	13	0,33	0,25	2,86	94,78	115,22	8,69	15	0,47	0,33	2,90	98,05	118,19	8,76	16
	p_{98}	0,35	0,32	3,96	170,26	155,42	11,95	18	0,41	0,37	6,51	232,36	212,45	15,91	22	0,58	0,50	6,22	210,83	205,89	17,27	32
TN4	p_{38}	0,06	0,06	0,14	7,09	5,22	1,74	8	0,06	0,08	0,16	8,16	6,06	1,77	10	0,06	0,07	0,13	6,83	4,75	1,80	8
	p_{68}	0,15	0,17	0,43	21,82	15,41	3,79	11	0,18	0,18	1,12	52,51	40,05	5,59	17	0,15	0,17	0,51	32,00	18,80	4,51	12
	p_{86}	0,41	0,29	3,65	171,35	151,58	11,82	17	0,43	0,38	6,80	258,97	282,38	21,37	28	0,46	0,35	5,68	229,45	266,33	18,78	21
	p_{95}	0,60	0,61	9,18	419,63	393,30	36,35	32	0,87	0,80	11,36	486,18	463,53	49,81	49	0,80	0,71	10,58	484,31	445,62	42,42	42
	p_{98}	1,37	1,32	17,78	883	612,24	72,87	49	1,65	1,50	17,02	932,33	776,66	79,49	49	1,42	1,25	17,08	816,36	789,18	81,89	49

thus the computational complexity is decreased drastically. This holds true especially if the overlap is small compared to the rest.

A smaller confidence interval reduces the range where potential correspondences could be found. This has positive effects if the error lies within this interval. In this case it does not happen that some of the potential good corresponding partners were clipped. However, if the current error is located outside of the confidence interval, points were clipped which could have a good corresponding partner in the other set. Thus the number of details, which can be used for the alignment, is reduced together with the overall possibility that ideal pairings can be found in general. This limits the alignment abilities drastically if the current initial error is high.

Thus we conclude that the uncertainty FoV and range clipping is especially useful for low tracking noise. There the number of points, which have to be aligned is reduced significantly and potential outliers were detected even before they could take any bad influence on the registration process. The confidence interval should not be chosen too small to avoid the clipping of potential useful point pairings. For higher tracking noise the uncertainty range clipping is less useful because occlusion effects can be detected just roughly. Therefore, outliers within the overlapping area were not thrown out. The FoV clipping on the other hand is always useful because the number of points, which have to be aligned, is decreased at low computational costs and potential wrong pairings were avoided.

5 Conclusions

The results show that an alignment of two data sets can be achieved with the help of a pose tracking system although they were captured from completely different perspectives. During the previous evaluation the contribution of certain system design aspects have been examined and the findings are summarized in the following sections.

The primary goal of the FoV clipping was to determine the overlapping area of the sensors FoV due to the pose estimation uncertainty. This reduces the number of points, which have to be aligned, significantly especially if the overlapping part is just a small subset of the complete data. Thus it further reduces the amount of possible outliers, which were not detected by the applied heuristics. That's why we conclude that the utilization of this method is always recommendable.

The idea of the uncertainty range clipping was to find the points, which could have a corresponding partner in the other data set, taking occlusion effects into account. This works fine if the tracking accuracy is already good because in this case the uncertainty ranges are small and a lot of potential outliers are ignored. Additionally the number of points, which have to be aligned is reduced further, which improves the overall performance. However, if the tracking noise is high, the contribution of the uncertainty range clipping is reduced drastically. That's why the uncertainty ranges are too big to detect occlusion effects, caused by small and medium sized objects like cars and trees, effectively. Nevertheless, occlusions caused by huge objects like buildings, were detected properly and thus points, which cannot have a corresponding point in the other data set are ignored. However, it is heavily situation dependent, whether the benefits justify the computational costs of this method.

According to Table 3 a rotation and translation accuracy below 0.5° and below 15 cm is attainable in most cases. Under good circumstances with a lot of details this falls even under 0.1° and 10 cm. This should be enough for a lot of practical applications. Using the improved transformation result it is now possible to fuse the complete data of both sensors.

5.1 Limitations of the presented approach

The presented approach is suitable for scenarios where two depth sensors observe partly the same surface of an arbitrary scene from completely different viewpoints. Additionally their relative pose have to be roughly known, due to some also known confidence interval. However some additional conditions must be met. First of all, enough details should be included in the overlapping part, such that an unambiguous registration result is possible. Therefore, it must contain at least 3 non parallel surface patches. This implies further that the point density in this region is sufficient to estimate proper normal features and point correspondences. Additionally the extent of this region has to be significantly higher than the measurement noise of the sensor data and the tracking error.

5.2 Further improvements and future work

So far there is no method applied which checks if the above mentioned requirements were fulfilled. For practical applications this have to be checked for each sensor data pair separately before the presented approach can be applied. If that is successfully done the plausibility of the registration result should be checked. Currently there is no method applied which is able to determine the quality of the registration result. A completely wrong alignment can be detected easily by evaluating, if the alignment result lies within the confidence interval of the applied tracking mechanism. For a further assessment of the achieved alignment accuracy it is necessary to develop a suitable metric. This could be done based on the remaining average distance between the correspondences found in the last ICP iteration. Further on the similarity of the normal features between the point pairs could be a helpful measurement.

References

Besl, P. and McKay, N. D.: A method for registration of 3-D shapes, Pattern Analysis and Machine Intelligence, IEEE Transactions, 14, 239–256, doi:10.1109/34.121791, 1992.

Chen, Y. and Medioni, G.: Object modeling by registration of multiple range images, in: Robotics and Automation, 1991, Proceedings, 1991 IEEE International Conference, Vol. 3, 2724–2729, doi:10.1109/ROBOT.1991.132043, 1991.

Gabriel, M.: LiDAR-Signaturberechnung von raeumlich ausgedehnten Zielen im Fahrzeugumfeld, Bachelor thesis, Chemnitz University of Technology, 2010.

Ibeo Automotive Systems GmbH: Operating Manual ibeo LUX® Laser scanner, ibeo Automobile Sensor GmbH, Merkurring 20, 22143 Hamburg, 2.5 Edn., 2008.

Jähn, B.: Fusion of multi-view point cloud data for cooperative object detection, Master thesis, Chemnitz University of Technology, 2014.

Klasing, K., Althoff, D., Wollherr, D., and Buss, M.: Comparison of surface normal estimation methods for range sensing applications, in: Robotics and Automation, 2009, ICRA '09, IEEE International Conference, 3206–3211, doi:10.1109/ROBOT.2009.5152493, 2009.

Pomerleau, F., Colas, F., Siegwart, R., and Magnenat, S.: Comparing ICP variants on real-world data sets, Auton. Robots, 34, 133–148, doi:10.1007/s10514-013-9327-2, 2013.

Pottmann, H., Leopoldseder, S., and Hofer, M.: Registration without ICP, Comput. Vis. Image Und., 95, 54–71, 2002.

Pulli, K.: Multiview registration for large data sets, in: 3-D Digital Imaging and Modeling, 1999, Proceedings, Second International Conference, 160–168, doi:10.1109/IM.1999.805346, 1999.

Rusinkiewicz, S. and Levoy, M.: Efficient variants of the ICP algorithm, in: 3-D Digital Imaging and Modeling, 2001, Proceedings, Third International Conference, 145–152, doi:10.1109/IM.2001.924423, 2001.

Permissions

All chapters in this book were first published in ARS, by Copernicus Publications; hereby published with permission under the Creative Commons Attribution License or equivalent. Every chapter published in this book has been scrutinized by our experts. Their significance has been extensively debated. The topics covered herein carry significant findings which will fuel the growth of the discipline. They may even be implemented as practical applications or may be referred to as a beginning point for another development.

The contributors of this book come from diverse backgrounds, making this book a truly international effort. This book will bring forth new frontiers with its revolutionizing research information and detailed analysis of the nascent developments around the world.

We would like to thank all the contributing authors for lending their expertise to make the book truly unique. They have played a crucial role in the development of this book. Without their invaluable contributions this book wouldn't have been possible. They have made vital efforts to compile up to date information on the varied aspects of this subject to make this book a valuable addition to the collection of many professionals and students.

This book was conceptualized with the vision of imparting up-to-date information and advanced data in this field. To ensure the same, a matchless editorial board was set up. Every individual on the board went through rigorous rounds of assessment to prove their worth. After which they invested a large part of their time researching and compiling the most relevant data for our readers.

The editorial board has been involved in producing this book since its inception. They have spent rigorous hours researching and exploring the diverse topics which have resulted in the successful publishing of this book. They have passed on their knowledge of decades through this book. To expedite this challenging task, the publisher supported the team at every step. A small team of assistant editors was also appointed to further simplify the editing procedure and attain best results for the readers.

Apart from the editorial board, the designing team has also invested a significant amount of their time in understanding the subject and creating the most relevant covers. They scrutinized every image to scout for the most suitable representation of the subject and create an appropriate cover for the book.

The publishing team has been an ardent support to the editorial, designing and production team. Their endless efforts to recruit the best for this project, has resulted in the accomplishment of this book. They are a veteran in the field of academics and their pool of knowledge is as vast as their experience in printing. Their expertise and guidance has proved useful at every step. Their uncompromising quality standards have made this book an exceptional effort. Their encouragement from time to time has been an inspiration for everyone.

The publisher and the editorial board hope that this book will prove to be a valuable piece of knowledge for researchers, students, practitioners and scholars across the globe.

List of Contributors

T. Vennemann, T. Frye, Z. Liu and W. Mathis
Institut für Theoretische Elektrotechnik (TET), Gottfried Wilhelm Leibniz Universität , Hannover, Germany

M. Kahmann
Physikalisch-Technische Bundesanstalt (PTB), Braunschweig, Germany

R. Rambousky
Bundeswehr Research Institute for Protective Technologies and NBC Protection (WIS), Munster, Germany

J. Nitsch and S. Tkachenko
Otto-von-Guericke University Magdeburg, Magdeburg, Germany

A. Fatemi and H. Klar
Institute of Microelectronics, Technical University of Berlin, Berlin, Germany

H. Gaul and U. Keil
FCI Deutschland GmbH, Berlin, Germany

David Krause
AUDI AG, 85045 Ingolstadt, Germany

Werner John
SiL System Integration Laboratory GmbH, Technologiepark 32, 33100 Paderborn, Germany

RobertWeigel
Lehrstuhl für Technische Elektronik, Friedrich-Alexander-Universität Erlangen-Nürnberg, Cauerstraße 9, 91058 Erlangen, Germany

A. Ascher, M. Lehner, M. Eberhardt and E. Biebl
Fachgebiet Höchstfrequenztechnik, Technische Universität München, Munich, Germany

S. F. Helfert
FernUniversität in Hagen, Hagen, Germany

F. Pfeiffer
perisens GmbH, Munich, Germany

M. Rashwan
Daimler AG, Sindelfingen, Germany

E. Biebl
Fachgebiet Höchstfrequenztechnik, Technische Universität München, Munich, Germany

B. Napholz
Daimler AG, Sindelfingen, Germany

I. Ali, U. Wasenmüller and N. Wehn
Microelectronic Systems Design Research Group, University of Kaiserslautern, 67663 Kaiserslautern, Germany

F. Ossevorth and H. G. Krauthäuser
TU Dresden, Elektrotechnisches Institut, Helmholtzstraße 9, 01069 Dresden, Germany

S. Tkachenko and J. Nitsch
Otto-von-Guericke-Universität Magdeburg, Lehrstuhl für Elektromagnetische Verträglichkeit, 39016 Magdeburg, Germany

R. Rambousky
Bundeswehr Research Institute for Protective Technologies and NBC Protection, Humboldtstraße 100, 29633 Munster, Germany

P. Jansen, D. Vergossen and D. Renner
Audi Electronics Venture GmbH, Gaimersheim, Germany

W. John
SiL GmbH – Paderborn/Leibniz Universität Hannover, Hanover, Germany

J. Götze
Technische Universität Dortmund (AG DAT), Dortmund, Germany

S. Parr, H. Karcoon and S. Dickmann
Faculty of Electrical Engineering, Helmut-Schmidt-University/University of the Federal Armed Forces Hamburg, Germany

R. Rambousky
Bundeswehr Research Institute for Protective Technologies and NBC Protection (WIS) Munster, Germany

A. Koenig, T. Rehg and R. Rasshofer
BMW Group Research and Technology, Munich, Germany

M. Kucharski and F. Herzel
IHP, Im Technologiepark 25, 15236 Frankfurt (Oder), Germany

O. Floch, A. Sommer, O. Farle and R. Dyczij-Edlinger
Chair of Electromagnetic Theory, Dept. of Physics and Mechatronics, Saarland University, Saarbrücken, Germany

M. Eberhardt, M. Lehner, A. Ascher, M. Allwang and E. M. Biebl
Fachgebiet Höchstfrequenztechnik, Technische Universität München, Munich, Germany

S. Pashmineh and D. Killat
Brandenburg University of Technology, Department of Microelectronics, Cottbus, Germany

J. Petzold, S. Tkachenko and R. Vick
Chair of Electromagnetic Compatibility, Otto-von-Guericke-University, Magdeburg, Germany

C. Adami, S. Chmel, M. Jöster, T. Pusch and M. Suhrke
Fraunhofer Institute for Technological Trend Analysis (INT), Euskirchen, Germany

D. Vollbracht
Faculty of Electrical Engineering and Information Technology, Technische Universität Chemnitz, 09126 Chemnitz, Germany

M. S. L. Mocker, F. Spinnler and T. F. Eibert
Technische Universität München, Lehrstuhl für Hochfrequenztechnik, Arcisstrasse 21, 80333 Munich, Germany

S. Hipp
CST AG, Bad Nauheimer Str. 19, 64289 Darmstadt, Germany

H. Tazi
Audi AG, August-Horch Str., 85055 Ingolstadt, Germany

M. Nawito, H. Richter and J. N. Burghartz
Institut für Mikroelektronik Stuttgart, Stuttgart, Germany

A. Stett
NMI Naturwissenschaftliches und Medizinisches Institut an der Universität Tübingen, Reutlingen, Germany

O. Kharshiladze
Ilia Vekua Institute of Applied Mathematics, Ivave Javakhishvili Tbilisi State University, Tbilisi, Georgia

K. Chargazia
Ilia Vekua Institute of Applied Mathematics, Ivave Javakhishvili Tbilisi State University, Tbilisi, Georgia
M. Nodia Institute of Geophysics, Tbilisi, Georgia

C. Schmidt, E. Lloret Fuentes and M. Buchholz
Research Group RI-ComET at the University of Applied Sciences Saarbrücken, Hochschul-Technologie-Zentrum, Altenkesselerstr. 17/D2, 66115 Saarbrücken, Germany

H. Hoffmann and T. Kürner
Deparmtent of Telecommunication Engineering, TU Braunschweig, Braunschweig, Germany

P. Ramachandra and F. Gunnarsson
Ericsson Research, Linköping, Sweden

I. Z. Kovács
Nokia, Aalborg, Denmark

L. Jorguseski
TNO, Delft, the Netherlands

M. Kreitlow and F. Sabath
Bundeswehr Research Institute for Protective Technologies and NBC Protection, Munster, 29633 Germany

H. Garbe
Institute of Electrical Engineering and Measurement Technology, Leibniz University Hannover, Hannover, 30167 Germany

T. Renkwitz, C. Schult, R. Latteck and G. Stober
Leibniz-Institute of Atmospheric Physics at the Rostock University, Schloss-Str. 6, 18225 Kühlungsborn, Germany

E. D. Schmitter
University of Applied Sciences Osnabrueck, 49076 Osnabrueck, Germany

B. Jaehn, P. Lindner and G. Wanielik
Professorship of Communications Engineering, Chemnitz University of Technology, Chemnitz, Germany

Index